However well the anatomy of the gastro-intestinal tracts of a wide range of animal species are described and quantified, there can be no real explanation of observed patterns without consideration of the mechanical and chemical properties of the food consumed and the digestive stages involved in its processing. This book aims to integrate findings from the many different types of investigation of mammalian digestive systems into a coherent whole.

Using the themes of food, form and function, researchers discuss models of digestive processes, linking this with evolutionary aspects of food use. Macroscopic and ultrastructural studies of the gastro-intestinal tract are also presented, as are physiological, ecological and biochemical aspects of the digestion of different food types. The book ends with an integrative chapter, bringing together the themes running through earlier sections.

The book will be of interest to researchers and graduate students of anatomy, zoology, physiology, ecology, evolution and animal nutrition.

THE DIGESTIVE SYSTEM IN MAMMALS: FOOD, FORM AND FUNCTION

Workshop participants by the main gate of Selwyn College, Cambridge, March 1992. *Back* (left to right): Peter Lucas, Bruno Simmen, Robert Snipes, Mike Perrin, Neill Alexander, Reg Moir, Göran Björnhag, Marcus Young Owl, Ian Hume, David Chivers.
Front (left to right): Marcel Hladik, Peter Langer, Carlos Martínez del Rio, Katie Milton, George Batzli, Steve Cork, Takashi Sakata.

THE DIGESTIVE SYSTEM IN MAMMALS: FOOD, FORM AND FUNCTION

Edited by
D. J. CHIVERS
University of Cambridge

P. LANGER
Justus-Liebig-Universität Giessen

CAMBRIDGE
UNIVERSITY PRESS

Published by the Press Syndicate of the University of Cambridge
The Pitt Building, Trumpington Street, Cambridge CB2 1RP
40 West 20th Street, New York, NY 10011-4211, USA
10 Stamford Road, Oakleigh, Melbourne 3166, Australia

First published 1994

Printed in Great Britain at the University Press, Cambridge

A catalogue record for this book is available from the British Library

Library of Congress cataloguing in publication data

The digestive system in mammals : food, form, and function / edited by
D.J. Chivers, P. Langer.
p. cm.
Includes bibliographical references and index.
ISBN 0-521-44016-5 (hardback)
1. Digestion. 2. Gastrointestinal system – Physiology.
3. Mammals – Physiology. I. Chivers, David John. II. Langer,
Peter, 1942– .
QP145.D565 1994
599′.0132 – dc20 93-32561 CIP

ISBN 0 521 44016 5 hardback

KW

Dedication

A quarter of a century ago, Professor Reg J. Moir of the University of Western Australia, Nedlands, wrote the following lines:

The development of the ruminant and other ruminant-like digestive systems is surely a progressive evolutionary pathway enabling life to be lived under wider, more difficult, and even inhospitable nutritional environments.
Moir, R. J. (1968). Ruminant digestion and evolution. In *Handbook of Physiology*, Section 6, Vol. 5, pp. 2673–2694. Washington DC: American Physiological Society.

These words, and the paper in which they were published, initiated the comparative type of investigation of the digestive system, not only of ruminant and ruminant-like herbivores but also of mammals in general. As this volume tries to give an overview of the state of the art, we wish to dedicate this book to its real initiator, whose presence at the Workshop, with his wife Shirley, was a special pleasure:

Professor Reg Moir

It was Reg who drew our attention to the words of Sir Geoffrey Vickers, as a guidance to our deliberations:

Even the dogs may eat the crumbs which fall from the rich man's table; and in these days, when the rich in knowledge eat such specialised food at such separate tables, only the dogs have a chance of a balanced diet.

Contents

Contributors

R. McNeill Alexander
Department of Pure and Applied Biology, University of Leeds, Leeds LS2 9JT, UK

G. O. Batzli
Department of Ecology, Ethology and Evolution, University of Illinois, 606 East Healey Street, Champaign, IL 61820, USA

G. Björnhag
Department of Animal Physiology, Swedish University of Agricultural Sciences, Box 7045, S-750 07 Uppsala, Sweden

The late A. D. Broussard
Formerly of Department of Ecology, Ethology and Evolution, University of Illinois, Champaign, IL 61820, USA

D. J. Chivers
Department of Anatomy, Downing Street, Cambridge CB2 3DY and Selwyn College, Cambridge CB3 9DQ, UK

S. J. Cork
Division of Wildlife and Ecology, CSIRO, Lyneham, ACT 2602, Australia

W. J. Foley
Department of Zoology, James Cook University, Townsville, Queensland 4811, Australia

C. M. Hladik
Laboratoire d'Ecologie Générale, Museum National d'Histoire Naturelle, F-91800 Brunoy, France

I. D. Hume
Zoology Building (AO8), University of Sydney, Sydney, NSW 2006, Australia

W. H. Karasov
Department of Wildlife Ecology, University of Wisconsin, Madison, WI 53705, USA

P. Langer
Institut für Anatomie und Zytobiologie, Justus-Liebig-Universität Giessen, Aulweg 123, D-35385 Giessen, Germany

P. W. Lucas
Department of Anatomy, Faculty of Medicine, University of Hong Kong, 5 Sassoon Road, Hong Kong

C. Martínez del Rio
Department of Ecology and Evolutionary Biology, Princeton University, Princeton, NJ 08544, USA

Clare McArthur
Department of Ecology and Evolutionary Biology, Monash University, Clayton, Vic. 3168, Australia

R. J. Moir
School of Agriculture, University of Western Australia, Nedlands, WA 6009, Australia

Eileen M. O'Brien
Department of Geography, Georgia State University, Atlanta, GA 30303, USA

R. J. Oliver
Department of Ecology, Ethology and Evolution, University of Illinois, 606 East Healey Street, Champaign IL 61820, USA

M. R. Perrin
Department of Zoology and Entomology, University of Natal, PO Box 375, Pietermaritzburg, 3200 South Africa

C. R. Peters
Department of Anthropology and Institute of Ecology, University of Georgia, Athens, GA 30602, USA

T. Sakata
Department of Basic Sciences, Ishinomaki Senshu University, Minamizakai Shinmito 1, Ishinomaki 986, Japan

B. Simmen
Laboratoire d'Ecologie Générale, Museum National d'Histoire Naturelle, F-91800 Brunoy, France

R. L. Snipes
Institut für Anatomie und Zytobiologie, Justus-Liebig-Universität Giessen, Aulweg 123, D-35385 Giessen, Germany

M. Young Owl
Department of Ecology, Ethology and Evolution, University of Illinois, Urbana, IL 61801, USA

Preface

We started corresponding early in 1988 with exchange of reprints and discovered that we had very similar anatomical interests and a shared desire to promote interdisciplinary discussion to resolve problems of mutual interest. However well the anatomy of the gastro-intestinal tract of a wide range of animal species is described and quantified, there can be no real explanation of observed patterns without consideration of the mechanical and chemical properties of the different foods consumed, and the digestive processes – mechanical and chemical – involved in their processing.

We met in Cambridge in May 1989 in the gardens of Selwyn College and continued discussions over dinner in Hall. We resolved to hold a workshop in Cambridge, from which a book would result, on the Form, Function and Evolution of the Digestive System in Mammals. Plans were developed, with the target of April 1991 once September 1990 proved unpopular among the 30 or so prospective participants – a blend of anatomists, physiologists, zoologists, botanists, ecologists and anthropologists. Lack of funds caused postponement for a further 12 months and it was only possible to hold the Workshop through the enthusiasm and determination of the participants in finding their travel (and some subsistence) money and with help from Dalgety plc and Cambridge University Press who covered most of the costs of the 4 days in Cambridge. We are indebted to them all.

Seventeen of us gathered in Selwyn College on 31 March 1992 and departed on the morning of 4 April. In between we worked through the food, form and function topics, hearing what each of us could contribute (based on manuscripts submitted in advance of the meeting), and then dividing into groups to start the synthesis that provides the basis for the final chapter. Throughout we were very well nourished by the College, and had a long lunch break to allow the participants to enjoy Cambridge and the good spring weather and to recharge their batteries for the intensive sessions. The evenings were relaxed and prolonged, contributing in some intangible way to the production of the book.

The impressive effort has been sustained since, with the 19 core chapters being revised and refereed and the five introductory or concluding chapters produced in 9 months – a very real team effort involving 23 contributors in all. Thus, the seed planted in 1988 germinated during the following 3 years to flower in 1992 and to fruit once Cambridge University Press complete their task, in 1994. Hopefully these fruits will be widely dispersed and digested chemically – by the brain rather than by the guts.

Finally, we reiterate our thanks to the contributors for their dedicated efforts throughout, to Dalgety plc and to Dr Alan Crowden and Cambridge University Press for agreeing to publish this volume and for advancing the fee to make the venture possible.

January 1993

D. J. Chivers
P. Langer

Part I

Introduction

1

Gut form and function: variations and terminology

DAVID J. CHIVERS and PETER LANGER

The aim in this volume is to present a synthesis of a wide range of data pertinent to the digestive system in mammals, focussing on three areas, food, form and function. There are introductions to each of these three sections and a final discussion encompassing all aspects but working from the view-points of the three groups of contributors. The essence of this volume is that it brings together very different lines of research, and that an integrating process is initiated to develop a broader understanding of the digestive system.

In this preliminary section we set the scene and discuss general, often problematic, issues and then Langer (Ch. 2) presents an evolutionary perspective. This is followed by introductions to modelling gut function by Martinez del Rio, Cork and Karasov (Ch. 3) and optimum gut structure for specified diets by McNeill Alexander (Ch. 4).

As we focus on the details of diets and digestive systems (structure and function), we need to think about the nature of the niches to which they are adapted: in particular, as to how broad or narrow these may be or how variable in the short (days) or long term (seasons, years). To what extent is form and function accounted for by preferred food and body size? To what extent are they moulded by critical events ('bottlenecks' in food availability). We need to think about them against an evolutionary background of increasing environmental diversity over the last 50 million years in relation to the dramatic radiation of angiosperm (flowering) plants. It is the increasingly dynamic nature of environments that encourages more dietary plasticity, varied adaptations and niche breadth – polytrophism or omnivory – but these may be discouraged by anatomical and/or physiological constraints of the digestive system.

Gut form and function

Gastro-intestinal tracts show dramatic variation from those adapted for processing animal matter (very nutritious and easy to digest) mainly in the small intestine, with small stomachs and large intestines, to those dominated by stomach and/or caecum and colon for the processing of plant cell walls (very common food but very difficult to digest) by fermentation. Here, then, is the contrast between specialist **faunivores** and **folivores**, with an intermediate morphology shown by **frugivores** (with a larger small intestine if supplementing the fruit staple with animal matter, or larger fermenting chambers if leaves are the main supplement). This dietary spectrum corresponds closely with body size and even more so with biomass density. This is because animal matter is scarce compared with plant material.

Herbivory (or **florivory**) – the consumption of plant material – needs to be sub-divided into **frugivory** and **folivory**, with numerous sub-divisions of each (see Ch. 5). Folivory ranges from the more selective 'browsers', through to the less selective bulk-feeders, 'grazers'. Chitin, the skeleton of invertebrates, can be digested in those mammals that have the enzyme chitinase. Its digestion requires at least a comparable dentition to folivores (although much smaller) to similarly maximise the surface area available for digestion.

The concept of **omnivory** is weakened by the anatomical and physiological difficulties of digesting significant quantities of animal matter *and* fruit *and* leaves. The supposed evolutionary pathway shown by mammals radiating from their insectivorous ancestor is reflected by evolving anatomy and physiology. As mammals increased in size (over about 1 kg for primates), so they were able to add plant material (mainly the more digestible fruit) to their insectivorous diet. Increasing body size allowed even more voluminous guts that allowed the fermentation of plant cell walls, giving nutritional access to foliage. Adaptations for digesting seed coats (by fermentation) is perhaps the link between fruit- and leaf-eating diets (Bodmer, 1989). Therefore, animal matter is swamped in a large gut, and foliage cannot be digested in a small gut. A compromise is not really feasible, although some rodents (dormice), pigs and opportunistic primates (macaques, baboons and chimpanzees) may be true omnivores. Humans are only omnivorous thanks to food processing and cookery; their guts have the dimensions of a (faunivore) carnivore but the taeniae, haustra and semi-lunar folds are characteristic of folivores. Among the so-called omnivores, most eat either mainly fruit and animal matter (if smaller) or fruit and foliage (if larger) but not all three.

The basic gut design persists clearly, as reflected by development, basic position, innervation and blood supply (Fig. 1.1), despite dramatic diversifi-

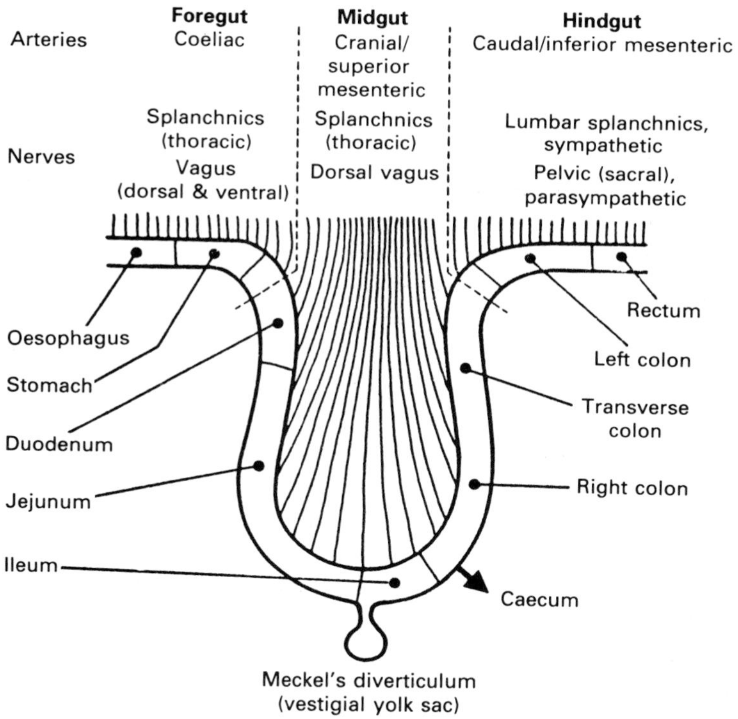

Fig. 1.1. The basic design of the gastro-intestinal tract showing the blood supply and innervation.

cation among different dietary types. Thus, the **stomach** and start of the **duodenum** (as far as the entrance of the bile duct) develop from the first relatively-fixed part of the **foregut** in the abdomen, crossing from left to right, and are innervated by the **dorsal** *and* **ventral vagus** (Xth cranial) nerves (parasympathetic) and the **splanchnic nerves** (sympathetic). They are supplied with blood by branches of the **coeliac artery** and blood is drained into the **hepatic portal vein** by gastro-splenic and gastro-duodenal veins. The **left (descending) colon** and **rectum** are relatively unvarying in all mammals, with little digestive function, and run caudally to the anus; this **hindgut** is innervated by **sacral (pelvic) nerves** (parasympathetic) and **lumbar splanchnic nerves** (sympathetic) and is supplied with blood by the **caudal mesenteric artery**, draining into the vein of the same name. Therefore, anatomically speaking, *hindgut* cannot be equated with large intestine as is so commonly done.

It is the mobile **midgut** loop that is so readily lengthened, proximally in faunivores and distally in folivores. It herniates through the umbilicus early in development and is later withdrawn into the peritoneal cavity by an anti-clockwise rotation (viewed from ventrally) at the base of the loop, with the proximal end preceding the distal end so that the **caeco-colic junction** comes to lie ventral to the duodenum (Fig. 1.2). The midgut is supplied by the **cranial mesenteric artery** and **vein** and is innervated by the **dorsal vagus** and **splanchnic nerves**. The duodenum, crossing from right to left, gives way to coils of **jejunum**, with the **small intestine** terminating as the **ileum** on the right. The **large intestine** starts with the **caecum** (with an appendix sometimes, where lymphatic tissue may be concentrated distally) which branches blindly off the large intestine. It continues with the **right (ascending) colon**, which can expand dramatically, and very variably in

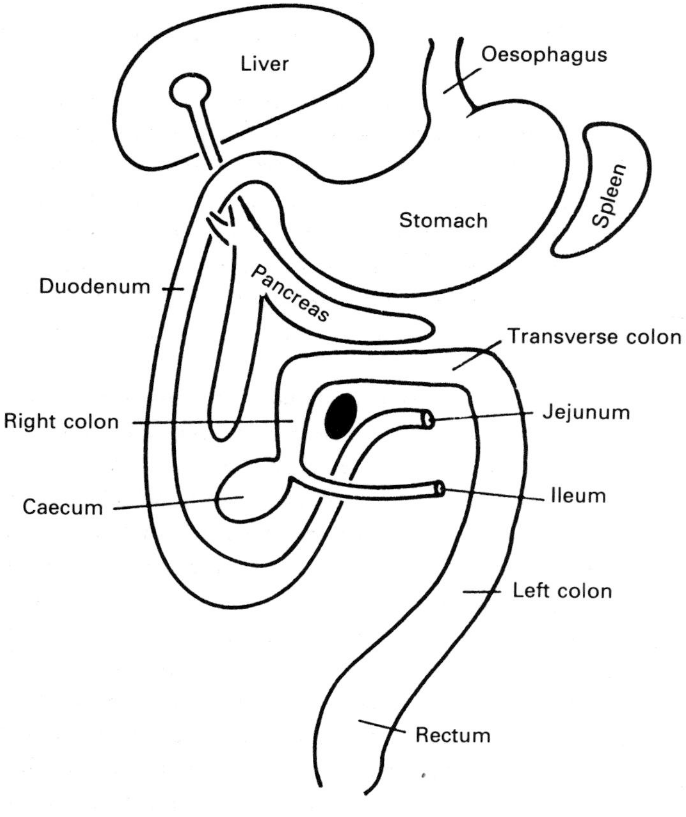

Fig. 1.2. Basic layout of the mamalian gut (ventral view) with the stomach crossing from left to right, duodenum on the right and the colon crossing from right to left.

mammals, into a fermenting chamber (instead of or as well as the caecum) in many folivores; this in turn gives way to the usually short **transverse colon**, near the stomach. Thus, we have the contrasting strategies for fermenting plant cell walls of **caeco-colic fermentation** (or caecal or colic), often called hindgut or post-gastric fermentation, and of **foregut fermentation**, often called pre-gastric or forestomach fermentation. The latter occurs in the keratinised (except in Camelidae and Macropodidae) enlarged first part of the stomach, before the true glandular part of the stomach that contains the gastric (cardial, fundic and pyloric) glands.

Terminology

We have already defined numerous terms. It remains to clarify others and give the synonyms that appear in this volume. The parts of the gut nearest to the head are proximal, cranial or oral, those nearer the tail are distal, caudal or aboral.

Stomachs are enlarged parts of the otherwise tubular gut; they are simple, one-chambered (uni-locular) or compound, multi-chambered (pluri-locular). The entrance is the **cardia**, the exit is the **pylorus**, with the body (or **fundus**) in between. Since it is now clear that they always develop from the simple gastric spindle, it is not correct to talk about monogastric or polygastric mammals – no mammal has more than one stomach! The confusion was generated by the contrast in ruminants between keratinised (like the oesophagus) and glandular parts of the stomach, implying that they might have had different origins.

Such enlargements of the stomach may be sac-like (sacculation, sacciform), diverticulated (diverticulum), or tube-like (tubiform). Enlargements in the large intestine, in particular, where the cylindrical shape is maintained, are sacculations when they are separated from each other by circular folds, or **plicae circulares**. These folds may occur in the small and large intestines. When longitudinal muscular bands, or **taeniae**, are differentiated in the outer layer of the muscular wall which ‘anchor’ so-called **semi-lunar folds** or **plicae semilunares**, the widened lumina between these folds are called **haustra** (singular haustrum). The two main regions of the large intestine, the colon and the caecum, can be haustrated. Only in the colobid monkeys, kangaroos and rat kangaroos are parts of the forestomach haustrated.

Histologically, the gastro-intestinal tract has (a) an inner epithelial layer of **tunica mucosa**, with a thin layer of muscle, the **lamina muscularis mucosa**; (b) a connective tissue layer of **tela sub-mucosa** containing vessels and nerves; (c) an inner circular and outer longitudinal layer of **smooth mus-**

cle (with an additional third – oblique – layer in the body of the stomach); (d) an outer connective tissue coat – **tunica serosa** – overlaid by (e) **visceral peritoneum** or, in retroperitoneal organs, by an **adventitia**. The tunica mucosa is of endodermal origin, the rest is derived from mesoderm.

An important parameter in understanding the relation between structure and function in the digestive system is the amount of time that digesta spends in the system. Terms commonly used are transit time, retention time, mean retention time, residence time, passage time, passage rate and turnover time (Martinez del Rio *et al*, Ch. 3). **Transit time** has been used for every possible measurement of **retention time**! It is usually used for the time of first appearance; transit time will only equal **mean retention time** if there is no mixing (otherwise it will underestimate mean retention time). **Residence time** refers to time spent in various gut compartments; **gut clearance time** refers to time to completely evacuate the gut under a starvation regime.

The digestive processes are modelled in terms of **plug flow** (PFR), a **piston** or **tubular reactor** and a **continuous stirred tank reactor** (CSTR); there are also **batch-stirred reactors** (BSR). Digestion may be **catalytic** or **autocatalytic, alloenzymatic** or **autoenzymatic**. Observations may be made **post-prandially** (after a meal) or **post-injectionally**.

This, then, provides the mainly structural background to the initial discussion of evolutionary and functional issues (Part I) and the detailed discussion of food (Part II), form (Part III) and function (Part IV). We aim to integrate very different lines of research to promote a better understanding of the digestive system in mammals.

Reference

Bodmer, I. (1989). Frugivory in Amazon ungulates. Ph.D. thesis, University of Cambridge.

2

Food and digestion of Cenozoic mammals in Europe

PETER LANGER

Fossil European mammalian species belonging to orders that have survived to the present day can – most probably – be allocated to the same *type of digestion* as their extant close relatives. The general habitus can be determined from fossil remains; the habitus – as well as the dentition – represents adaptations to a certain mode of life and a certain type of digestion that is similar to conditions in a closely-related species from the Recent (Holocene). Of course, the type of digestion in an extinct taxon can only be considered as highly probable, and not as absolutely certain.

It is also probable that the *type of food* is similar in extinct and recent species of the same order and/or family. For example, it is plausible that a perissodactyl ingested plant material and was certainly not zoophagous. Also, it was most probably a caecum-colon fermenter, as are its present-day relatives, and did not ferment the ingested plant material in a forestomach. The above-mentioned assumptions accepted, it is possible to classify mammals from the Cenozoic era (fossil mammals) according to the food they ate and according to their type of digestion – as long as they belong to orders that still survive today!

The following questions will be covered in the present study:

1. What were the steps – and when were these steps undertaken – to widen the range of *food* that could be used by European fossil mammals?
2. When did major steps in the differentiation of new types of *digestion* take place?
3. What can be said about the combinations between types of food and types of digestion in European mammalian orders and families?
4. Which answers to the previous questions present insights into the relationships between body weight, class of food and type of digestion during mammalian evolution in Europe?

Materials and methods

Extinct species, as well as their families and orders, are listed by Savage and Russell (1983). From textbooks of paleontology (Romer, 1966; Carroll, 1988) the approximate body size of a species can be obtained by comparing fossil species with well-known recent ones (Flindt, 1985). 'We cannot measure body mass directly for fossil species and must derive estimates from skeletal remains that are usually fragmentary and incomplete' (Damuth and MacFadden, 1990, p. 3). Difficulties in the determination of body size from fossil material were emphasized by Heinrich (1991): changes in *proportions* of body parts have to be identified to avoid confusion with changes in body *size*. Although it is not possible to obtain direct biological measurements for members of an extinct fauna, the variables of skeletal elements 'have Gaussian distribution, and thus averages are biologically valid species parameters' (Martin, 1990, p. 63).

For the present study, five classes of animals based on differing mean body weights were compiled. In the following diagrams the weight groups will be identified by their numbers (1 through 5), thus sub-dividing the continuous mammalian weight range between the pygmy white-toothed shrew (*Suncus etruscus*) weighing 2 g and the African elephant (*Loxodonta africana*) weighing 6 t into just five steps. The following **groups of weight** were differentiated:

1 (W1). Minute mammals: weight below 5 kg
2 (W2). Small mammals: weight between 5 and 50 kg
3 (W3). Medium-heavy mammals: weight between 50 and 500 kg
4 (W4). Heavy mammals: weight between 500 and 5000 kg
5 (W5). Very heavy mammals: weight above 5000 kg.

In a further endeavour to differentiate fossil species, the dietary categories of mammals were attributed to the species listed by Savage and Russell (1983) according to the actuo-zoological approach. To avoid the misleading impression of over-exactness, four food or **dietary classes** were determined here, considering and modifying data from Eisenberg (1981) and Langer (1991).

1 (F1). **Zoophages** (faunivores): piscivores and squid-eaters, crustacivores, carnivores (eating vertebrates), sanguinivores, planktonivores, myrmecophages and insectivores
2 (F2). **Polyphages** (omnivores): species that eat foods of animal *and* of plant origin

3 (F3). **Phytophages** (florivores): nectarivores, gumivores, frugivores, granivores, herbivores and browsers
4 (F4). **Poëphages** (grazers, graminivores): the grazers.

Graminivores is based on the Greek name for grass (Werner, 1961); plant taxonomists apply the name Poaceae as well as the Latin Gramineae for the grass family (Jones and Luchsinger, 1987).

The different **strategies of digestion**, as they can be found in mammals, were also grouped into four main types, following Langer (1988) and Langer and Snipes (1991).

1 (D1). Autoenzymatic digestion: digestion with the mammal's own set of digestive enzymes
2 (D2). Alloenzymatic digestion in the caecum-colon complex or large intestine: fermention is performed by microbes
3 (D3). Alloenzymatic digestion in the forestomach without rumination: again microbes perform the fermentation of the food
4 (D4). Alloenzymatic digestion in the forestomach plus repeated chewing of the cud (rumination): microbial fermentation.

The differentiation of groups of body weight, food classes and types of digestion was made for fossil European mammalian species listed by Savage and Russell (1983) for 13 geological units, of varying ages given in million years before the Present (Ma BP): five for the European Eocene (Sparnacian, approximately 54; Cuisian, 48; Lutetian, 46; Rubiacian, 44; Headonian, 38), three for the Oligocene (Early Oli, 35; Middle Oli, 30; Late Oli, 24) and five for the Miocene (Agenian, 22; Orleanian, 18; Astaracian, 14; Vallesian, 11; Turolian, 8). Because of Quaternary extinctions and extirpations under the influence of man during the Pleistocene era (Barnosky, 1991), especially of forms with higher body weights (Edwards, 1967; Caughley and Krebs, 1983; Caughley, 1987), only the Tertiary will be dealt with in this paper.

Three-dimensional diagrams were drawn to help to clarify the interrelationships (Figs. 2.1 and 2.2). The first type (*a*) compares dietary category or food class, group of body weight and the number of species that show the respective combination of both criteria. In the second type of diagram (*b*) the connection between the type of digestion, the weight group and the number of species is given. Finally, the third type of diagram (*c*) illustrates dietary category or food class, type of digestion and the number of species with the respective combination.

Setting the diagrams in chronological order, it was possible to obtain a visual impression as to how European mammals during the Tertiary era occupied niches that were characterised by different classes of food and types of

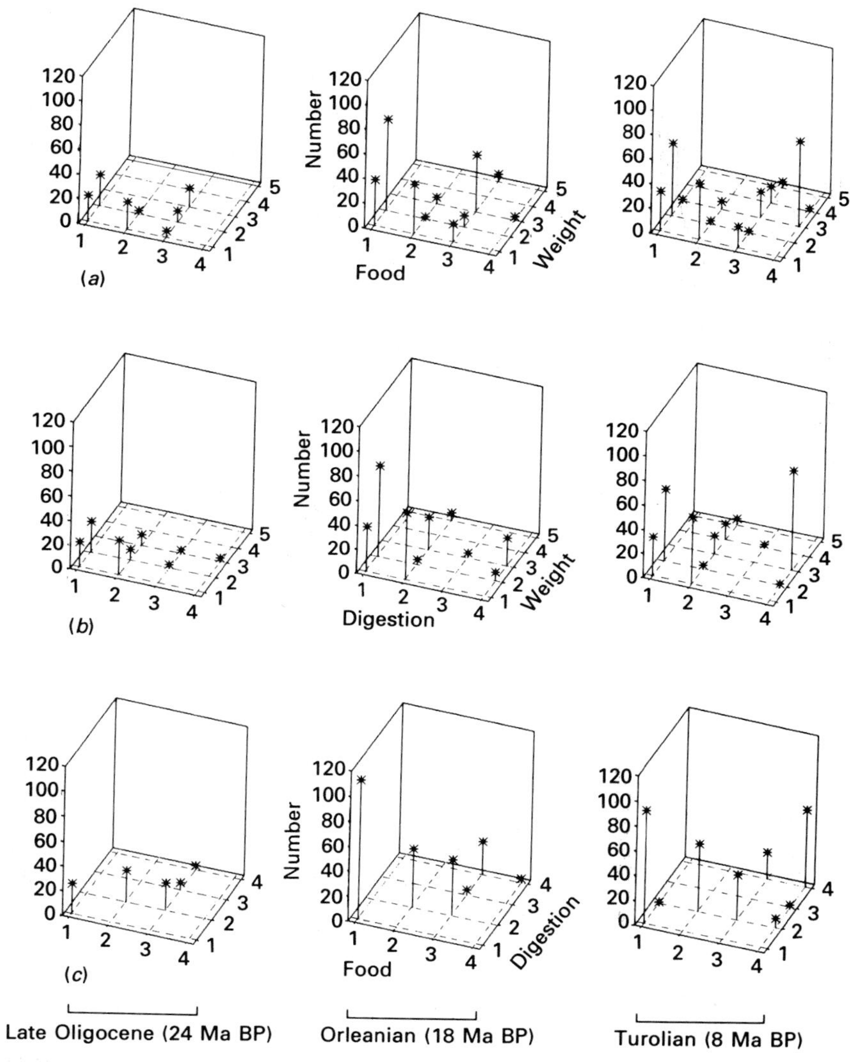

Fig. 2.1. Three sets of three-dimensional diagrams for three geological time units, Sparnacian, Cuisian and Early Oligocene. Their time scale is given in parentheses as millions of years before Present. In each horizontal line only the central diagram is marked but in the right and left ones the axes have the same designations. (*a*) Dietary categories or food classes of mammals versus weight group. The vertical axis gives the number of species (from data published by Savage and Russell, 1983.) (*b*) Types of digestion versus weight groups and number of species. (*c*) Food classes versus types of digestion and number of species.

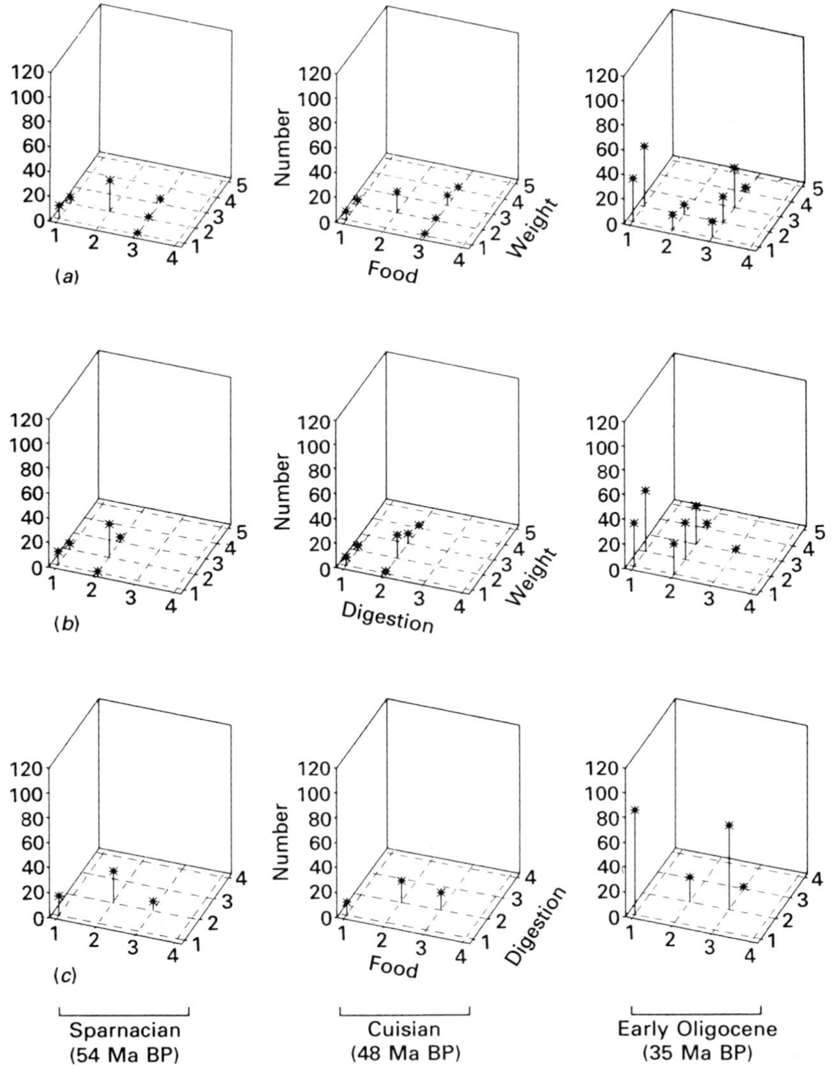

Fig. 2.2. This set of diagrams is the chronological sequel to Fig. 2.1, Late Oligocene, Orleanian and Turolian, and has the same designations.

digestion. As the whole number of 13 geological units gives a confusing number of details, only six of them are considered in Figs. 2.1 and 2.2, namely the Sparnacian and Cuisian for the Eocene, the Early and Late Oligocene and the Orleanian and Turolian for the Miocene.

Results

Widening of the range of food used by European fossil mammals during the Tertiary

The widening of the range of ingested food can be followed in Figs. 2.1*a* and 2.2*a*: 54 Ma BP. In the Sparnacian, only zoophages (F1), polyphages (F2) and phytophages (F3) can be found as (maximally) medium-heavy mammals. Already in the Cuisian (48 Ma BP), heavy phytophages (W4, F3) can also be found. This tendency of increase in body weight of phytophages (F3) can be followed over the Orleanian to the Turolian (8 Ma BP) during the Miocene. In the Orleanian (18 Ma BP) phytophagy (F3) is extended to graminivory (F4). The number of 'sure' and 'probable' fossil grass remains drastically increased in number during the Miocene (Thomasson, 1987), indicating the formation of grasslands. It is obvious that the number of grazing species (F4) increased before the Turolian, especially in medium-heavy mammals (W3), but not in very large ones! It is interesting that both zoophages (F1) and polyphages (F2) increased their number of small (W2) and minute-size (W1) species.

Widening of the types of digestion in European fossil mammals of the Tertiary

As can be seen in Fig. 2.1*b* for the Sparnacian, only minute (W1) and small (W2) or at most medium-heavy mammals (W3) that digest autoenzymatically (D1) or alloenzymatically in the large intestine (D2) were present in Europe. However, the following diagrams of Figs. 2.1*b* and 2.2*b* show an increase in body size in the group of alloenzymatic caecum-colon digestors (D2)(Cuisian), as well as the appearance of a new type of digestion, namely, alloenzymatic digestion in the forestomach of non-ruminants (D3)(Early Oligocene 35 Ma BP) and further to ruminants (D4)(Late Oligocene, 24 Ma BP). In the Miocene (Orleanian and Turolian), the ruminants (D4) did not occupy the upper range of body sizes (W4 and W5) but their number of species increased considerably. Only caecum-colon digestors (D2) extended their body weights to very large mammals (W5) but the highest numbers of species representing this type of digestion can be found in very small (W1) representatives.

Combinations of food classes and types of digestion in European fossil mammals of the Tertiary

Figure 2.1*c* shows only three types of combination during the Sparnacian: mammals were either autoenzymatically-digesting zoophages (D1, F1), polyphages (F2) or phytophages (F3) digesting their food alloenzymatically in the large intestine (D2). In the Early Oligocene, the diversity of types of digestion extended; however, it was not before the Orleanian (Fig. 2.2*c*) that the first ruminants (D4, F4) appeared. This combination of food and digestion was very successful, as evidenced by the high number of species present during the Turolian. Large numbers of species can also be found among autoenzymatically digesting zoophages (D1, F1) as well as polyphages (D2, F2) and phytophagous caecum-colon digestors (D2, F3) in the Turolian.

Discussion

From this presentation it can be generalised that, starting from the autoenzymatically digesting zoophages (D1, F1) and the large-intestinal fermenting polyphages (D2, F2) and phytophages (F3), all of which are either of medium (W3), small (W2) or minute (W1) weight, the food range was extended towards grazing (F4, already attained at the end of the Late Oligocene, 24 Ma BP), as well as towards forestomach fermentation in non-ruminants (D3, attained in the Early Oligocene, 35 Ma BP) and rumination (D4, in the Late Oligocene). Body weight showed a general tendency to increase, a tendency that was also noted by Wing and Tiffney (1987).

It is also interesting (Figs. 2.1*a* and 2.2*a*) that grass-eating forestomach fermenters, non-ruminants (D3, F4) as well as ruminants (D4, F4), are never found in the very heavy range but are mostly medium-weight (W3) mammals. The physiological aspects of this problem have been discussed by Case (1979), van Soest (1982), Demment and van Soest (1983, 1985), Peters (1986), Owen-Smith (1988) and Prins and Kreulen (1990). The nutritional problem that has to be overcome by herbivores (F3 and F4) is the reduction of food *quality* and, in some cases, also the reduction of the *quantity* of food that is available for ingestion. These reductions in quality and quantity are the result of climatic seasonality starting during the Late Oligocene (Furon, 1972; Snaydon, 1981; Janis, 1984; Stiles, 1989). Some herbivores are able to 'increase' the quality of food postprandially by using the plant material as a culture medium for microbes. Bacteria occur in all mammals, protozoa are prominent in many and even fungi may be important (Clarke, 1977;

Hungate, 1988; Prins and Kreulen, 1990; Fonty and Joblin, 1991). The mammal makes use of products of microbial synthesis, mainly short-chain fatty acids but also vitamins and essential amino acids. Ammonia, which is released by bacterial urease, can be absorbed and used for amino acid synthesis (Mitvedt, 1982). Gut microorganisms are able to retain nitrogen (Shirley, 1986). 'Furthermore, this flora plays an important role in preventing foreign organisms from becoming established in the gut' (Mitvedt, 1982, p. 40).

A 'disadvantage' of microbial fermentation lies in the fact that, compared to autoenzymatic digestion, it takes a relatively long time. Because of this, the rate of passage of digesta through the gastro-intestinal tract is reduced. To solve this basic dilemma in herbivores, different strategies can be used. One solution, mainly applied in caecum-colon fermenters (D2) but also in some forestomach fermenters (D3), is the differentiation of taeniae, haustra and semi-lunar folds, all of which most probably represent adaptations for the regulation of digesta transit (Langer, 1984, 1988). In forestomach fermenters rumination increases the passage of digesta through the ruminoreticulum. Pearce and Moir (1964) showed that sheep experimentally prevented from rumination retained food for an extended time in their forestomachs. Under these circumstances room for new food by passage of gastric contents could not be realised. On the other hand, intake in ruminants will be markedly slower as compared to non-ruminating mammals. Ruminants have an effective fermentation; but on very low-digestible food the ruminants cannot speed up the passage, so that they can even starve with a full stomach (Björnhag, personal communication).

Returning to the problem of body weight in herbivores with different types of digestion, we should concentrate on the nutritionally 'narrow' situation when food quality is low, e.g. in the dry or cold season. (The alternative adaptation, to select for a higher quality food, is not considered in this discussion.) Food of low quality is generally abundant (Case, 1979) and there is, most often, no *quantitative* limitation concerning available plant material. However, the voluminous forestomachs, as well as the complex process of gastric empyting (Baumont and Deswysen, 1991) reduce the transit through the digestive tract considerably. For example, in cattle and sheep more than 50% of the total mean retention time of the complete digestive tract takes place in the ruminoreticulum (Langer, 1987). In the caecum-colon fermenters, e.g. pig and horse, the same can be said for the large intestine. Recent studies on four wild species of Artiodactyla presented feeding time expressed as a percentage of total daylight activities. These studies considered wild boar, *Sus scrofa* (Gerald *et al.*, 1991), roe deer, *Capreolus capreolus* (Maublanc *et al.*, 1991), chamois, *Rupicapra rupicapra* (Pépin *et al.*, 1991) and mouflon, *Ovis ammon musimon* (Bon *et al.*, 1991). These four species are either caecum-colon fermenters (wild boar) or ruminating forestomach fermenters. A decrease in food quality from that of the polyphagous wild boar via that of the concentrate selector roe deer and the intermediate type chamois to that of the poëphagous (grass- and roughage-eating, Hofmann, 1988) mouflon is connected with the increase in time that is necessary for nutrition and, most probably, also with the increase in quantity of ingested food. In sheep, intake of organic matter decreases with cell wall content of foodstuffs (van Soest, 1982). High cell-wall content is found in low-quality food. Feeding time is longer in winter than in the other three seasons in the pig but not so in the ruminant chamois and mouflon. The reason for these differences could be an increased uptake by the wild boar during winter of food that is of lower quality than that eaten in spring, summer and autumn during the vegetation period. Transit through the pig's digestive tract can be 'speeded-up'. On the other hand, the feeding time is practically identical in the chamois and the mouflon in winter as during the vegetation period because the voluminous ruminoreticulum in both species prevents the transit from being 'speeded-up'.

Considering the availability of large amounts of low-quality plant food which can be microbially fermented (Prins and Kreulen, 1990) and consequently 'improved' postprandially, the question why *only* large intestinal fermenters (D2) and herbivores (F3) and *not* forestomach fermenters (D3 and D4) and grazers (poëphages) (F4) reach very heavy body weights (W5) can now be discussed.

Large intestinal fermenters

The large intestinal fermenter is able to ingest large amounts of low-quality food, which pass through the digestive tract relatively quickly. This is very variable. Hill (1952) presented data showing that the large intestine of the horse retained digesta for 70% of the total digestive tract retention time, while the values published by van Soest (1982) represent about 32% (Langer, 1987). The remarkable differences in digesta transit could be considerably influenced by different qualities of food during the experimental periods in different laboratories. Mason (1983) emphasised in a study on microbial digestion in the large intestine of the pig that transit through this region of the digestive tract can vary from about 20 h to about 38 h. When the pig is fed with feed-components of low digestibility, large-intestinal digestion can increase considerably (Drochner and Meyer, 1991). The requirement for large food quantities because of the short times of retention in the digestive tract of caecum-colon fermenters compensates for the low quality of food. As the metabolic need per unit body weight decreases with increasing body mass (Demment and van Soest, 1983, 1985), very large caeco-colic-fermenting herbivores should be able to tolerate a lower minimum dietary quality than smaller species (Bell, 1970, 1971; Jarman, 1974; Owen-Smith, 1988). The African and the Indian elephant, which digest their plant food in the large intestine, belong to the group of very heavy mammals (W5). The largest known land mammal, *Baluchitherium* from the Late Oligocene and early Miocene periods in Central Asia, stood 5 m tall (Economos, 1981), was a representative of the odd-toed Rhinocerotoidea (Carroll, 1988) and was, most probably, a large-intestinal fermenter.

Non-ruminating forestomach fermenters

The non-ruminating forestomach fermenters (D3) have difficulties in increasing the quantity of ingested food by 'speeding-up' digesta transit time. Australian Macropodidae (kangaroos) probably are able to regulate digesta passage with the help of their semi-lunar gastric folds (Langer *et al.*, 1980; Langer, 1984) but a large forestomach volume without such functionally variable folds retains digesta for a more extended time than a small one. For example, 'hippos exhibit an exceptionally long retention time, amounting to almost four days for the grass hay diet' (Owen-Smith, 1988). This is probably because of their voluminous forestomach of 200 litres (Langer, 1975, 1976). Microbial fermentation in the forestomach makes products of microbial activity available and thus 'improves' food quality postprandially. An example of a non-ruminating forestomach-fermenting grazer is the river hippo (*Hippopotamus amphibius*), which belongs to the group of heavy mammals (W4). Other mammals that ferment in the forestomach but do not ruminate are much smaller and belong to weight group 2 (W2), such as the tree sloths (Bradypodidae) or the colobine monkeys (Colobinae). The number of species using forestomach fermentation without rumination has always been, and still is, very low.

Ruminating forestomach fermenters

The ruminants (D4) modify the situation of the non-ruminant types. They 'improve' food quality postprandially by rechewing their forestomach digesta and thus opening cell walls and making the nutritious cell contents accessible. This allows them 'to profit fully from the abundantly available but slowly released energy in cell wall polysaccharides' (Prins and Kreulen, 1990,

p. 114) because 'grinding of forage increases the likelihood of passage' (van Soest, 1982, p. 223). In Fig. 2.2*b* most of the ruminants (D4) of the Turolian (8 Ma BP) were medium-heavy mammals (W3) with a remarkable number of species.

Summary

In Fig. 2.3, the different processes that led to a diversification in the range of food, types of digestion and body weight in mammals within less than 50 million years are compiled schematically. Increase in the number of species during different time periods and in different lines of development is represented by irregular stars and the letter 'N'; increase in body weight is denoted by a different type of irregular star and the letter 'W'. Physiological differentiations are represented by various simple geometrical shapes.

Zoophages that digest autoenzymatically and polyphages that are caecum-colon fermenters (large-intestinal fermenters) showed a considerable increase in the number of species. In the zoophages this quantitative growth took place before the Early Oligocene, i.e. before 35 Ma BP; the polyphagous large-intestinal fermenters became more numerous in the Early Miocene. What could be the reason for the quantitative increase in polyphagous caecum-colon fermenters? Many of them, besides eating food of animal origin, expanded their food range to seeds, which are of high caloric value compared to vegetative plant material (DLG, 1984). To reach these seeds, mammals very often have to eat fruit pulp, which is characterised by a high variability in nutrient content and often represents a poor nutrient source (Stiles, 1989). In other cases, the seeds will be in dry fruits and without pulp. Buds are also eaten by polyphages; they represent a plant material with slightly higher caloric value than ripe vegetative parts.

During the Oligocene, the climate in Europe became seasonal (Furon, 1972). This resulted in the spread of annual or biennial angiosperms with buried or near-ground buds that were easily accessible to small polyphages. According to Wing and Tiffney (1987), the general size of many fruits and seeds decreased under the influence of seasonality during the Oligocene. The conditions in the Oligocene induced a slow increase in the number of small polyphagous large-intestinal fermenters. In addition, the slightly increased humidity of the Early Miocene (Schwarzbach, 1974) was partly responsible for high productivity of both phytomass and insect life in the sub-tropical type of forest in Europe (Probst, 1986).

In the phytophagous mammals we find different types of modifying steps that produced a considerable morpho-functional variability in plant-digesting

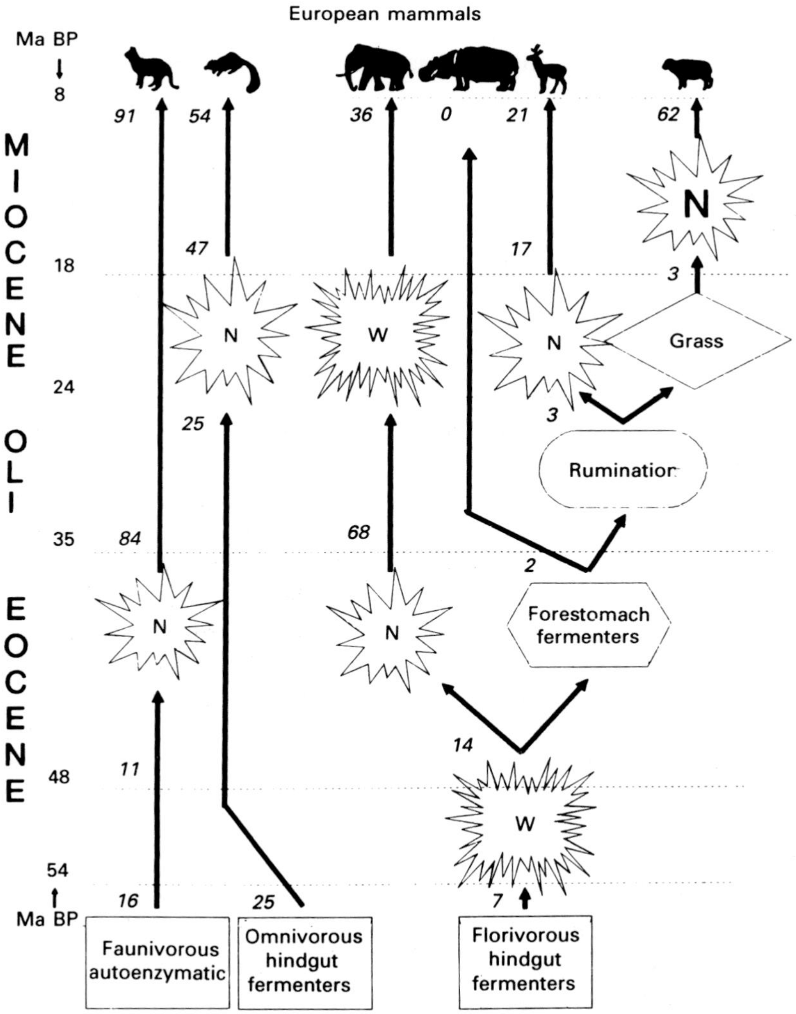

Fig. 2.3. Processes that diversified the range of food classes, the types of digestion and body weight in European fossil mammals. Data are from Savage and Russell (1983). Three types of differentiation, from which diversification started, are given at the foot of the diagram. 'Hindgut-fermenters' is used to indicate large intestine fermenters. Increase in species number is illustrated by irregular stars and the letter 'N'. Numbers of species for different morpho-functional differentiations are given in italics.

Increase in body weight is symbolised by a second type of star and the letter 'W'. Other types of differentiation are marked by different polygons and on top of the diagram six representative recent mammalian species are depicted.

From the original three groups in the early Eocene, six groups have developed by the end of the Miocene.

mammals: already, before the Cuisian (48 Ma BP), there was a considerable increase in heavy large-intestinal fermenters. In this group there was a considerable increase in the number of species before the Early Oligocene period (35 Ma BP), followed by a weight increase to very heavy mammals (Probscidea) before the Orleanian period (18 Ma BP). Phytophagous large-intestinal fermenters are further differentiated by the formation of a plurilocular (multichambered) forestomach (Langer, 1991) before the Early Oligocene. This type of differentiation is not combined with a considerable increase in species number, and the size oscillates around medium-heavy forms. Therefore, there are no symbols representing increase in species number or weight increase in the lines leading to the Upper Tertiary in Fig. 2.3.

Forestomach fermenters belonging to the Pecora (Cervoidea) very probably started rumination during the Oligocene (before 24 Ma BP). This functional differentiation was followed by a rise in the number of species in the Early Miocene. On the other hand, the differentiation of forestomach fermentation obviously 'opened a door' towards grazing. 'The first grasses were spreading globally in the late Cretaceous and were both widespread and abundant by Miocene times (20 million years B.P.)' (Williams, 1981, p. 3; Thomasson, 1987). A reasonable increase in species number followed soon after this adaptation to the ingestion of grass and before the Turolian (8 Ma BP). Grass eating can also be found in a few species of non-ruminating forestomach fermenters and in a few caecum-colon fermenters, but these differentiations did not cause development of high numbers of species.

References

Barnosky, A.D. (1991) The Late Pleistocene Event as a paradigm for widespread mammal extinction. In *Mass Extinctions: Processes and Evidence*. ed. S. K. Donovan, pp. 235–254. London: Belhaven Press.

Baumont, R. & Deswysen, A. G. (1991). Mélange et propulsion du contenu du réticulo-rumen. *Reproduction, Nutrition Dévelopement*, **31**, 335–359.

Bell, R. H. V. (1970). The use of the herb layer by grazing ungulates in the Serengeti. In *Animal Populations in Relation to their Food Resources*, ed. A. Watson, pp. 111–124. Oxford: Blackwell Scientific.

Bell, R. H. V. (1971). A grazing ecosystem in the Serengeti. *Scientific American*, **225**(1), 86–93.

Bon, R., Cugnasse, J. M., Dubray, D., Houard, P. G. T. & Rigaud, P. (1991). Le mouflon de Corse. *Revue d'Ecologie* (*Terre et Vie*), Suppl. 6, 67–110.

Carroll, R. L. (1988). *Vertebrate Paleontology and Evolution*. New York: W. H. Freeman.

Case, T. J. (1979). Optimal body size and an animal's diet. *Acta Biotheoretica*, **28**, 54–69.

Caughley, G. (1987). The distribution of eutherian body weights. *Oecologia* (Berlin), **74**, 319–320.

Caughley, G. & Krebs, C. J. (1983). Are big mammals simply little mammals writ large? *Oecologia* (Berlin), **59**, 7–17.
Clarke, R. T. J. (1977). The gut and its micro-organisms. In *Microbial Ecology of the Gut*. ed. R. T. J. Clarke & T. Bauchop, pp. 35–71. London: Academic Press.
Damuth, J. & MacFadden, B. J. (1990). Body size and its estimation. In *Body Size in Mammalian Paleobiology*. ed. J. Damuth & B. J. MacFadden, pp. 1–10. Cambridge: Cambridge University Press.
Demment, M. W. & van Soest, P. J. (1983). *Body Size, Digestive Capacity, and Feeding Strategies of Herbivores*. Morrilton, AR: Winrock International.
Demment, M. W. & van Soest, P. J. (1985). A nutritional explanation for body-size patterns of ruminant and nonruminant herbivores. *American Naturalist*, **125**, 641–672.
DLG (1984). *DLG-Futterwerttabellen fur Schweine*, 5. Aufl. Frankfurt: DLG-Verlag.
Drochner, W. & Meyer, H. (1991). Verdauung organischer Substanzen im Dickdarm verschiedener Haustierarten. *Fortschritte der Tierphysiologie und Tierernährung*, **22**, 18–40.
Economos, A. C. (1981). The largest land mammal. *Journal of Theoretical Biology*, **89**, 211–215.
Edwards, W. E. (1967). The Late-Pleistocene extinction and diminution in size of many mammalian species. In *Pleistocene Extinctions. The Search for a Cause*. ed. P. S. Martin & H. E. Wright, pp. 141–154. New Haven: Yale University Press.
Eisenberg, J. F. (1981). *The Mammalian Radiations. An Analysis of Trends in Evolution, Adaptation, and Behavior*. Chicago: University of Chicago Press.
Flindt, R. (1985). *Biologie in Zahlen*. Stuttgart: Gustav Fischer Verlag.
Fonty, G. & Joblin, K. N. (1991). Rumen anaerobic fungi: Their role and interactions with other rumen microorganisms in relation to fiber digestion. In *Physiological Aspects of Digestion and Metabolism in Ruminants*. ed. T. Tsuda, Y. Sasaki & R. Kawashima, pp. 655–680. San Diego: Academic Press.
Furon, R. (1972). *Eléments de Paléoclimatologie*. Paris: Librairie Vuibert.
Gerard, J. F., Teillaud, P., Spitz, F., Mauget, R. & Campan, R. (1991). Le sanglier. *Revue d'Ecologie (Terre et Vie)*, Suppl. 6, 11–66.
Heinrich, D. (1991) *Untersuchungen an Skelettresten wildlebender Säugetiere aus dem mittelalterlichen Schleswig. Ausgrabung Schild 1971–1975*. Neumünster: Karl Wachholtz Verlag.
Hill, H. (1952). Die Motorik des Verdauungskanals bei den Equiden mit besonderer Berücksichtigung des Röntgenbildes. *Archiv für Tierernährung*, **3**, 1–78.
Hofmann, R. R. (1988). Anatomy of the gastro-intestinal tract. In *The Ruminant Animal. Digestive Physiology and Nutrition*. ed. D. C. Church, pp. 14–43. Englewood Cliffs, NJ: Prentice Hall.
Hungate, R. E. (1988). The ruminant and the rumen. In *The Rumen Microbial Ecosystem*. ed. P. N. Hobson, pp. 1–19. London: Elsevier Applied Science.
Janis, C. M. (1984). The use of fossil ungulate communities as indicators of climate and environment. In *Fossils and Climate*. ed. P. Brenchley, pp. 85–104. Chichester: John Wiley.
Jarman, P. J. (1974). The social organisation of antelope in relation to their ecology. *Behaviour*, **48**, 215–267.
Jones, S. B. & Luchsinger, A. E. (1987). *Plant Systematics*, 2nd edn. New York: McGraw-Hill.

Langer, P. (1975). Macroscopic anatomy of the stomach of the Hippopotamidae GRAY, 1821. *Zentralblatt für Veterinärmedizin, Anatomia, Histologia, Embryologia*, **C4**, 334–359.
Langer, P. (1976). Functional anatomy of the stomach of *Hippopotamus amphibius* L. 1758. *South African Journal of Science*, **72**, 12–16.
Langer, P. (1984). Comparative anatomy of the stomach in mammalian herbivores. *Quarterly Journal of Experimental Physiology*, **69**, 615–625.
Langer, P. (1987). Evolutionary patterns of Perissodactyla and Artiodactyla (Mammalia) with different types of digestion. *Zeitschrift für Zoologische Systematik und Evolutionsforschung*, **25**, 212–236.
Langer, P. (1988). *The Mammalian Herbivore Stomach. Comparative Anatomy, Function and Evolution*. Stuttgart: Gustav Fischer Verlag.
Langer, P. (1991). Evolution of the digestive tract in mammals. *Verhandlungen der Deutschen Zoologischen Gesellschaft*, **84**, 169–193.
Langer, P., Dellow, D. W. & Hume, I. D. (1980). Stomach structure and function in three species of macropodine marsupials. *Australian Journal of Zoology*, **28**, 1–18.
Langer, P. & Snipes, R. L. (1991). Adaptations of gut structure to function in herbivores. In *Physiological Aspects of Digestion and Metabolism in Ruminants*. ed. T. Tsuda, Y. Sasaki & R. Kawashima, pp. 349–384. San Diego: Academic Press.
Martin, R. A. (1990). Estimating body mass and correlated variables in extinct mammals: travels in the fourth dimension. In *Body Size in Mammalian Paleobiology*. ed. J. Damuth & B. J. MacFadden, pp. 49–68. Cambridge: Cambridge University Press.
Mason, V. C. (1983). Microbial digestion in the hind-gut of the pig. In *Digestion and Absorption of Nutrients*. ed. H. Bickel, & Y. Schutz, pp. 27–38. Bern: Hans Huber.
Maublanc, M. L., Cibien, C., Gaillard, J. M., Maizeret, C., Bideau, E. & Vincent, J. P. (1991). Le chevreuil. *Revue d'Ecologie (Terre et Vie)*, Suppl. 6, 155–183.
Mitvedt, T. (1982). What are the main functions of a normal gut flora? *Fortschritte der Veterinärmedizin*, **33**, 39–40.
Owen-Smith, N. (1988). *Megaherbivores. The Influence of very large Body Size on Ecology*. Cambridge: Cambridge University Press.
Pearce, G. R. & Moir, R. J. (1964). Rumination in sheep I. The influence of rumination and grinding upon passage and digestion. *Australian Journal of Agricultural Research*, **15**, 635–644.
Pépin, D., Gonzalez, G. & Bon, R. (1991). Le chamois et l'isard. *Revue d'Ecologie (Terre et Vie)*, Suppl. 6, 111–153.
Peters, R. H. (1986). *The Ecological Implications of Body Size*. Cambridge: Cambridge University Press.
Prins, R. A. & Kreulen, D. A. (1990). Comparative aspects of plant cell wall digestion in mammals. In *The Rumen Ecosystem*. ed. S. Hoshino, R. Onodera, H. Minato & H. Itabashi, pp. 109–120. Tokyo: Japan Scientific Societies Press.
Probst, E. (1986). *Deutschland in der Urzeit. Von den Entstehung des Lebens bis zum Ende der Eiszeit.* München: Bertelsmann.
Romer, A. S. (1966). *Vertebrate paleontology*, 3rd edn. Chicago: University of Chicago Press.
Savage, D. E. & Russell, D. E. (1983). *Mammalian Paleofaunas of the World.* Reading, MA: Addison-Wesley.
Schwarzbach, M. (1974). *Das Klima der Vorzeit.* Stuttgart: Ferdinand Enke Verlag.

Shirley, R. L. (1986). *Nitrogen and Energy Nutrition of Ruminants*. Orlando: Academic Press.

Snaydon, R. W. (1981). The ecology of grazed pastures. In *Grazing Animals*. ed. F. H. W. Morley, pp. 13–31. Amsterdam: Elsevier Scientific.

Stiles, E. W. (1989). Fruits, seeds, and dispersal agents. In *Plant–Animal Interactions*. ed. W. G. Abrahamson, pp. 87–122. New York: McGraw-Hill.

Thomasson, J. R. (1987). Fossil grasses: 1820–1986 and beyond. In *Grass Systematics and Evolution*. ed. T. R. Soderstrom, K. W. Hilu, C. S. Campbell & M. E. Backworth, pp. 159–167. Washington, DC: Smithsonian Institution Press.

van Soest, P. J. (1982). *Nutritional Ecology of the Ruminant*. Corvallis, OR: O. & B. Books.

Werner, C. F. (1961). *Wortelemente lateinisch-griechischer Fachausdrücke in den biologischen Wissenschaften*, 2 Aufl. Leipzig: Akademische Verlagsgesellschaft Geest und Portig.

Williams, O. B. (1981). Evolution of grazing systems. In *Grazing Animals*. ed. F. H. W. Morley, pp. 1–12. Amsterdam: Elsevier Scientific.

Wing, S. L. & Tiffney, B. H. (1987). Interactions of angiosperms and herbivorous tetrapods through time. In *The Origins of Angiosperms and their Biological Consequences*. eds. E. M. Friis, W. G. Chaloner, & P. R. Crane, pp. 203–224. Cambridge: Cambridge University Press.

3

Modelling gut function: an introduction

CARLOS MARTINEZ DEL RIO,
STEVEN J. CORK and
WILLIAM H. KARASOV

A cautious man should above all be on his guard against resemblances: they are a very slippery sort of thing.
Plato (in Gordon *et al.* 1972)

Digestive physiologists either focus narrowly, use a reductionist approach and work at the cellular, biochemical and molecular levels, or analyse digestive performance (measured usually as digestive efficiency and retention time of food in the gut) at the whole organism level. Consequently digestive physiology encompasses two more or less exclusive bodies of data: on one hand we have adequate knowledge of the molecular and biochemical characteristics of digestive enzymes and nutrient transport systems (Desnuelle *et al.*, 1986); on the other we have a large catalogue of retention times, assimilation efficiencies and digestive morphologies for a variety of animal taxa fed on an assortment of food types (e.g. Chivers and Hladik, 1980; Warner, 1981; Karasov, 1990). A consequence of the slightly schizophrenic nature of the trade is that we know very little about how the fine details of nutrient digestion and uptake function are integrated to affect whole organism digestive efficiency and food intake rates. Until very recently, digestive physiology has lacked a theoretical framework that integrates digestive processes with gut morphology and the chemical properties of food. Because nutrient assimilation is a remarkably complex phenomenon involving a variety of enzymatic and transport pathways that take place in a variety of organs within the gastro-intestinal tract, it is perhaps not surprising that attempts to analyse digestion from an integrated perspective have been few (Sibly, 1981).

Mertens and Ely (1982) and Illius and Gordon (1992), for example, have developed detailed mathematical models that integrate intake, digestion and food characteristics for ungulates. These models emphasize realism and precision; they are complex, rich in details, rely on numerical simulation and apply very specifically to fermenting herbivores. They are 'tactical' approaches to the modelling of digestion (*sensu* Levins, 1966) and have limited generality. Here we will describe a different approach to modelling digestive function. The models that we will describe sacrifice precision to realism and generality and can be labelled 'strategical' (Levins, 1966). The models described here provide primarily qualitative prediction and are very flexible, thus allowing broad comparisons. Alexander's model presented in this volume is a good example of strategical modelling of gut morphology and function.

The approach described here has been advocated by Penry and Jumars (1986, 1987). They argued that modelling the digestive process is analogous to modelling the behaviour of chemical reactors. The question faced by the digestive physiologist is to discover how various gut morphologies and digestive reactions may maximize an animal's rate of energy and nutrient gain given a distribution of food chemical compositions and animal energetic costs as boundary conditions. This challenge is very similar to the task of the chemical engineer who has to evaluate the performance of reactors with different designs (gut functional morphologies) with the goal of maximizing yield or yield rate (energy or nutrient assimilation), and is given a series of chemical reactions (digestive hydrolysis and nutrient uptake) taking place in these reactors.

Penry and Jumars' proposal follows the fruitful tradition in comparative physiology of establishing technological analogies to physiological processes. Distinguished examples are blood vessels and renal tubules as countercurrent exchangers and multipliers (Scholander and Krog, 1957; Kokko and Tisher, 1976) and Scholander and Irving's model of energy exchange based on Newton's cooling law (originally developed for industrial furnaces, Scholander *et al.*, 1950). Technological analogies show how something complex and poorly understood (e.g. gut function) resembles something that is understood (e.g. human-made reactors) and are therefore extremely useful as explanatory and clarifying tools in physiology (Hardison, 1989). Biologists can borrow the mathematical and conceptual tools of engineers to understand how animals function and why they have particular morphologies (Alexander, 1982).

Before reactor theory

The chemical reactor approach shares many features with other previous modelling efforts for digestive function and the borders between reactor-based and other more 'tactical' models are blurred. Non-reactor models fall into two broad categories: compartmental and physiological (Usry *et al.*,

1991). Physiological models start with knowledge of physiological processes, such as the kinetics of enzymatic reactions and bacterial fermentation or the electrophysiology of muscular contraction in the gut wall, and build complex mechanistic models that allow precise and detailed predictions of changes in gut function under various physiological conditions (Mertens and Ely, 1982; Usry *et al.*, 1991; Illius and Gordon, 1992).

Numerous compartmental models have been developed to describe results from marker-passage studies and have been in common use and evolving since the 1950s (e.g. Balch, 1950; Brandt and Thacker, 1958; Grovum and Phillips, 1973; Faichney, 1975; Pond *et al.*, 1989). These models attempt to provide mathematical descriptions of marker excretion curves that can be related to biologically realistic processes. The assumption underlying most of these models is that the gut consists of one or more mixing compartments in series linked by tubes.

Compartment models make predictions about the general shapes of marker output distribution in faeces that are identical to those of reactor-based models (see p. 35). The main difference between compartment and reactor models is in the interpretation of these curves. Compartmental modelling deals with tubular (non-mixing) sections of the gut as 'time delays' (Blaxter *et al.*, 1956), whereas reactor-based models consider the tubular portions of the gut and recognize them as functional units of digestion. The treatment of tubular portions of the gut as 'delays' probably arose because compartmental modelling was first employed in studies of domesticated herbivores in which fermentative digestion in mixing compartments was of prime interest and other forms of digestion taking place in non-mixing compartments was of secondary interest.

The application of reactor theory to digestive processes has caused us to recognize two things. First, that it is possible (and extremely interesting) to view the tubular portions of the gut not just as connections between the 'important bits' but as reaction vessels in their own right where important biological processes take place; second, that considering animal guts as combinations of different kinds of 'reaction vessels' (tubular, mixing, batch) opens up many important questions. What are the relative functions of the different types of digestive 'vessels'? How do the reactions occurring in these vessels interact? What combination of these types of vessel is optimal under different physiological and ecological conditions? (See Alexander, Ch. 4.) Reactor theory has broadened our perspective and has allowed us to appreciate the significance not only of guts dominated by mixing compartments but also of guts with widely ranging proportions of mixing to non-mixing compartments. Because reactor theory is general and relatively flexible, it permits the study of the guts of animals with widely diverging diets and digestive modes under a common paradigm. To date, reactor theory has been used to interpret the digestive processes of marine invertebrates (Penry and Jumars, 1990; Plante *et al.*, 1990), herbivorous mammals (Hume, 1989), nectar- and fruit-eating birds (Martínez del Rio and Karasov, 1990), insectivorous birds (Dykstra and Karasov, 1992), and marine herbivorous fishes (Horn and Messer, 1992).

Building mathematical models of the digestive process using a reactor engineering approach relies on two steps: (a) establishing explicit analogies between different kinds of gut morphologies and different types of reactor designs and (b) using physiological measurements (e.g. affinities and maximal rates of enzymes and transporters) as parameters of performance equations to predict reactor configurations and/or digestive behaviours that maximize nutrient and/or energy uptake rates (Hume, 1989; Dade *et al.*, 1990). We tackle these two steps in turn. First, we describe animal guts in terms of chemical reactors and describe some elementary principles in reactor design that allow evaluation of the performance of different gut configurations. Then, we provide a methodology that uses digesta flow patterns and output distributions of inert markers to identify gut functional morphologies with reactors of different configurations. Finally, we construct and analyse a mathematical model of digestive function for nectar- and fruit-eating animals. This last section highlights two aspects of 'guts-as-reactors' modelling that we find especially important: (a) modelling guts as chemical reactors requires making, and hence eventually validating, explicit assumptions about the digestive process and (b) reactor theory allows making crisp falsifiable predictions about whole organism digestive outcomes from measurements typically done *in vitro* at the organ and tissue level (Martínez del Rio and Karasov, 1990).

Guts as reactors: morphological analogies

Chemical engineers recognize three basic types of reactors: batch reactors, continuous-flow stirred tank reactors (CSTR) and plug-flow reactors (PFR). Reactors of complex configurations can be constructed by setting two or more reactors in series or parallel (Levenspiel, 1972). Modelling guts as reactors is facilitated by establishing an analogy between a digestive organ and one or several of these reactors.

Kinds of reactors

Batch reactors process reactants in discrete portions. Reagents are introduced into the chamber and, after a given reaction time, reaction products and the unreacted fraction are retrieved. In batch reactors, reaction periods alternate with 'down time' periods during which the reaction is interrupted to empty and fill the reactor. An important feature of batch reactors is that the concentration of reactants changes over time. Figure. 3.1*a* illustrates the changes in concentration of reagents and products as a function of time in a typical batch reactor in which a simple catalytic reaction takes place. The blind coelenteron of cnidarians and the enteron of turbellarians probably function like batch reactors (Yonge, 1937). The blind compartments of the vertebrate gut that fill and empty at discrete intervals probably also function as batch reactors (e.g. the fundic caecum of vampire bats, the caecae of herbivorous birds and the pyloric stomach of fish; Forman, 1972; Duke, 1989; Kapoor *et al.*, 1975).

Continuous-flow stirred tank reactors (CSTR) are usually large tanks into which reagents flow and out of which products and unreacted materials flow continuously. Reagent concentration is diluted immediately upon entry into the reaction chamber. Because flow is continuous and the contents of the reactor are well mixed, both the concentration of reagents and products in the reaction chamber and the reaction rate remain constant. Figure 3.1*b* illustrates the influence of flow rate on the efficiency and rate of product formation in a CSTR in which a simple catalytic reaction takes place. At high flow rates, the rate of conversion (amount of product formed per unit time) is high, but the efficiency of conversion (fraction of reactant transformed into product) is low. CSTR-like conditions probably occur in the camelid forestomach, the sacchiform region of the kangaroo forestomach and the reticulorumen of ruminants (Hume, 1989).

Plug-flow reactors (PFR) are tubular reaction vessels through which materials flow continuously and in an orderly fashion. In PFRs there is little or no mixing along the length of the reactor, and consequently reaction levels and reagent concentrations are high at the reactor's entrance and decline along the length of the tube. In a PFR functioning at steady state, the gradient in the concentration of reagents and products remains constant along the length of the reactor. Figure 3.1*c* portrays the influence of flow rate on the efficiency and rate of product formation in a PFR in which a simple catalytic reaction takes place. The simple tubular guts of many deposit feeders and the small intestines of many vertebrates seem to be analogous to PFRs.

With few exceptions (e.g. euphonias and mistletoe birds that have simple tubular guts, see Martínez del Rio, Ch. 8), most vertebrate digestive tracts are complex and contain several elements that may function like one of the chemical reactors identified above. Dellow *et al.* (1983), for example, treated the tubiform stomach of the wallabies *Thylogale thetis* and *Macropus eugenii* as a series of mixing chambers (formally equivalent to CSTRs). Penry and

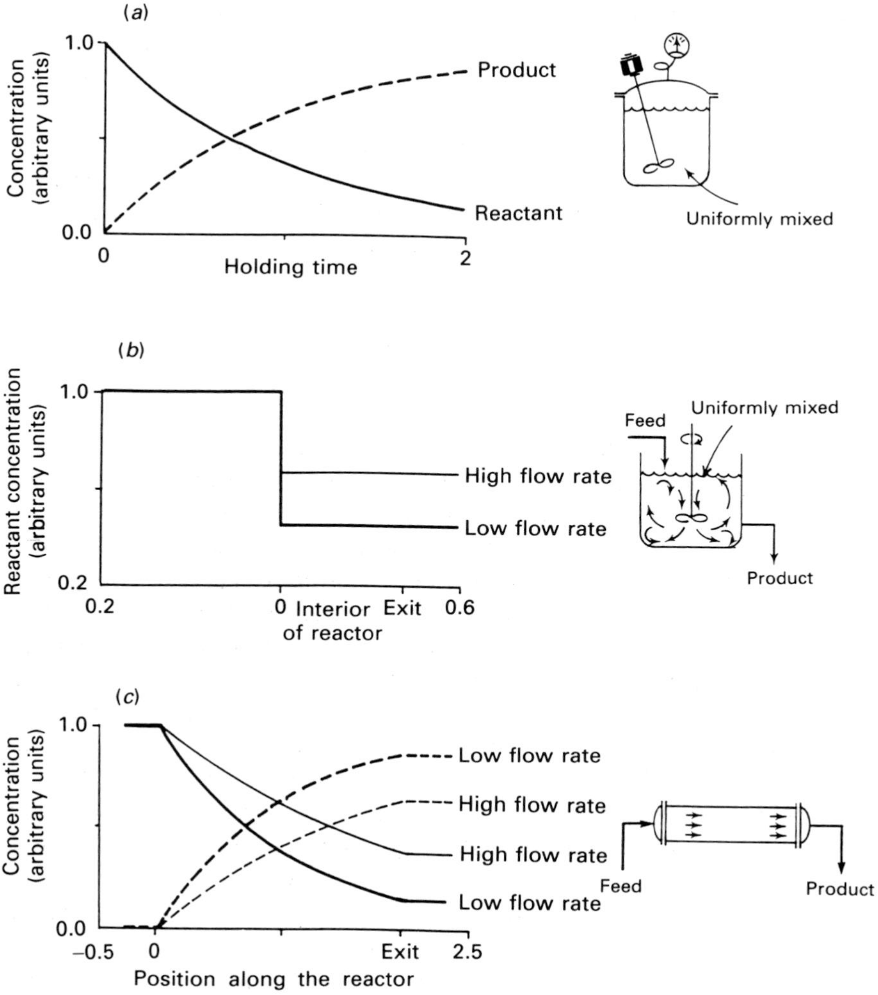

Fig. 3.1. Three types of chemical reactors analogous to gastro-intestinal organs. (*a*) In a batch reactor, reactants and reagents are mixed in a container. The concentration of reactants and products changes as reaction time progresses. (*b*) In continuous stirred tank reactors (CSTR), reactants and reagents flow in and out constantly maintaining both their concentration and the rate of reaction constant. Reaction rate and product output rate are functions of flow rate through the reactor. (*c*) Plug-flow reactors (PFR) are characterized by continuous constant flow in and out of the reactor and a steady state gradient in the concentration of reagents, reactants, products and reaction rates along the length of the reactor.

Jumars (1987) have argued that digestion in ruminants involves a fermentation CSTR vat (the ruminoreticulum) in series with a posterior PFR (the small intestine). In caecum fermenters, the PFR (small intestine) is followed in series by a modified CSTR (the caecum, Hume, 1989).

Clearly the three 'ideal' reactor types described above represent ends in a continuum. A long series of CSTRs behaves like a PFR, and the behaviour of a PFR with intense recycling resembles that of a CSTR (Levenspiel, 1972). The use of each reactor type as an analogy for different kinds of digestive organs depends on how closely the flow and kinetic behaviours of these structures resemble those of ideal reactors or reactor configurations.

Performance of different kinds of reactor

A sufficient description of the assimilation process requires two elements: we need to know the kinetics of enzymatic, microbial and transport processes, and we need a mass balance equation that relates how these result in changes in nutrient and product composition in the elements of the gut. The kinetics of digestion, or at least *in vitro* approximations to them, are relatively well studied. Although we do not have adequate mass balance equations for animal guts, considering the performance of ideal reactors for which the mass balance equations are known can lead to several intriguing insights (Penry and Jumars, 1987). The discussion of reactor performance presented here is very simplified. We include only the few mathematical details necessary to justify some of the results of reactor performance analysis that we believe are interesting for the digestive physiologist. Penry and Jumars (1987) and Horn and Messer (1992) provide analytical derivations for mass-balance equations for the ideal reactor types described above and interested readers can consult any chemical engineering textbook for a formal exposition of reactor performance analysis (e.g. Aris, 1970).

Performance is easily calculated in batch reactors where reactants are initially loaded, allowed to react, and then the mixture of reactants and products is removed. A measure of performance in a batch reactor is the holding time needed to achieve a given conversion, where conversion is the fraction of original reactant that is transformed into product. For the sake of simplicity we will assume that the reaction of interest involves the transformation of a reactant A into product P. Figure 3.2 shows the time required to achieve a given conversion assuming that the conversion from A to P occurs through first-order kinetics. Note that in a batch reactor in addition to holding time we have to consider the 'down time' required to empty and fill the reactor in each reaction cycle.

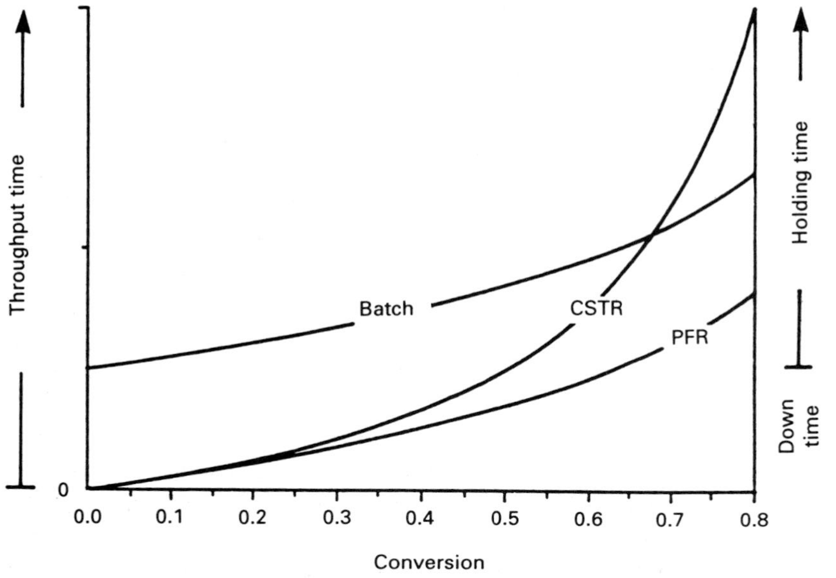

Fig. 3.2. Comparison of performance among the three chemical reactor types. Performance is estimated by the throughput or holding time required to obtain a given conversion of reactants into products assuming a first-order chemical reaction (i.e. the rate of change in reactant disappearance is proportional to reactant concentration). For any given conversion, PFR reactors require shorter times than either batch or CSTR reactors.

A slightly more complicated situation arises when one tries to estimate the performance of a PFR where materials flow in and out of the vessel continually. In a PFR, conversion is the fraction of reactant that is transformed into product measured at the exit of the reactor. An estimate of performance that allows comparison of PFRs with batch reactors is the throughput time needed to achieve a given conversion. Throughput time is the time needed to completely turnover the contents of a continuously flowing reactor and can be calculated as the ratio of the reactor volume and its flow rate. At high flow rates, throughput time is short and conversion is low. As shown in Fig. 3.2, the relationship between conversion and throughput time in a PFR is identical to the relationship between conversion and holding time in a batch reactor of equal volume. PFRs perform better (i.e. achieve the same conversion in less time) than batch reactors because they function continuously and thus no down time has to be added to the throughput time required to process a volume of reactor.

As in PFRs, we can estimate the performance of CSTRs as the throughput time needed to achieve a given conversion. If reaction rates increase with reactant concentration (i.e. if the order of reaction > 0) then a PFR always attains the same conversion in a shorter throughput time than an equivalent CSTR (Fig. 3.2). The reason is that in a CSTR the reactant is diluted to a lower constant concentration immediately upon entry into the reaction chamber (Fig. 3.1*b*). In a PFR, reactant concentrations, and hence reaction rates, decrease from the entrance to the end of the reactor but they are always higher than the concentration in a CSTR with equal conversion. The superiority of PFRs over CSTRs is accentuated at high conversions (Fig. 3.2). An additional potential advantage of PFRs is that they are by definition tubular and therefore have higher area/volume ratios than CSTRs. High area/volume ratios are preferable for reactions occurring at the surface of the reactor (as is the case in many digestive hydrolyses, Alpers, 1987).

The preceding argument suggests that, for most digestive reactions, animal guts should function as PFRs. Then why do we often find digestive structures that appear to work as CSTRs (e.g. the ruminoreticulars, some caecae)? We have dealt only with catalytic reactions where reaction rates are monotonically increasing functions of reactant concentration. This kinetic condition is not valid in general. If the digestive reaction is of order zero (rate is independent of concentration) for example, the performances of CSTRs and PFRs are identical. We will now suggest that for a particular class of chemical reactions (autocatalytic reactions) CSTRs may have superior performance to PFRs.

In autocatalytic reactions, the rate of reaction depends on both the concentration of the reactant *and the concentration of the product*. At the start of autocatalytic reactions, rates are low because little product is present; rates increase to a maximum as product is formed and then drop again to a low value as reactant is consumed (see Fig. 3.3). At low conversions, when reactant concentrations are to the right of the hump of the reaction curve, CSTRs perform better than PFRs. The reason is that the average concentration (and hence the average reaction rate) at the PFR is lower than at the CSTR. At low conversions, the rate of reaction in a PFR is highest at the exit of the reactor and lowest at the entrance (Fig. 3.3). In a CSTR with exactly the same conversion, the rate of reaction throughout the reactor is the same as the rate at the exit of the PFR. As conversion increases, the advantages of a CSTR become smaller and eventually at high conversion rates the PFR becomes superior. In conclusion, for catalytic reactions, PFRs show higher performance than both CSTRs and batch reactors. For autocatalytic reactions, however, a CSTR is superior to a PFR provided that reactant conversion is low.

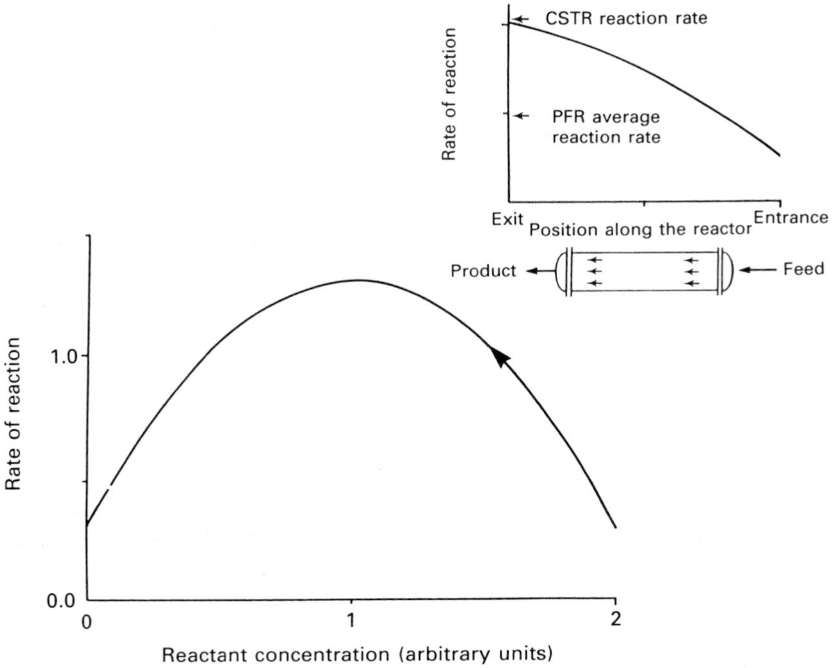

Fig. 3.3. Autocatalytic reactions are characterized by 'humped' rate of reaction versus reactant concentration curves. If the rate of conversion is low (i.e. the reactant concentration is to the right of the hump) CSTRs perform better than PFRs. The reason is that the concentration of reactant can be kept at intermediate levels in the CSTR and thus reaction rates can be maximized. In a PFR, reaction rates are low at the entrance of the reactor and increase towards the end giving lower average reaction rates.

The digestive process includes both autocatalytic and catalytic reactions occurring concurrently and/or in series. The optimal reactor design for complex reactions often contains more than one reactor or requires a single reactor with varying degrees of recirculation. Place (1990) has argued that the guts of seabirds function as plug-flow reactors with recycle, and that this design probably allows efficient assimilation of dietary lipids. Penry and Jumars (1987) believe that the optimal design for animals relying on autocatalytic microbial fermentation requires that a portion of the gut is a CSTR. Reactor theory highlights the significance of digestive kinetics for optimal gut design. Most modelling efforts for the digestive process assume that all digestive reactions are first order. Incorporating complex kinetics into models reduces mathematical tractability but may lead to added realism and increased understanding (see Dade *et al.*, 1990).

'Diagnosing' reactor (gut) function with input–output experiments

The geometry of some gastro-intestinal structures immediately suggests an analogy with one of the three main types of ideal reactors. How can these analogies be validated? The behaviour of human-made reactors often departs from that expected of ideal reactors of similar design. Chemical engineers use a variety of input–output experiments to explore the behaviour and misbehaviour of reactors. These experiments rely on introducing an inert (non-reactive) marker into the reactor and then analysing its output distribution (Levenspiel and Turner, 1970). These experiments are exactly analogous to the experiments used by digestive physiologists to estimate the temporal parameters that characterize digestive function (i.e. transit time, turnover time, turnover rate and retention time; Warner, 1981). In this section we first describe the marker retention time distributions of ideal reactors, then we describe output distribution of a reactor consisting of a CSTR and a PFR in series and use this distribution to clarify the meaning of several terms that are very commonly used (and misused) in digestive physiology (transit time, throughput time, turnover time, turnover rate and mean retention time).

Retention time distributions

To characterize the flow of materials in a reactor, we need to know how long individual particles or molecules stay in the vessel or, more precisely, the distribution of residence or **retention times** of the flowing materials (we will use the terms 'residence' and retention time interchangeably). This information can be determined by stimulus–response experiments. The stimulus is a tracer input into the fluid entering the reactor. Ideally, the tracer travels through the reactor in the same fashion as the component of the feed whose flow pattern we are interested in tracking. More than one marker can be used simultaneously to trace the fate of different digesta components. A wide range of markers have been used in animal studies (Kotb and Luckey, 1972; Udén *et al.*, 1980) and there has been much discussion about the relationship of the behaviour of various markers to the flow pattern of the various components of digesta (Ellis and Huston, 1967).

The response is a time record of the tracer as it leaves the vessel. The two simplest possible inputs of tracers are pulse signals and step signals (Fig. 3.4). Pulse inputs consist of a certain amount of tracer placed instantaneously at the entrance of the vessel and are equivalent to the 'single dosage' experiments of digestive physiologists (Faichney, 1975). The time course of tracer output after a pulse can be analysed in two ways. We can plot the concentration of

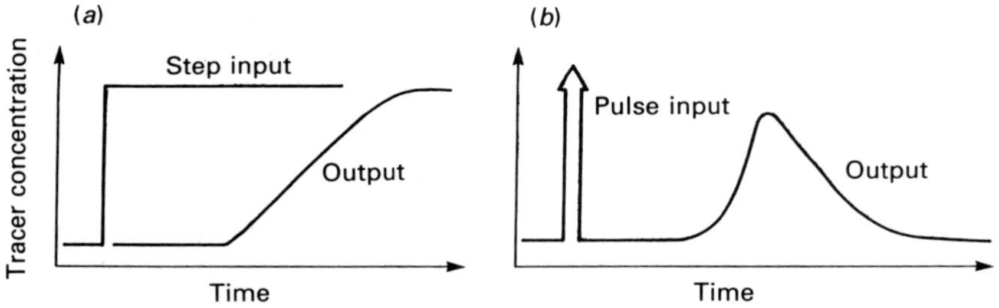

Fig. 3.4. Reactor geometry can be analysed by input–output experiments in which an inert marker is introduced to the reactor and the output signal is analysed. The two most commonly used types of input–output experiment analyse the output of marker after either (*a*) a step input or (*b*) a pulse input of marker.

tracer in the outflow of the reactor (**C** curves), or we can plot the fraction of the total amount of tracer that came out in each small time interval (**E** curves) (Levenspiel, 1972). In closed reactors (i.e. those in which fluid enters and leaves the reactor by plug-flow and is uncontaminated), **C** curves can be easily transformed into **E** curves by dividing each concentration by the area under the **C** curve. If the outflow of the reactor does not satisfy these conditions (i.e. if outflow is turbulent or is contaminated), outflow samples may not be representative and hence concentration values cannot be transformed into the fraction of total marker defecated. In animals, such as birds and reptiles, where urine and faecal materials are mixed in the cloaca, for example, the concentration curve cannot be directly transformed into a frequency distribution of retention times. The mean residence time of a particle can be calculated from residence time distribution curves from the formula

$$\text{mean residence time} = \int_0^\infty tE(t)\mathrm{d}t = \sum_{i=1}^{\infty} t_i E_i\ (t_i)$$

where E is the density function for residence time and $E_i(t_i)$ is the discrete form of this density function. Warner (1981) provides alternative forms to these general formulae that can be applied to a variety of experimental situations.

Step input experiments are equivalent to the 'continuous infusion' experiments of digestive physiologists and rely on suddenly introducing fluid containing a given concentration of marker into the reactor and maintaining this input of constant concentration (Fig. 3.4). In closed reactors (defined above), the time record of tracer concentration in the exit stream of the vessel is called the **F** curve. The information provided by pulse or step input experiments is identical. In closed reactors, the **F** curve differs from the *cumulative* distribution of residence time distributions by a constant. Physiologists have been using and interpreting retention time distributions (equivalent to **C** and **E** curves) and cumulative marker output distributions (equivalent to **F** curves) for years (Warner, 1981). Here we emphasize again that concentration curves (**F** and/or **C** curves) can only be used to estimate residence time distributions, and hence mean retention time, in animals with guts in which there is no mixing of faecal and urinary output. For animals in which mixing occurs in the cloaca, the concentration curve is not equivalent to the residence time distribution (**E** curve). The retention time distribution has to be obtained from complete collection of excreta. The usefulness of continuous infusion experiments to determine retention time distribution in animals with cloacal mixing is limited.

Ideal PFRs and CSTRs exhibit characteristic **E** and **F** curves. In ideal PFRs there is nil axial mixing and, hence, a pulse input of marker yields a pulse output of marker after a period given by the time required for the marker to travel throughout the reactor (Fig. 3.5*b*). A step input yields a step output (Fig. 3.5*a*). The time between marker input and the first appearance of marker at the end of the reactor is the time that a particle stays in the reactor and an estimate of the time required for the complete turnover of one volume of feed. More precisely, in reactors with a single homogeneous phase, the mean retention time of a particle is equal to the ratio of the volume of the reactor and the flow rate. In PFR reactors with modest amounts of longitudinal mixing the residence time distribution after a pulse is leptokurtic (tall and skinny) and symmetrical (Firmer and Cutler, 1988). Increased mixing, and/or 'dead spaces' (regions of stagnant material), flatten the distribution and tend to make it asymmetrical to the right (Levenspiel and Smith, 1957). Indeed, the output distribution of a PFR with large amounts of longitudinal mixing is exactly equivalent to that of a CSTR (described below).

In ideal CSTRs, a pulse of marker is immediately diluted as it penetrates the reactor and continues to be diluted at a constant fractional rate as new unlabelled material enters the reactor (Fig. 3.1). Consequently, the output residence time distribution is a negative exponential (Fig. 3.5*d*). If the reactor has a single homogeneous phase, the fractional rate of dilution equals the ratio of flow rate and reactor volume; mean retention time of a marker particle equals the reciprocal of this fractional rate. The **F** curve increases in a decelerating fashion and tends asymptotically to the concentration of marker in the input fluid (Fig. 3.5*c*). When many CSTRs are placed in series, the

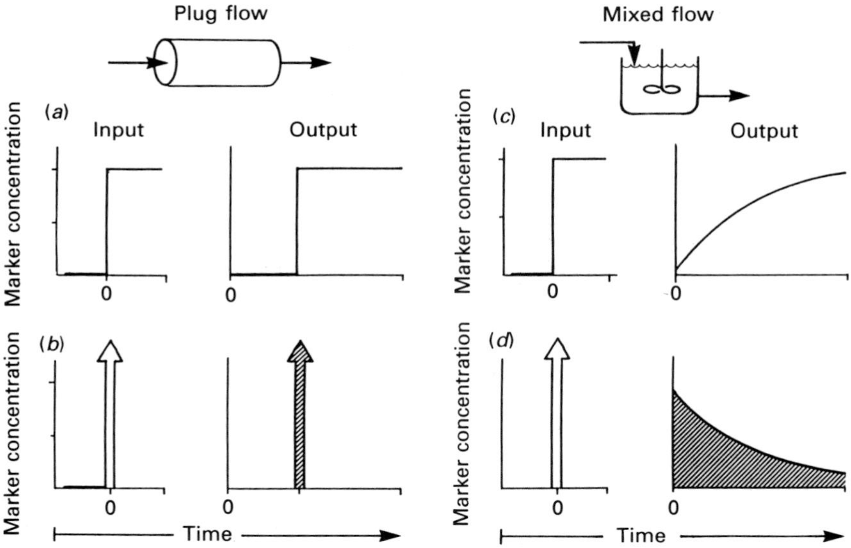

Fig. 3.5. Ideal CSTR and PFR reactors have characteristic ouput distribution of markers. (*a*) PFRs show step function outputs if the marker is introduced as a step signal. (*b*) If the marker is introduced as a pulse, the output signal is a pulse. The time required for the pulse to appear at the end of the reactor is equal to the throughput time of the PFR. CSTRs are characterized by (*c*) coexponential output curves after a step input of marker and (*d*) negative exponential output curves after pulse inputs.

retention time distribution of the series resembles that of a PFR with longitudinal mixing. Indeed, a series containing a very large number of very small CSTRs has a retention time distribution that approaches that of a PFR.

Figure 3.6 shows the residence time distribution for a reactor consisting of a CSTR in series with a PFR (analogous to a stomach and a small intestine). The **E** curve is a negative exponential shifted to the right. This curve can be used to illustrate an important property of reactors in series and to clarify some commonly misused terms. First, under fairly non-restrictive conditions, the marker mean retention time of an output of several reactors in series equals the sum of the means of the distributions of the individual reactors (Levenspiel, 1972). In the reactor illustrated in Fig. 3.6, mean retention time equals the sum of the mean retention in the PFR and the CSTR.

Physiologists use a variety of terms derived from the retention time distribution of markers to estimate the time that food particles remain inside digestive structures. Many of these terms are used interchangeably to mean several possible things (e.g. transit time, passage rate). Although the likelihood that a unified terminology will be adopted is slim (digestive physiologists are a stubborn lot), here we use Fig. 3.6 to develop a clearly defined set of terms

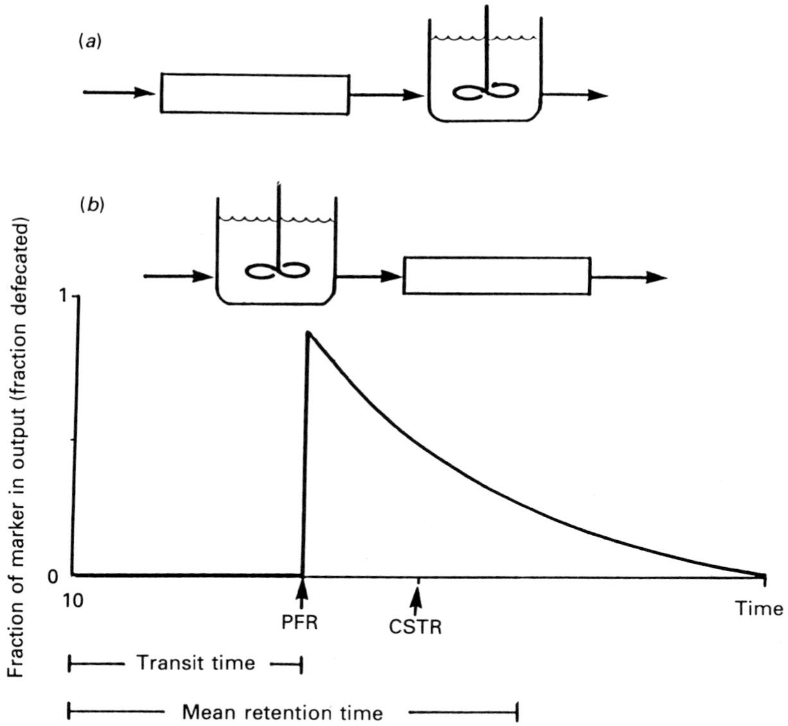

Fig. 3.6. Distribution of marker residence time for an ideal reactor consisting of a PFR and CSTR in series. Note that the order of the reactors does not affect the form of the retention time distribution. The time at which the marker first appears (transit time) equals the throughput time of the PFR. Mean retention time equals the sum of transit time and the mean of the retention time of the marker in the CSTR.

to characterize retention time distributions. We follow the terminologies of van Soest (1983) and Warner (1981) with small modifications that allow explicit analogies with reactors.

Probably the most misused expression in digestive physiology is transit time. The term **transit time** has been applied to almost every possible measurement of retention time (Warner, 1981). We will define transit time as synonymous to 'time of first appearance', namely the time at which marker particles first appear in the output flow of the reactor. In PFRs with no mixing, transit time equals mean retention time (Smith, 1981). Transit time, however, underestimates mean retention time in PFRs with mixing (Balch, 1950). Mean retention time is the arithmetic mean of the residence time distribution. In Fig. 3.6, mean retention time equals the sum of transit time and the mean residence time of the CSTR. We emphasize the point made earlier that for animals with gastro-intestinal tracts consisting of two or more compartments in series, mean residence time is an additive function of the residence in each compartment. This point is especially important for animals in which a significant fraction of the volume of the gastro-intestinal tract is used for storage (e.g. the crop in some birds). Mean retention time can include a fraction of time during which food *is not* in contact with digestive secretions and surfaces. The terms gut-passage time, gut-passage rate and turnover time are all synonymous with mean retention (or residence) time. Because mean retention time has a clear statistical meaning and a direct biological interpretation, we suggest that it is adopted as an estimator of the time ingested materials spend from mouth to anus (or cloaca). Throughput time is defined as the time required to process one volume of reactor (Levenspiel, 1972) and in closed reactors (defined above) with a single reacting phase it is equal to mean retention time. Gut-clearance time is the time required for an animal entering starvation to evacuate its gut completely. Gut-clearance time is often used as an estimator of retention time (Jackson, 1992). As Penry and Jumars (1987) have pointed out, gut-clearance time is measured under conditions that are difficult to model and very often unnatural.

A large number of studies calculate estimators of residence times in idiosyncratic ways (e.g. time to *p*% of total marker excretion with *p*% ranging from 0–100%, median retention time, modal retention time . . . etc.; Koopman and Kennis, 1977; Hinton *et al.*, 1969). While there is nothing intrinsically wrong with these estimates, for standardization we suggest that mean retention time is always reported. For statistical purposes it is probably better to use well known standard distribution parameters – such as variance, coefficient of variation, kurtosis and skewness (Kendall and Stewart, 1966) – rather than parameters with unknown statistical properties (e.g. retention time defined as time between appearance of 5% and 80% of marker, Balch and Campling, 1965; van Soest, 1983). We also encourage reporting the full retention time distribution curves or, at the very least, concentration output curves. As our discussion indicates, it is often the form of these curves that is informative rather than the few statistical parameters that summarize them.

Most of the work calculating the rate of passage in gut segments from residence time distribution data has been done in ruminants (Spalinger and Robbins, 1992). Most of this body of work implicitly relies on modelling the gut as a series of reactors with continuous flow (Grovum and Phillips, 1973). Modelling residence time distributions in simple stomached non-fermenters and in caecal and large intestine fermenters has received much less attention (Laplace, 1972). Consequently, we do not know if animals within 'digestive strategy' classes sharing common gastro-intestinal morphologies exhibit similar retention time distribution curves. It is safe to predict that similar particle sorting mechanisms and flow patterns within the gut will yield similar retention time distributions (see Cork and Warner, 1983; Chilcott and Hume, 1984). Gastro-intestinal tracts are complex and it is unlikely that interpreting retention time distributions will be a straightforward exercise. Many animals exhibit antiperistaltic movements (Björnhag, 1987), others alternate discrete meals with long fasting intervals and hence some sections of the gut may function continuously (i.e. the intestine) whereas others function in a more discontinuous fashion (i.e. the stomach and the rectum). All these factors influence the form of retention time distributions. The interpretation of whole animal retention time distributions will be aided by experiments that examine retention time distributions of markers in individual organs and by modelling efforts that focus on specific gut morphologies and temporal patterns of feeding and digestion. Explaining retention time distributions and linking them to the varied morphologies of animals is a worthwhile exercise that can provide a link between gut morphology and digestive function and hence establish the basis for the quantitative functional morphology of the gut.

Some examples from the literature: things are rarely as we expect them

In Fig. 3.7 we have included some examples of whole-gut marker excretion curves from various animals. After describing the behaviour of markers in ideal reactors, we wish to make two points about the behaviour of markers in real animals: (a) marker behaviour is rarely exactly as predicted and (b) there are substantial benefits to be gained from plotting the curves and interpreting them in relation to the structure of the gut. Marker behaviour can only be properly interpreted in the light of the functional morphology of the gut.

Our first example (Fig. 3.7*a*) is a frugivorous bird, the cedar waxwing (*Bombycilla cedrorum*) eating fruit mash marked with polyethylene glycol, molecular weight 4000 (Martínez del Rio *et al.*, 1989). The first value for marker excretion was for an excreta collection made at 15 min; none were made earlier. Though early time points are lacking, the pattern of marker excretion suggests a gut consisting of a single mixing compartment (a CSTR) followed by a tube, i.e. an **E** curve shifted to the right (see Fig. 3.6). This interpretation requires the mixing compartment to occupy a large fraction of the gut. Inspection of the waxwings' gross morphology reveals a small stomach (Fig. 3.7*a*). The gut of this bird is probably best interpreted as a plug-

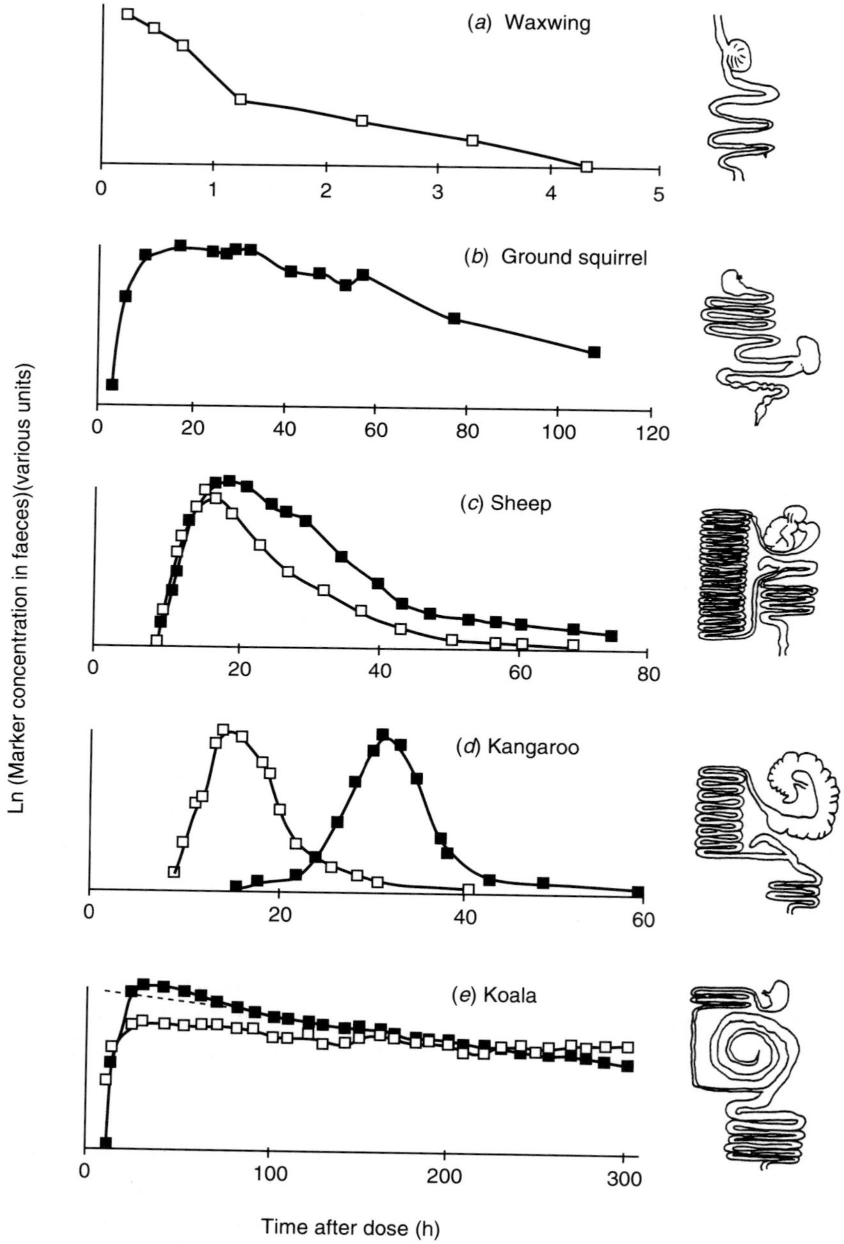

Fig. 3.7. Retention time distribution of markers for several species. (□)Fluid marker ([^{3}H] polyethylene glycol for *a, b* and [^{51}Cr] labelled EDTA for *c, d* and *e*). (■) Particulate marker ([^{103}RU] labelled phenanthroline (^{103}Ru-phen). Details of the interpretation of these distribution are provided in the text.

flow system with significant axial mixing. Most of the components of a meal stay together but some mixing occurs causing the expected pulse of marker to be skewed to the right. Radiological studies (Levey and Duke, 1992) revealed that meal components pass through the small intestine together but that mixing can occur due to prolonged retention and antiperistalsis in the rectum. The mouth to anus mean retention time was 41 min (Martínez del Rio *et al.*, 1989) and the half time for empyting the gizzard was 8 min (D. J. Levey, personal communication), so the best estimate of mean residence in this modified PFR intestine is 35 min. Transit time is about 10 min (Levey and Grajal, 1991).

The pattern of marker concentration in the faeces of the golden-mantled ground squirrel (Fig. 3.7*b*) resembles that in the frugivorous bird. In the squirrel, however, there is a clear single mixing chamber (the caecum) following a relatively short tube (the small intestine, Fig. 3.7*b*). The short interval between dosing and the first appearance of marker is probably a consequence of a short intestine (i.e. a small PFR). The rising phase of the curve suggests that mixing in the caecum is rapid but not instantaneous as it would be in an ideal CSTR.

Excretion of orally administered particulate and solute markers in sheep also appears to be a near-perfect example of excretion from a single mixing compartment (CSTR) followed by a non-mixing tube (PFR)(Fig. 3.7*c*). This behaviour is deceptively simple. Faichney and Griffiths (1978) have suggested that the forestomach alone probably behaves as a series of two to three mixing chambers. Because retention time in the ruminoreticulum of the forestomach is very long relative to time in other compartments of the stomach, digesta retention patterns in other components are not easily resolvable. Mouth–anus retention time distribution curves may not provide the fine detail needed for some studies.

The marker excretion curves from kangaroos (Fig. 3.7*d*) illustrate how the distinction between CSTRs and PFRs can become blurred. Dellow *et al.* (1984) and Hume (1989) postulated that the tubiform forestomach of kangaroos (Fig. 3.7*d*) functions as a series of mixing chambers (CSTRs). Consequently, it is not surprising that kangaroo marker excretion curves behave as expected from a series of CSTRs plugged to a PFR. The concentration of marker in faeces as a function of time rises more slowly in kangaroos than in sheep, as predicted for several CSTRs in the series (a single CSTR has an instantaneous increase). Because a series of CSTRs approaches PFR behaviour, the declining phase of the concentration curve for kangaroos is relatively steep. In a single CSTR this phase would decrease as a negative exponential. The marker excretion curves for kangaroos are intermediate between the pulse output expected from a PFR and the instantaneously rising exponentially falling output expected from a CSTR.

The faecal concentration curve for the particulate marker ^{103}Ru-phen in the koala (Fig. 3.7*e*) does not conform with that predicted for any combination of PFRs and CSTRs in series, because it has more than one descending slope (i.e. there is a 'bump' in the peak). Cork and Warner (1983) argued that the only model consistent with this curve has two (or more) mixing compartments in parallel. This situation could arise in koalas because large and small particles of digesta (both labelled with the same marker) are separated in the caecum/proximal colon and seem to have very different rates of passage through those organs (Cork and Warner, 1983). Koalas provide an example of how gut *function*, in addition to gut morphology, must be considered when interpreting excretion curves. The koala exhibits slower excretion of solute than particulate markers. This pattern may be characteristic of selective retention mechanism in some small mammals adapted to utilize high-fibre diets (Björnhag, Ch. 17; Cork, Ch. 21). Penry (1989) provides additional examples of the use of gut-reactor retention time distributions to describe and interpret patterns of particle movement and digestive function in deposit-feeding invertebrates.

Modelling from the bottom-up: an example with nectar- and fruit-feeding birds

Martínez del Rio and Karasov (1990) modelled the guts of nectar- and fruit-eating birds as PFRs because all digestion of sugars (the primary nutrients in nectar and many fruits) takes place in the intestine. The performance equations they chose were based on current understanding of the pathways of sugar absorption: (a) carrier-mediated absorption (facilitated diffusion for fructose, active transport for glucose) (Karasov and Diamond, 1983; Alpers, 1987) is described by Michaelis–Menten kinetics with parameters of maximal absorption rate (V_g) and Michaelis constant (k_g, the sugar concentration at which absorption is equal to $V_g/2$); (b) simple diffusive absorption is described by the product of lumenal concentration (G) and a permeability coefficient (k_d).

The rate of intestinal absorption of a single hexose such as glucose can be modelled as:

$$r_g = GV_g/(k_g+G) + k_dG.$$

The units of V_g, k_g, and k_d are μmol (μl min)$^{-1}$, μmol μl^{-1}, and min^{-1} respectively.

If lumenal concentration is initially high (i.e. > 5 times k_g so that mediated absorption is near maximal) then the hexose absorption rate is V_g+k_gG. As hexoses are removed from the lumen and their concentration drops so does the absorption rate, at an accelerating rate once in the range of k_g (Fig. 3.8a). These kinetic features determine the shape of the relation between hexose absorbed as a function of residence time (T) of the intestine. The asymptote of the absorption curve (A(t)) is the initial value of G. Modelling the digestion and absorption of sucrose, which has to be hydrolysed prior to absorption of its monosaccharide monomers, glucose and fructose, requires an additional equation for the hydrolysis step. The overall hexose absorption curve depends on the relative rates of hydrolysis and absorption (Martínez del Rio and Karasov, 1990). For now we will restrict our attention to the simpler model of absorption of hexose meals, which is ecologically relevant because hexoses predominate in the fruits consumed by these birds (see Martínez del Rio, Ch. 8).

The kinetic parameters determining the shape of the curves in Fig. 3.8 can be seen as boundary conditions in an optimization model where the control variable is T. We can ask what value of T maximizes the net rate of energy assimilation. To do so we must transform glucose absorbed from moles to joules, and specify the costs of processing a volume of intestine $C(T)$. Martínez del Rio and Karasov (1990) assumed that $C(T)$ equalled the sum of the cost of obtaining the food required to fill the intestine (C_o) plus the costs associated with surviving during the T units of time required to process this amount of food ($C_m \cdot T$). Because net rate of assimilation equals the difference between gains (the energy equivalent of A(T) and costs ($C(T)$)), the graphical result of including costs in the model is to move the curve down C_o units and to give it a hump (Fig. 3.8c). The T^* that maximizes the net rate of energy intake is that in which a straight line passing through the origin is tangent to the benefit-cost curve (Fig. 3.8*c*).

In this model, both the concentration of hexose in food and the cost of feeding influence T^*. Optimal retention time increases with increased cost of food acquisition and decreases with increased sugar concentration in food (Figs. 3.9*a*,*b*). The fraction of sugars extracted from food depends on the time that food spends in contact with digestive surfaces. A corollary of the model is that the efficiency of sugar extraction (1−[amount defecated]/[amount eaten]) decreases with increased sugar concentration in food and increases with the cost of acquiring food.

To this point, we have used explicit assumptions about the digestive process along with measurements done *in vitro* at the organ or tissue level to generate crisp falsifiable predictions about whole organism digestive out-

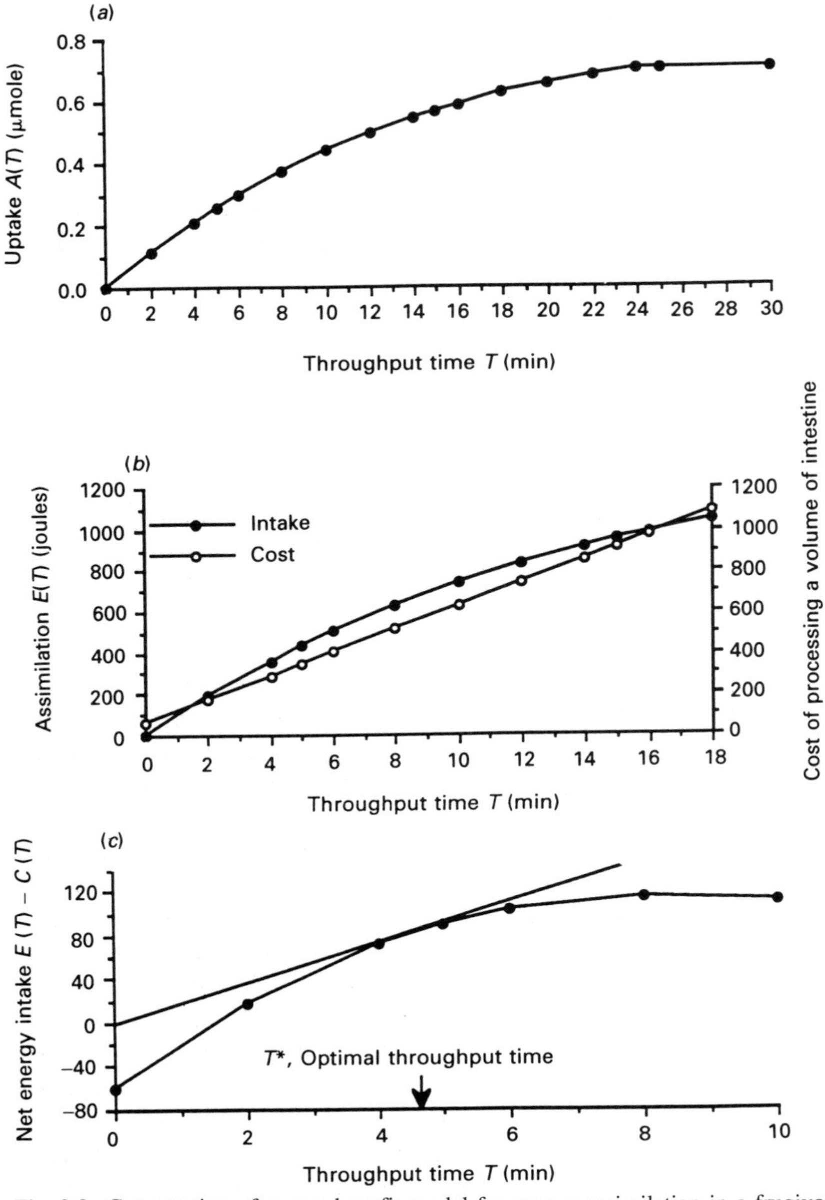

Fig. 3.8. Construction of a cost–benefit model for energy assimilation in a frugivorous bird as a function of retention time. (*a*) Nutrient uptake increases in a decelerating fashion with time that food spends in the intestine. (*b*) Net energy obtained is constructed as the difference between net energy assimilation minus the cost of processing a volume of intestine. Note that this cost increases as a linear function (with a positive intercept) of retention time. The retention time that maximizes the rate of energy gains can be obtained by finding the point of intersection of the cost–benefit curve and a straight line through the origin.

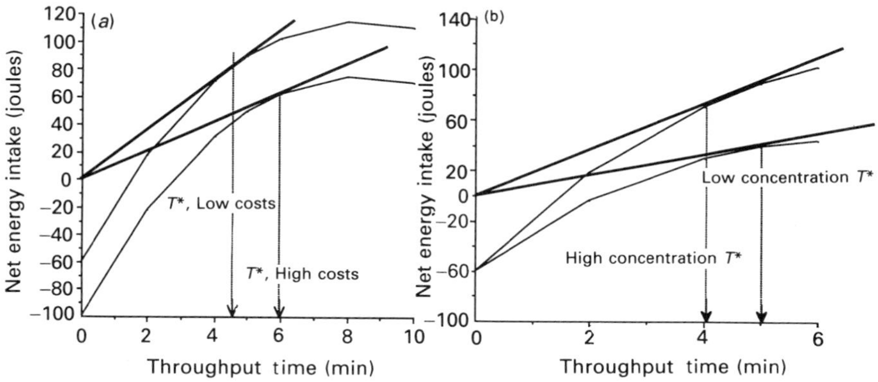

Fig. 3.9. Effect of (*a*) cost of food acquisition and (*b*) energy concentration on the optimal retention time. Increasing the cost of food acquisition (e.g. by increasing the distance between food patches) has the effect of shifting the cost–benefit curve downwards and increasing optimal *T*. Increased food concentration has the effect of increasing the steepness of the cost–benefit curve and lowering optimal *T*. Note that these predictions are strongly dependent on the form of the cost–benefit curve.

comes. The predictions would differ had we assumed that there was no passive hexose absorption or had we modelled sucrose hydrolysis/absorption for the case of sucrose hydrolysis being rate-limiting (the alternative models are described in detail in Martínez del Rio and Karasov, 1990). How well do the models and their predictions conform to reality?

The prediction that T^*, and consequently extraction efficiency, is decreased for more concentrated foods appears contrary to the conventional view in digestive physiology. Generally, in mammals and birds, gastric emptying is inhibited by negative feedback arising from duodenal receptors stimulated by the products of food digestion (Malagelda and Azpiroz, 1989; Duke, 1989). An analogous negative feedback on transit in the intestine of mammals has been termed the **ileal brake** (Spiller *et al.*, 1984). In dogs, for example, ileal infusions of glucose inhibited motility in and slowed transit through the proximal and distal small intestine (Fich *et al.*, 1990; Siegle *et al.*, 1990). These features seem to describe a system in which a particular extraction efficiency would be maintained when ingesting more concentrated foods by increasing (not decreasing) intestinal residence time.

Two recent empirical studies yielded results inconsistent with the predictions based on rate maximization. D. J. Levey (personal communication) fed cedar waxwings artificial fruits with sugar concentrations ranging from 3–30% and observed no change in either transit time or mouth to anus mean retention time. W. H. Karasov and S. J. Cork (unpublished data) observed increased mouth to anus retention time and the same extraction efficiency when rainbow lorikeets were fed glucose solution of higher concentrations. In both of these studies tissue-level experiments first characterized the aforementioned pathways of glucose absorption and, thus, validated some of the model's major assumptions. Is the match between predictions and observations poor because the wrong optimization criterion was chosen or because a crucial assumption was ignored?

This question highlights the usefulness of models. A primary argument for their use is that they provide a critical component of the feedback loop that includes hypothesis generation, data collection, analysis, hypothesis refinement, more data collection ... etc. Mathematical models are especially important for the study of the digestive system which is an interlocking complex of processes characterized by several reciprocal cause–effect pathways whose overall function we can better appreciate by viewing it as a whole. The insight and direction one gains by taking a systems perspective rather than a one factor at-at-time view is illustrated in the following application.

The model described above is mechanistic and permits an explicit estimate of glucose absorption when the performance equations, lumenal glucose concentration and residence time are specified. Martínez del Rio *et al.* (1989) fed freely-feeding cedar waxwings a meal of 192 μmol of radiolabelled D-glucose of which 92% (176 μmol) was apparently absorbed. Lumenal concentration was not measured but, had it been saturating (>50 mM in birds; Levey and Karasov, 1992), could active absorption explain all the observed absorption? Maximal active uptake (V_g) over the length of the small intestine (i.e. pyloric valve to the minute caecae) was 2.12 μmol min^{-1} when measured *in vitro* (Karasov and Levey, 1990) plus an additional 20% uptake in the colon (Levey and Duke, 1992) yielding 89 μmol min^{-1} over the 35 min intestinal residence time. Therefore, active glucose uptake cannot account for more than about 50% of the observed glucose absorption. Passive uptake is known to occur and the measured k_d in vitro appears high enough to explain the rest of the total glucose absorption (Karasov and Levey, 1990) if one makes reasonable assumptions about lumenal glucose concentrations. To make more explicit predictions and more accurate estimations we need measurements of glucose lumenal concentrations.

Future challenges for digestive tract models?

All current models of digestive tract function fall far short of being perfect. The field, however, is in the midst of rapid advances, especially with respect to understanding the ecological and evolutionary implications of digestion.

We believe that this is in part a result of the insights provided by the application of reactor theory. Although it is clear that reactor-based models have many deficiencies, we believe that the underlying theory is flexible and robust enough to allow progress.

The first generation of reactor-based models sought to be general enough to permit interspecific comparisons and consideration of optimal design in a broad comparative sense. A consequence of this aim was that reactor-based gut models sacrificed some realism for generality. Current reactor-based gut models of mammalian guts, for example, do not take into account several important physiological and ecological traits and many subtle complexities that should be taken into account when interpreting interspecific differences in small herbivores (see Cork, Ch. 21). There seems to be no reason, however, why these factors cannot be incorporated into reactor-based models in the future. This new generation of models will probably be much more mechanistic and detailed. They will permit more precise prediction and estimation but will have lost generality. Modelling is subject to tradeoffs. Some reactor models will probably increase in complexity and richness of detail, will lose generality and analytical tractability, and will probably converge with previous computer simulation models of gut function.

It would be unrealistic to suggest that the future of digestive physiology and nutritional ecology depends exclusively, or even predominantly, on reactor-based theoretical models. We believe, however, that this theoretical framework will greatly benefit our field. By thinking of guts as reactors and attempting to create reactor models of guts, we force ourselves to make assumptions explicit and hence to examine how much (or how little) we really know about the digestive process. A theoretical framework can help us by providing clear falsifiable predictions, organizing knowledge and locating areas where research is needed. Although we believe that a good dose of theory is useful for progress at this stage of the development of digestive physiology and nutritional ecology, we also believe that our theoretical efforts have to be tempered by the continuous examination of theoretical developments under the light of empirical research. Digestive physiologists would be wise to follow Whitehead's (1925) wise counsel to theoreticians about always referring to the concrete in search of inspiration. Chemical reactors are fruitful analogies for digestive systems and can tell us much about how animals process nutrients. Regardless of how convenient and useful analogies are, however, we should use them prudently. Analogies, as Plato warned philosophers, and Gordon *et al.* (1972) warned comparative physiologists, are a very slippery sort of thing.

References

Alexander, R. M. (1982). *Optima for Animals*. London: Edward Arnold.

Alpers, D. H. (1987). Digestion and absorption of carbohydrates and proteins, In *Physiology of the Gastrointestinal Tract*, vol. 2, ed. L. R. Johnson, pp. 1469–1486. New York: Raven Press.

Aris, R. (1970). *Elementary Chemical Reactor Analysis*. Englewood Cliffs, NJ: Prentice-Hall.

Balch, C. C. (1950). Factors affecting the utilization of food by dairy cows. 1. The rate of passage of food through the digestive tract. *British Journal of Nutrition*, **4**, 361–388.

Balch, C. C. & Campling, R. C. (1965). Rate of passage of digesta through the ruminant digestive tract. In *Physiology of Digestion in the Ruminant*, pp. 108–146. ed. R. W. Dougherty, Washington DC: Butterworths.

Björnhag, G. (1987). Comparative aspects of digestion in the hindgut of mammals: the colonic separation mechanism (CSM, a review). *Deutsche tierärztliche Wochenschrift*, **94**, 33–36.

Blaxter, K. L., Graham, N. & Wainman, F. W. (1956). Some observations on the digestibility of food by sheep and on related problems. *British Journal of Nutrition*, **10**, 69–91.

Brandt, C. S. & Thacker, E. J. (1958). A concept of rate of food passage through the gastrointestinal tract. *Journal of Animal Science*, **17**, 218–223.

Chilcott, M. J. & Hume, I. D. (1984). Digestion of *Eucalyptus andrewsii* foliage by the common ringtail possum, *Pseudocheirus peregrinus*. *Australian Journal of Zoology*, **32**, 605–613.

Chivers, D. J. & Hladik, C. M. (1980). Morphology of the gastrointestinal tract in primates: comparison with other mammals in relation to diet. *Journal of Morphology* **166**, 337–386.

Cork, S. J. & Warner, A. C. I. (1983). The passage of digesta markers through the gut of a folivorous marsupial, the koala *Phascolarctos cinereus*. *Journal of Comparative Physiology*, **152**, 43–51.

Dade, W. B., Jumars, P. A. & Penry, D. L. (1990). Supply-side optimization: maximizing absorptive rates. In *Behavioural Mechanisms of Food Selection*, ed. R. N. Hughes, pp. 531–556. Berlin: Springer-Verlag.

Dellow, D. W., Nolan, J. V. & Hume, I. D. (1984). Studies on the nutrition of macropodine marsupials. V. Microbial fermentation in the forestomach of *Thylogale thetis* and *Macropus eugenii*. *Australian Journal of Zoology*, **31**, 433–443.

Desnuelle, P., Sjöstrom, H. & Norén, A. (eds.) (1986). *Molecular and Cellular Basis of Digestion*. New York: Elsevier Science.

Duke, G. E. (1989). Relationship of cecal and colonic motility to diet, habitat, and cecal anatomy in several avian species. *Journal of Experimental Biology*, Suppl. 3, 38–47.

Dykstra, C. R. & Karasov, W. H. (1992). Changes in gut structure and function of house wrens (*Troglodytes aedon*) in response to increased energy demands. *Physiological Zoology*, **65**, 422–442.

Ellis, W. C & Huston, J. E. (1967). Caution concerning the stained particle technique for determining gastrointestinal retention time of dietary residues. *Journal of Dairy Science*, **50**, 1996–1999.

Faichney, G. J. (1975). The use of markers to partition digestion within the gastrointestinal tract of ruminants. In *Digestion and Metabolism in the*

Ruminant, ed. I. W. McDonald & A. C. I. Warner, pp. 277–291. Armidale NSW Australia, University of New England Publishing.

Faichney, G. J. & Griffiths, D. A. (1978). Behavior of solute and particle markers in the stomach of sheep given a concentrated diet. *British Journal of Nutrition*, **40**, 71–82.

Fich, A., Phillips, S. F., Neri, M., Hanson, R. B. & Zinsmeister, A. R. (1990). Regulation of postprandial motility in the canine ileum. *American Journal of Physiology*, **259**, G767–G774.

Firmer, S. J. & Cutler, D. J. (1988). Simulation of gastrointestinal drug absorption I. Longitudinal transport in the small intestine. *International Journal of Pharmacology*, **48**, 231–246.

Forman, G. L. (1972). Comparative morphological and histochemical studies of stomachs of selected American bats. *University of Kansas Science Bulletin*, **49**, 591–729.

Gordon, M. S., Bartholomew, G. A., Grinnell, A. D., Jorgensen, C. B. & White F. N. (1972). *Animal Physiology: Principles and Adaptations*. 2nd edn. New York: Macmillan.

Grovum, W. L. & Phillips, G. D. (1973). Rate of passage of digesta in sheep. 5. Theoretical considerations based on a physical model and computer simulations. *British Journal of Nutrition*, **30**, 313–329.

Hardison, O. B. (1989). *Disappearing through the Skylight: Culture and Technology in the Twentieth Century*. London: Penguin Books.

Hinton J. M., Lennard-Jones, J. E. & Young, A. C. (1969). A new method for studying gut transit times using radioopaque markers. *Gut* **10**, 842–847.

Horn, M. H. & Messer, K. S. (1992). Fish guts as chemical reactors: a model of the alimentary canals of marine herbivorous fishes. *Marine Biology* **113,** 527–535.

Hume, I. D. (1989). Optimal digestive strategies in mammalian herbivores. *Physiological Zoology*, **62**, 1145–1163.

Illius, A. W. & Gordon, I. J. (1992). Modelling the nutritional ecology of ungulate herbivores: evolution of body size and competitive interactions. *Oecologia* **89**, 428–434.

Jackson, S. (1992). Do seabird gut sizes and mean retention times reflect adaptation to diet and foraging method. *Physiological Zoology*, **65**, 674–697.

Kapoor, B. G., Smith, H. & Verighina, I. A. (1975). The alimentary canal and digestion in teleosts. *Advances in Marine Biology*, **13**, 109–239.

Karasov, W. H. (1990). Digestion in birds: chemical and physiological determinants, and ecological implications. In *Avian Foraging: Theory, Methodology, and Applicatons*, ed. M. L. Morrison, C. J. Ralph, J. Verner & J. R. Jehl, Studies in Avian Biology No. 13, 391–415. Lawrence, KS: Cooper Ornithological Society.

Karasov, W. H. & Diamond, J. M. (1983). A simple method for measuring intestinal solute uptake in vitro. *Journal of Comparative Physiology*, **B152**, 105–116.

Karasov, W. H. & Levey, D. J. (1990). Digestive system trade-offs and adaptations of frugivorous passerine birds. *Physiological Zoology*, **63**, 1248–1270.

Kendall, M. G. & Stuart, A. (1966). *The Advanced Theory of Statistics*, Vol. 2. New York: Hafner.

Kokko, J. P. & Tisher, C. C. (1976). Water movement across nephron segments involved with the countercurrent multiplication system. *Kidney International*, **10**, 64–81.

Koopman, J. P. & Kennis, H. M. (1977). Two methods to assess gastrointestinal transit time in mice. *Zeitschrift für Versuchstierkunde*, **19**, 298–303.

Kotb, A. K. & Luckey, T. D. (1972). Markers in nutrition. *Nutrition Abstracts and Reviews*, **42**, 813–845.

Laplace, J. P. (1972). Le transit digestiff chez les monogastriques, I. Les technique d'etude. *Annales de Zootechnie* **21**, 83–105.

Levenspiel, O. (1972). *Chemical reactor engineering*. 2nd edn. New York: Wiley.

Levenspiel, O. & Smith, W. K. (1957). Notes on the diffusion-type model for the longitudinal mixing of fluids in flow. *Chemical Engineering Science*, **6**, 227–233.

Levenspiel, O. & Turner, J. C. R. (1970). The interpretation of residence-time experiments. *Chemical Engineering Science*, **25**, 1605–1609.

Levey, D. J. & Duke, G. E. (1992). How do frugivores process fruit? Gastrointestinal transit and glucose absorption in cedar waxwings (*Bombycilla cedrorum*). *Auk* (in press).

Levey, D. J. & Grajal, A. (1991). Evolutionary implications of fruit processing and intake limitation in cedar waxwings. *American Naturalist*, **138**, 171–189.

Levey, D. J. & Karasov, W. H. (1992). Digestive modulation in a seasonal frugivore, the American robin (*Turdus migratorius*). *American Journal of Physiology*, **262**, G711–G718.

Levins, R. (1966). The strategy of model building in population biology. *American Science*, **54**, 421–431.

Malagelda, J. R. & Azpiroz F. (1989). Determinants of gastric emptying and transit time in the small intestine. In *The handbook of Physiology*, Section 6, Vol. 1, ed. S. G. Schultz, pp. 409–437. Bethesda, MD: American Physiological Society.

Martínez del Rio, C. & Karasov, W. H. (1990). Digestion strategies in nectar- and fruit-eating birds and the composition of plant rewards. *American Naturalist*, **136**, 618–637.

Martínez del Rio, C., Karasov, W. H. & Levey, D. J. (1989). Physiological basis and ecological consequences of sugar preferences in cedar waxwings. *Auk*, **106**, 64–71.

Mertens, D. R. & Ely, L. O. (1982). Relationship of rate and extent of digestion to forage utilization – a dynamic model evaluation. *Journal of Animal Science*, **54**, 895–905.

Penry, D. L. (1989). Tests of kinematic models for deposit-feeder's guts: patterns of sediment processing by *Parastichopus californicus* (Stimpson) (Holothuridae) and *Amphicteis scaphobranchiata* Moore (Polychaeta). *Journal of Experimental Marine Biology and Ecology*, **128**, 127–146.

Penry, D. L. & Jumars, P. A. (1986). Chemical reactor analysis and optimal digestion. *Bioscience*, **36**, 310–315.

Penry, D. L. & Jumars, P. A. (1987). Modelling animal guts as chemical reactors. *American Naturalist*, **129**, 69–96.

Penry, D. L. & Jumars, P. A. (1990). Gut architecture, digestive constraints and feeding ecology of deposit-feeding and carnivorous polychaetes. *Oecologia*, **82**, 1–11.

Place, A. R. (1990). The avian digestive system – an optimally designed plug-flow chemical reactor with recycle? *Memoirs of the Digestive Strategies of Animals Symposium*, pp. 53–59. Washington, DC: The Center for Biological Parks, National Zoological Park.

Plante, C. J., Jumars, P. A. & Baross, J. A. (1990). Digestive associations between

marine detritivores and bacteria. *Annual Review of Ecology and Systematics*, **21**, 93–127.
Pond, K. R., Ellis, W. C., Matis J. H. & Deswysen A. G. (1989). Passage of chromium-mordanted and rare earth labeled fiber: time of dosing kinetics. *Journal of Animal Science*, **67**, 1020–1028.
Scholander, P. F. & Krog, J. (1957). Countercurrent heat exchange and vascular bundles in sloths. *Journal of Applied Physiology*, **10**, 405–411.
Scholander, P. F., Walters, V., Hock, R. & Irving, L. (1950). Body insulation of some arctic and tropical mammals and birds. *Biological Bulletins*, **99**, 225–236.
Sibly, R. M. (1981). Strategies of digestion and defecation. In *Physiological Ecology: an Evolutionary Approach to Resource Use*, ed. C. R. Townsend & P. Calow, pp. 109–141. Oxford: Blackwell Scientific.
Siegle, M. L., Schmid, H. R. & Ehrlein, H. J. (1990). Effects of ileal infusions of nutrients on motor patterns of canine small intestine. *American Journal of Physiology*, **259**, G78–G85.
Smith, J. M. (1981). *Chemical Engineering Kinetics*. New York: McGraw Hill.
Spalinger, D. E. & Robbins, C. T. (1992). The dynamics of particle flow in the rumen of mule deer (*Odocoileus hemionus hemionus*) and elk (*Cervus elaphus nelsoni*). *Physiological Zoology*, **65**, 379–402.
Spiller, R. C., Trotman, I. F., Higgins, B. E. *et al.* (1984). The ileal brake–inhibition of ileal motility after ileal fat perfusion in man. *Gut*, **25**, 365–374.
Udén, P., Colucci, P. E. & van Soest P. J. (1980). Investigation of chromium, cerium and cobalt as markers in digesta rate of passage studies. *Journal of Science and Food Agriculture*, **31**, 625–632.
Usry, J. L., Turner, L. W., Stahly, T. S., Bridges, T. C. & Gates, R. S. (1991). GI tract simulation model of the growing pig. *Transactions of the American Society of Agricultural Engineers*, **34**, 1879–1890.
van Soest, P. J. (1983). *Nutritional Ecology of the Ruminant*. Corvallis, OR: O & B Books.
Warner, A. C. I. (1981). Rate of passage of digesta through the gut of mammals and birds. *Nutritional Abstracts and Reviews*, **51B**, 789–820.
Whitehead, A. N. (1925). *Science and the Modern World*. New York, Macmillan.
Yonge, C. M. (1937). Evolution and adaptation in the digestive system of metazoa. *Biology Reviews*, **12**, 87–115.

4

Optimum gut structure for specified diets

R. MCNEILL ALEXANDER

Mammals all have their guts built of the same units – oesophagus, stomach and small and large intestines – but there are many variants on the basic design (Stevens, 1988). Not only are carnivore guts very different from those of herbivores but there are marked differences between one herbivore and another. For example, cattle have huge, complex stomachs but horses have small stomachs and enormous large intestines. This paper tries to explain the differences by means of a mathematical model that predicts optimum gut structures for specified diets. The details of the model have been presented elsewhere (Alexander, 1991). We will be concerned principally with herbivores, because plant food presents particularly interesting problems to the animals that eat it.

Digestion and fermentation

The foodstuffs in plant cells, like those in animals, are easily digested by the enzymes of vertebrates, which break them down into compounds such as simple sugars and amino acids that are easily absorbed from the gut into the bloodstream. Plant cells, however, are enclosed in fibrous walls that consist largely of compounds that cannot be broken down by any of the enzymes that vertebrates produce. Some of these compounds (for example, lignin) are useless to herbivores, passing through their guts unchanged. Others (notably cellulose) are equally resistant to the herbivore's own enzymes, but may be broken down by microbes living in the herbivore's gut.

The microbes cannot oxidize the food because partial pressures of oxygen in the gut are very low. Instead, they ferment it, converting cellulose anaerobically to fatty acids, carbon dioxide and methane. The herbivore absorbs and uses the fatty acids, which typically retain 75% of the heat of combustion of the original cellulose (Blaxter, 1962).

Types of reactor

Some gut components (for example, all small intestines) are long, slender tubes. Others (including stomachs) are wide bags. Penry and Jumars (1987) made the significance of this difference clear by applying to guts the theory of chemical reactors.

We have to consider two types of reactor, both of which are used in industry. We will compare them, using as an example the simple case of a reaction that proceeds according to first-order kinetics with rate constant r. A reagent (for example, a foodstuff) enters with concentration C_{in} and leaves with concentration C_{out}. The average time taken to pass through the reactor is t.

A continuous-flow stirred tank reactor (CSTR) is a tank whose contents are kept well mixed. Because of this mixing, the reagent entering the reactor is immediately diluted to the concentration C_{out} at which it will leave, and the reaction proceeds throughout at a rate rC_{out}. For each sample of reagent this continues, on average, for time t so

$$C_{in} - C_{out} = rC_{out}\,t \qquad (1)$$
$$C_{out} = C_{in}/(1 + rt)$$

A plug-flow reactor (PFR) is a slender tube along which reagents flow without opportunity for mixing. At the entry end, the concentration is still C_{in} and the reaction proceeds at a rate rC_{in}. The concentration falls along the length of the reactor until it reaches C_{out} at the exit end, where the reaction proceeds at the lower rate rC_{out}. The fall in concentration is exponential

$$C_{out} = C_{in} \exp(-rt) \qquad (2)$$

These equations tell us that reactions proceed faster in PFRs. For example, when $rt = 1$, C_{out}/C_{in} is 0.50 for a CSTR but 0.37 for a PFR: 50% of the reagent is broken down in the CSTR and 63% in the PFR. This advantage makes the PFR the preferable reactor for digestion, but it is unsuitable for fermentation because the flow of reagents through the reactor would soon wash all the microbes out. A population of microbes can, however, be maintained indefinitely in a CSTR because it is kept well stirred. Consequently we can expect animals to digest food in slender tubes that will function as PFRs, unless (like some carnivores) they eat food in large chunks that need a large stomach to accommodate them. However, fermentation requires chambers wide enough to ensure the mixing that will enable them to function as CSTRs. Fermentation in herbivores does indeed occur in wide chambers, such as the rumen of cattle and the large intestine of horses.

Fore- and hindgut fermenters

Cattle and other ruminants have the rumen, at the anterior end of the gut, as their principal fermentation chamber but also have a small fermentation chamber in the hindgut. The same basic arrangement (a large fore fermentation chamber and a small hind one) is also found in kangaroos, sloths and colobus monkeys. Horses, elephants and wombats, however, have no foregut fermentation chamber. Their stomachs are used for digestion, as in carnivores, and the enlarged large intestine is the only fermentation chamber. Rodents, rabbits and howler monkeys are also hindgut fermenters, but their fermentation chamber is the caecum (an outgrowth from the junction of the large and small intestine) rather than the large intestine itself (Hume, 1989). What are the relative merits of fore- and hindgut fermentation chambers?

Each brings advantages and disadvantages. A foregut fermentation chamber has the advantage that the microbes washed out of it with the remains of the food pass into the small intestine where they are digested. Thus energy and materials used for microbial growth eventually become available to the herbivore. Against this, cell contents that could have been digested instead get fermented in the foregut and so some of their energy is lost to the herbivore. The converse is true of hindgut fermentation: energy loss due to fermentation of food cell contents is avoided if these are largely digested by the time they reach the hindgut, but the microbes are lost in the faeces. The relative merits of fore- and hindgut fermentation must depend on the composition of the food.

A model

Therefore, it seems clear that PFRs should generally be preferred for digestion and CSTRs should always be preferred for fermentation. However, we have found advantages and disadvantages both for fore- and for hindgut fermentation and, indeed, in the use of fermentation anywhere in the gut. It seems dangerous to try to decide the merits of different gut designs by verbal argument and better to use a mathematical model.

A model formulated by Alexander (1991) is shown in Fig. 4.1. It has three chambers each of which may be present or absent and can be varied in size. The total volume V of the gut is made up of a foregut fermentation chamber (a CSTR) of volume v_1V; a digestive tube (a PFR) of volume v_2V; and a hindgut fermentation chamber (another CSTR) of volume v_3V. Plainly, $v_1 + v_2 + v_3 = 1$. This is the simplest arrangement capable of modelling all

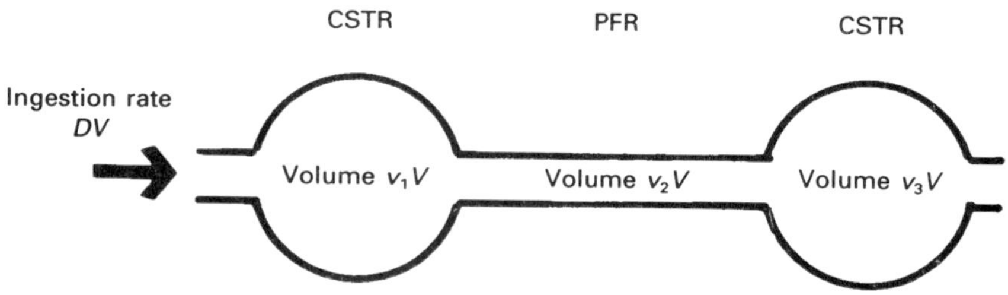

Fig. 4.1. The gut model described in the text, from Alexander (1991).

mammalian guts. Food passes through the gut at a constant rate DV: thus D is the feeding rate in gut volumes per unit time.

Penry and Jumars (1987) used the Michaelis–Menten equation to predict rates of digestion and the Monod equation for rates of fermentation. For this model, I use the simpler assumption of first-order kinetics, which seems reasonable for a steady-state model (but would be unsatisfactory if the model were taking widely-spaced meals and having to build up the microbial population after each). The rate constants and other parameters required for the model are given by Alexander (1991) who shows that they can be changed substantially with little effect on the main conclusions. Cell-wall materials are assumed to be fermented more slowly than cell contents are fermented or digested (van Soest *et al.*, 1988). A microcomputer was used to simulate the passage of food of different compositions at different rates through guts of various structures. In each case, the energy gained by the herbivore was calculated.

Results are shown in Fig. 4.2. Each triangle represents the entire range of gut structures allowed by the model. For example, the bottom left hand corner represents the gut for which $v_1 = 1$, $v_2 = v_3 = 0$; that is, a gut consisting of the foregut CSTR and nothing else. A point half way up the left edge of the triangle represents a gut with $v_1 = v_2 = 0.5$, $v_3 = 0$; that is, a gut consisting of a foregut CSTR and a PFR of equal volumes, and no hindgut CSTR. Points inside the triangle represent guts possessing all three components. The contours show for each gut structure the fraction of the potentially available energy that the herbivore assimilates.

Figure 4.2*a*, *d* refer to food containing a very low proportion of digestible cell contents, lower even than would be found in mature grass. If the feeding rate is low (Fig. 4.2*a*), the gut structure that maximizes energy gain (indicated by the star) consists of two large CSTRs and a small intervening PFR

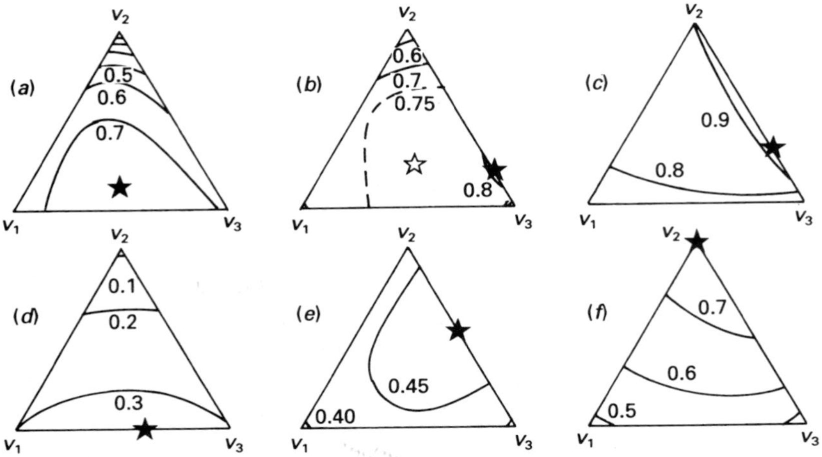

Fig. 4.2. Graphs showing how gut design affects the rate of energy gain from different diets. Each triangle represents the entire range of gut structures permitted by the model. The contours represent rates of energy gain and stars indicate maxima. (*a*) and (*d*) refer to a poor diet, containing a low proportion of digestible cell contents; (*b*) and (*e*) refer to a moderate diet; and (*c*) and (*f*) to a rich one. The food intake rate is low in (*a*), (*b*) and (*c*), and ten times faster in (*d*), (*e*) and (*f*). (From Alexander, 1991.)

($v_1 = 0.45$, $v_2 = 0.1$, $v_3 = 0.45$). At higher feeding rates (Fig. 4.2*d*) the optimum gut has no PFR but consists simply of two CSTRs in series.

As the proportion of cell contents in the food increases, the optimum shifts abruptly to a gut consisting of a PFR and a hindgut CSTR only. This is seen in Fig. 4.2*b*, *e*, which refers to food of moderate quality in which digestible cell contents represent half the energy content. Figure 4.2*b* is very close to the boundary at which the optimum shifts: the global maximum rate of energy gain (filled star) is at the new position but a local maximum (hollow star) remains at the old one. Figure 4.2*c*, *f* refers to food with a very high proportion of cell contents, higher even than mangolds or grain. If the food intake rate is low (Fig. 4.2*c*), the optimum gut still consists of a PFR and a hindgut CSTR, but if it is high (Fig. 4.2*f*) it consists of a PFR alone. A herbivore that eats rich food fast should abandon fermentation, allowing cell wall materials to escape, but use the largest possible PFR to digest as much as possible of the cell contents.

A strict carnivore would derive no benefit from fermentation chambers, but an omnivore might well eat digestible materials and plant cell walls in the proportion assumed in Fig. 4.2*c*, *f*. For such an omnivore, the optimum gut structure would depend on the rate of food intake, as for a herbivore eating rich food.

The optimum gut indicated by Fig. 4.2*a* resembles the guts of ruminants in having both a fore- and a hindgut fermentation chamber, with a small digestive tube between. However, ruminants have very large foregut fermentation chambers and only small hind ones, whereas the indicated optimum has fore and hind chambers of approximately equal volume. Two plausible modifications of the model predict an optimum gut for this diet, with a larger foregut chamber. In one, water is absorbed from the gut contents as they travel along the gut. In the other, the foregut fermentation chamber is replaced by several chambers of the same total volume, connected in series. A paper presenting these modifications of the model is in press (Alexander, 1993b).

Though the optimum in Fig. 4.2*a* resembles a ruminant gut, Fig. 4.2*b*, *e* shows a horse-like gut as optimal. The model suggests that foregut fermenters such as cattle are adapted to poorer diets (with a lower proportion of cell contents) than hindgut fermenters such as horses. Janis (1976) reached the opposite conclusion.

The howler monkey *Alouatta* eats leaves (food of moderate quality) slowly but the spider monkey *Ateles* eats fruit (with a high proportion of cell contents) fast (Milton, 1981). Both are hindgut fermenters, but the fermentation chamber is much larger in *Alouatta* than in *Ateles*, as Fig. 4.2 suggests it should be.

Coprophagy

Microbes produced in foregut fermentation chambers are digested further along the gut, so their energy content becomes available to the herbivore, but those produced in the hindgut are lost in the faeces. Rabbits, some rodents and a few other hindgut fermenters recover the energy from microbes by eating faeces during the part of the day when they are inactive and passing the faeces through the gut for a second time.

The habit of coprophagy is likely to be beneficial only during rest periods: when the animal is active it will generally do better to eat fresh food than faeces that, by their nature, can be expected to have a lower energy content. A modification of the model (Alexander, 1993a) compares two styles of feeding in which equal quantities of food are eaten. In both cases, feeding is restricted to 12 consecutive hours per day. In one case, the food is left stationary in the gut during the other 12 h but in the other it continues to pass through the gut and the faeces are eaten. This model shows, as expected, that foregut fermenters can gain little by coprophagy but that hindgut fermenters can make substantial gains, especially when the food contains low proportions of cell contents: coprophagy seems capable of increasing energy yields by up to 10–15%.

Mastication

The more a mammal chews the less it can eat, for it cannot eat and chew simultaneously. Chewed food is presumably digested and fermented faster because its particles are smaller and its cells damaged. How long should an animal chew?

Alexander (1991) tackled this question, modelling the effect of chewing as a progressive increase in the rate constants for digestion and fermentation. The conclusions were as might have been expected. Foods that break down slowly when chewed should be chewed for longer than those that break down fast. Foods that can be eaten rapidly should be chewed for longer than those that can be eaten only slowly. Fibrous foods, containing a lot of cell wall, should be chewed for longer than foods rich in cell contents. This result helps to explain why cattle spend so much time chewing. However, the model implies that chewing occurs before swallowing, so is not directly applicable to chewing the cud.

Optimum diets

Given the choice of equal quantities of food of different qualities, a herbivore or omnivore should choose the richer. This is true, whatever the structure of the gut, because cell contents can be digested or fermented faster than cell wall materials can be fermented, so will be more completely broken down. However, animals often have to choose between a smaller quantity of richer food and a larger quantity of poorer. Should a grazer select the richer young shoots or take larger mouthfuls of grass indiscriminately? Should a browser negotiate thorns to eat the rich leaves of a well-protected tree or content itself with poorer leaves that can be eaten faster from trees without thorns? The model has been applied to such problems (Alexander, 1991).

Again the results are not surprising. In what follows, feeding rate means the rate of intake of food in terms of (digestible or fermentable) energy content. If the maximum possible feeding rate is quite low even for the poorer food, the better quality of the rich one will compensate only for a slightly smaller feeding rate. If the poor food can be eaten fast however, a richer food may be preferable even at the expense of a considerably reduced feeding rate. (This is because poor food eaten fast is very incompletely broken down.) This helps to explain why grazing antelopes are less selective on low-biomass pastures (Murray, 1991). It also seems to explain why larger bovids (which have lower mass-specific metabolic rates and, presumably, lower feeding rates per unit gut volume, D) tend to be less selective than smaller ones (Jarman, 1974). The optimum diet depends, of course, on the structure of the gut as well as on the choice available. A mammal with small fermentation chambers that can break down cellulose only slowly should be more selective than one which can ferment faster.

Alexander (1991) also considers how diet choice should be affected by the time and energy used in travelling between patches of food.

Conclusion

The model that we have been discussing is crude but seems powerful as an aid to understanding gut design and diet choice. It predicts optimum gut structure for given herbivorous or omnivorous diets and optimum diets for a given gut structure. It is capable of refinement and may need to be refined for some further applications. One obvious limitation is that in its present form it takes no account of the fluctuations of gut microbe population that occur when there are intervals between meals.

References

Alexander, R. McN. (1991). Optimization of gut structure and diet for higher vertebrate herbivores. *Philosophical Transactions of the Royal Society, Series B*, **333**, 249–255.

Alexander, R. McN. (1993a). The energetics of coprophagy: a theoretical analysis. *Journal of Zoology* (Lond.), **230**, 629–637.

Alexander, R. McN. (1993b). The relative merits of foregut and hindgut fermentation. *Journal of Zoology* (Lond.), **231**, in press.

Blaxter, K. L. (1962). *The Energy Metabolism of Ruminants*. London: Hutchinson.

Hume, I. D. (1989). Optimal digestive strategies in mammalian herbivores. *Physiological Zoology*, **62**, 1145–1163.

Janis, C. (1976). The evolutionary strategy of the Equidae and the origins of rumen and cecal digestion. *Evolution*, **30**, 757–774.

Jarman, P. J. (1974). The social organisation of antelope in relation to their ecology. *Behaviour*, **48**, 215–267.

Milton, K. (1981). Food choice and digestive strategies of two sympatric primate species. *American Naturalist*, **117**, 496–505.

Murray, M. G. (1991). Maximising energy retention in grazing ruminants. *Journal of Animal Ecology*, **60**, 1029–1045.

Penry, D. L. & Jumars, P. A. (1987). Modelling animal guts as chemical reactors. *American Naturalist*, **129**, 69–96.

Stevens, C. E. (1988). *Comparative Physiology of the Vertebrate Digestive System.* Cambridge: Cambridge University Press.

van Soest, P. J., Sniffen, C. J. & Allen, M. A. (1988). Rumen dynamics. In *Aspects of Digestive Physiology in Ruminants*. ed. A. Dobson & M. J. Dobson, pp. 21–42. Ithaca: Comstock.

Part II

Food

5

Foods and the digestive system

C.M. HLADIK and D.J. CHIVERS

The structure and composition of the different parts of animals or plants (and, in a few instances, minerals) that are actually fed on by animal species result from long processes of interaction in the networks of past ecosystems where these species or their ancestors were present. Analysing the present status of 'food species' therefore involves taking a glimpse at this past, as does interpreting the adaptation of the form and function of the digestive tract of the consumers (Langer, Ch. 2).

Primate species and primate foods provide several examples in which recent investigations (presented in different chapters of this volume) have enabled us to understand these long-term processes, which also apply to other mammalian species. During the Mezozoic, when the continental plates were drifting apart (South and North America on one side, Africa and Eurasia on the other one), most plant species were quite different from those that we presently know, and, if we think of them as food, they probably also differed in terms of composition and taste. In parallel with the evolution of the Platyrrhini and Catarrhini primates, occurring on the west and east sides of the primitive Atlantic Ocean, the flowering plant species evolved in response to consumers and tended, simultaneously, to develop edible parts and non-edible parts.

Food edibility

Edible plant parts include leaves, exudates, tubers, fruits and seeds; the productive plant machinery (leaves), although needing to contain a diverse mixture of enzymes and co-enzymes and, therefore, containing a broad mix of essential amino acids and minerals, has most of its energy locked up in cellulose. We will come back to several important aspects of leaf-eating in primates and other mammals that involve the possibility of breaking the long-chain structural carbohydrates. However, most of the potential foods that require a minimum of processing are concentrated into the reproductive plant parts. Starch and fats play a major role, especially in seeds where constraints imposed by the need for dispersal make calorie-dense storage compounds a necessity.

The fleshy fruits of the phanerogamous plant species evolved with the most unspecialized primate species, together with birds and bats (as discussed in Ch. 8 by Martínez del Rio). They have been shaped by selection to be attractive to potential consumers whose positive taste response to the sweetness of sugars is generally adapted to find foods with high energy content. Efficient seed dispersal and mechanisms to promote outcrossing when the distance between conspecific individuals is high both fit well with the life-style hypothesized for early angiosperms as colonists of resource-rich gaps in low-diversity gymnosperm forests (Estrada and Fleming, 1986). In contemporary tropical forest, plants adapted for dispersal of seeds by primates and other frugivorous vertebrates account for a very large percentage of species and individuals of trees and lianas (Hladik, 1993).

Because plant–frugivore interactions are not species-specific, there is potential competition, not only among frugivores, but also among plant species for the services of shared dispersal agents. This competition has driven the evolutionary increases in the reward offered to seed dispersers up to limits set by a balance between the costs and benefits of dispersal. How concentrated the nutritional reward for frugivores is may depend on the intensity of the competition for the services of these frugivores. Variation in the concentration of the reward may also reflect the 'packaging' problem inherent in fleshy fruits. Larger seeds have more reserves and a better chance of survival; however, larger seeds mean larger diaspores. The larger the diaspore, the smaller the number of animals large enough to exploit the fruit. Therefore, when seed size is near the upper limit for a plant's disperser, there is selection for a concentrated reward that minimizes further increases in diaspore size (Herrera, 1985). This may explain why Myristicaceae evolved a fruit with a thin (but very fatty) aril surrounding a large seed and why the large fruits of several species have a pulp particularly concentrated in sugars.

For this kind of edible material, the taste buds and the associated taste response of the consumers can be considered as the first part of the digestive system. This aspect of chemical perception, which has been explored by Simmen (Ch. 10) in a comparative approach to the two major radiations of Platyrrhini primates, reflects the necessary successive steps in the coevolutionary processes occurring between plants and primates or other frugivorous species.

Exudates, especially gums, were probably the first type of plant food involved in this process, because the prosimian species that preceded the radiations of Old World and New World primates had several primitive characteristics (especially the dental comb) that are considered as efficient tools for collecting gums along tree trunks (Charles-Dominique and Martin, 1970). The composition of most gums – long-chains of C5 β-linked carbohydrates – that are presently eaten by Platyrrhini primates (Simmen, Ch. 10) and by several prosimian species is closer to that of leaves than to sweet fruit flesh. This necessitates, for the specialized gum-eating primates, a digestive system with a large caecum, such as that described for *Euoticus elegantulus* (Chivers and Hladik, 1980). Nevertheless, the composition of the gums of various plant species of Madagascar (*Terminalia* spp.) presently eaten by prosimian species with *and without* specialized gut morphology also includes large amounts of reducing sugars that can be digested as easily as fruit pulp (Hladik *et al.*, 1980). These might be comparable to the first types of gums eaten by the primate ancestors, together with other foods such as insects and other small invertebrates and vertebrates.

With reference to those items (gums, fruits, insects) that are likely to have been foods for early primates and other mammals, the produce of different environments is one of the most important parameters to consider. As we discussed in a previous paper (Hladik and Chivers, 1978), body size and density (the biomass) are pivotal factors affecting the composition of the diet of the different primate species. There are limited amounts of invertebrates, compared to fruit produce and other primary produce such as leaves in forests. In the Gabon rain forest, for instance, the annual production of leaf litter is 14 t dry weight with 500 kg for fruits and only 23 kg for invertebrates (Hladik, 1990b). There is no estimate for gum production but the drastic reduction in the secondary produce, as compared to primary, implies that species feeding on small prey and fruits have low densities and low body weight (100 g to 1 kg). The digestive tract of these species, which does not require a specialized shape, is probably close to that of primitive forms and its mode of functioning might be comparable to that of most young animals (Moir, Ch. 7).

Conversely, primate species able to feed on large amounts of leaves can supplement a partly frugivorous diet with protein, without using animal foods, and thus reach higher biomasses and body masses (generally between 2 and 10 kg), and the more folivorous the diet, the more biomass and body weight can be achieved (a mountain gorilla can weight about 200 kg) in relation to the actual carrying capacity of the environment. The diets of most primate species are composed of fruits, leaves and/or invertebrates (Fig. 5.1) and must necessarily fit with these ecological principles. The diversity of the

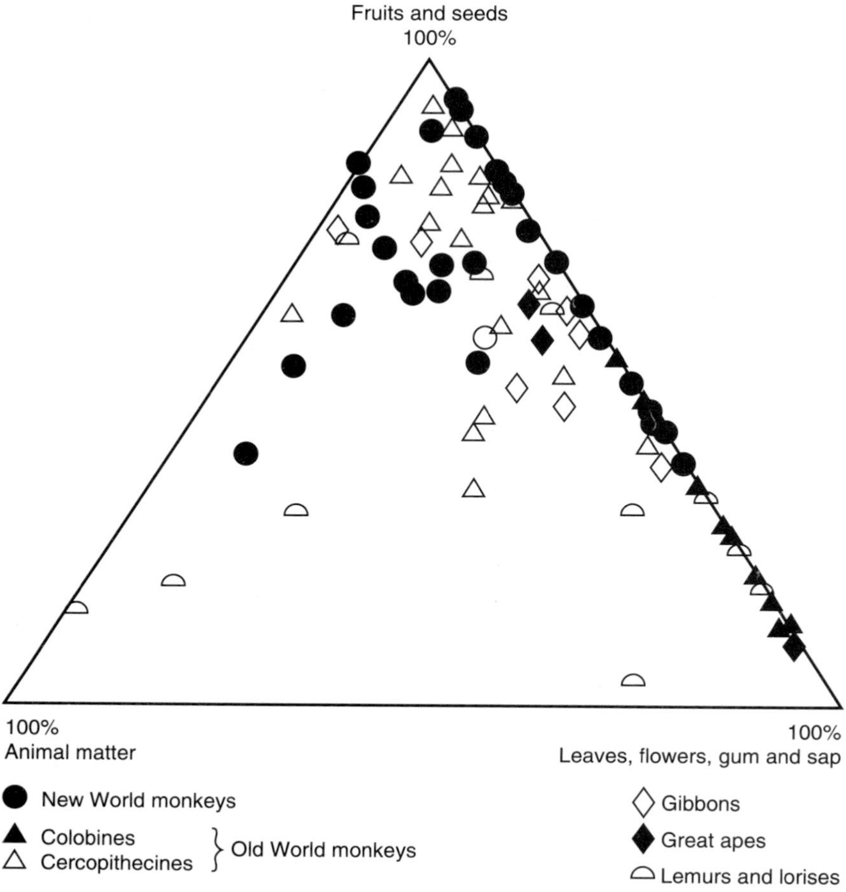

Fig. 5.1. Mean annual diets of some primates in the form of a triangular projection from a three-dimensional graph (all points lie in the same plane, since the three scores for each species are percentages totalling 100%), with 100% faunivory, frugivory and folivory at the three corners and 0 for each food type on the opposite side. Hence, fauni-frugivores are to the left, and foli-frugivores to the right. (From *The Cambridge Encyclopaedia of Human Evolution*, p. 61.)

diets reflects the diversity of ecological niches. A further discussion on the ecological niche concept of the folivorous mammals is presented by Perrin (Ch. 9) who shows the respective importance of the specialized morphological and physiological features necessary to cope with the particular composition of most leafy diets, and the diverging spread of pre-gastric fermenters and simple-stomached species with digestive modifications that increase the digestive efficiency but limit the flexibility of the diet.

The inventory of the different food items available to mammal species, presented by Langer and Chivers (Ch. 6), includes those different plant parts in which edibility can be due to a response to the consumers, and those fibrous parts that are edible only to consumers that themselves evolved to allow their utilization (the fibrous parts have a tendency not to become edible, as discussed below). The edible animal species can also be subject to evolutionary trends in response to consumers. These food items can be presented along a wide spectrum involving increasing difficulty in processing in the gut: animal flesh – fatty arils – oil-rich nuts – fruit pulps – tubers – flowers – sugary arils – immature leaves – gums – oil rich seeds – invertebrate cuticles – mature leaves – stems and other woodier parts. The physical characteristics of these food items are not directly dependent on their composition and the problems related to shearing to obtain small pieces, presented in the next section (Ch. 13, Lucas). In terms of gut morphology, a linear spectrum seems inapplicable. While eating animal matter produces a gut dominated by the small intestine (with reduced stomach and colon) and fruit-eaters have an intermediate morphology, leaves are either digested in a much expanded stomach or in a greatly enlarged caecum and/or colon (Fig. 5.2). The pattern of physiological variation may be linear, that of anatomical variation is not.

A special mention must be made of tubers because these underground plant parts, particularly rich in starch, play, and may have played in the past, an important role in the subsistence of a well-known primate species *Homo sapiens* (Hladik, 1985). In the present time the edible wild tubers of the rain forests are mostly yam (*Dioscorea* spp.). Their standing crop, recently measured during large scale surveys (Hladik and Dounias, 1993), demonstrated that the populations of hunter-gatherers can use – and can have used in the past – these wild tubers as a staple food.

Outside the rain forest, for instance in sub-tropical Africa where Peters and O'Brien have conducted their survey on potential foods for humans and non-human primates (Ch. 11), some tubers might be present. But many tuberous species of the dry areas are toxic and require, to become edible, a long preparation including steeping in water for several days. This technical skill partly replicates what occurs in the fermenting chambers of the mammal species able to ingest 'toxic' products without being harmed.

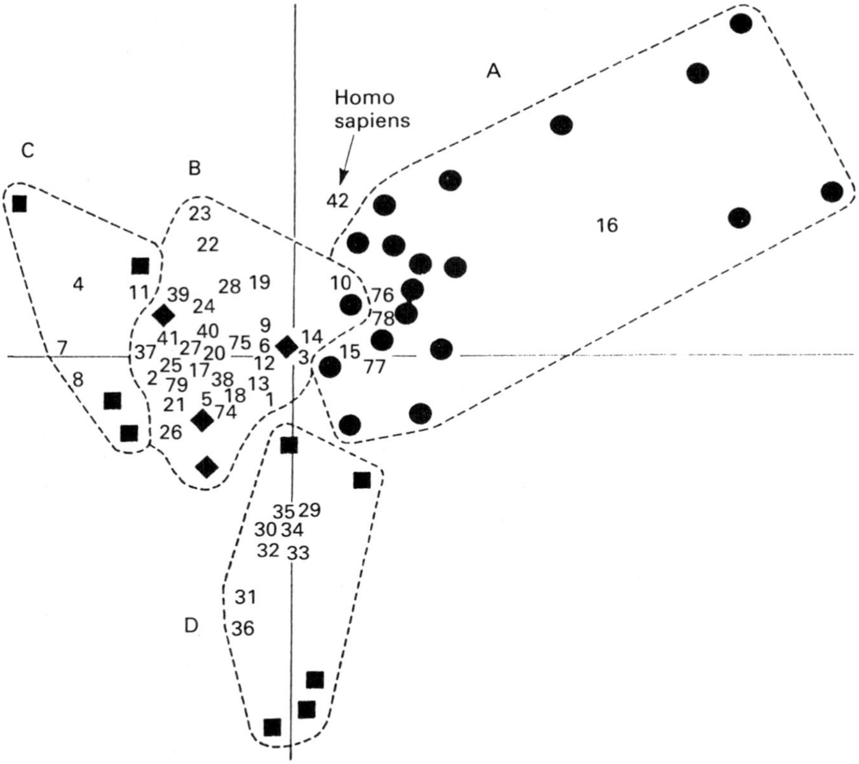

Fig. 5.2. Multi-dimensional plot of indices for surface areas of stomach, small intestine and caecum plus colon for about 80 primate and other mammal species. A, Faunivores, spreading to the upper right, with 'insectivores' and cetaceans more extreme than 'carnivores'; B, frugivores, in the central cluster, mostly primates; C, caeco-colic fermenting folivores, with primates nearer the central cluster and the horse most extreme; D, foregut-fermenting folivores, with ruminants more extreme than colobine monkeys. ●, Carnivores (Carnivora), whales and dolphins (cetacea), insectivores (Insectivora), seals (Pinnipedia), pangolins (Pholidota). ■, Even-toed ungulates (Artiodactyla), anteaters (Edentata), hyraxes (Hydrocoidea), odd-toed ungulates (Perissodactyla), rabbits (hagomorpha), marsupials (Marsupialia). ◆, Rodents: squirrels (Sciuridae). 1,2,3, etc., Primates. (From *The Cambridge Encyclopaedia of Human Evolution*, p. 61.)

Food toxicity

In fact, the toxicity of several potential food items is a major issue reflecting the long evolutionary processes and is a present necessity for many animal species to overcome. The toxic yam species are found exclusively in savannas or at forest edges because most plants of the rain forest ecosystem are so scattered among the other species that the risk of being eaten and destroyed in large numbers is low. In dry environments where biodiversity is not as high, plant species growing in aggregate among a smaller number of other species risk being destroyed by the animal consumers if they have not developed toxicity. The survey of alkaloid frequencies in plant species inside and outside the rain forest of Gabon (Hladik and Hladik, 1977) demonstrated this differential long-term response.

This evolutionary trend towards non-edible plant parts applies particularly to the leaves because the loss of this producing machinery reduces plant fitness. The chemical defences of leaves are generally digestibility-reducing, toxic or otherwise harmful and the herbivores have evolved physiological equipment to neutralize the defences occurring in the plants they select as food. Several aspects of these evolutionary processes resulting from the confrontation between plant eaters and plant secondary compounds are discussed by Waterman *et al.* (1988) and Foley and McArthur (Ch. 22), together with the adaptations to cope with the non-toxic but otherwise difficult to digest plant wall materials.

Among the most surprising materials frequently ingested by mammals (and that we consider as foods) are minerals of various origins. Although the most documented studies refer to the sodium content of the earth ingested by large herbivores, we have some evidence of another important role of these materials, when they are consumed with leaves, in relation to toxicity and/or low digestibility. When the content of sodium and other mineral elements in the earth ingested by colobine monkeys was compared to that of the whole diet (Hladik and Gueguen, 1974), it appeared that all the necessary dietary minerals were available in leaves and fruits regularly consumed, and the soluble mineral content of the earth was lower than that of the rest of the diet. Clay is frequently eaten by folivorous primates (and eaten in larger amounts when mature leaves are consumed); this material, in contact with particles of leaves in the stomach, can adsorb tannins and prevent their combination with protein which would decrease digestive efficiency. In this context, geophagy can enable the various fermenting systems of folivorous mammals to cope with leaf toxicity.

With regard to edible plant parts, the sensory response is the first adaptation of the digestive system to toxic plant parts (or to the insects that concentrate the substances of the plants). The high sensitivity of the taste buds to bitter substances and the negative responses elicited by astringent and bitter tastes can be considered as an adaptation to avoid toxic or nutritionally inefficient potential foods. The 'gusto-facial reflex' of primates is a genetically-programmed response which prevents even a newborn swallowing a bitter substance (Steiner and Glaser, 1984). These taste responses, shared by all human populations (Hladik, 1990a), are more flexible than any other adaptations to food composition. Tolerance to large quantities of bitter substances might become a necessity, for example for a primate species feeding on the exudates of trees with a bitter bark. This aspect, investigated by Simmen (Ch. 10) shows the importance of the above-threshold responses.

The digestive system of mammals appears as an integrated complete system responding to an environment where potential foods are the result of sucessive adaptations of both the plant and animal species.

References

Charles-Dominique, P. & Martin, R. D. (1970). Evolution of lorises and lemurs. *Nature*, **227**, 257–260.

Chivers, D. J. & Hladik, C. M. (1980). Morphology of the gastro-intestinal tract in primates: comparisons with other mammals in relation to diet. *Journal of Morphology*, **166**, 337–386.

Estrada, A. & Fleming, T. H. (eds.) (1986). *Frugivore and seed dispersal.* Dordrecht: Dr W. Junk Publisher.

Herrera, C. M. (1985). Determinants of plant–animal coevolution: the case of mutualistic dispersal of seeds by vertabrates. *Oikos*, **44**, 132–141.

Hladik, A. & Hladik, C. M. (1977). Significations écologiques des teneurs en alcaloïdes des végétauz de la forêt dense: Résultats des tests préliminaires effectués au Gabon. *Revue d'Ecologie (Terre et Vie)*, **31**, 515–555.

Hladik, A. & Dounias, E. (1993). Wild yams of the African rain forest as potential food resources. In *Tropical Forests, People and Food: Biocultural Interactions and Applications to Development*, Man and the Biosphere series, Vol. 15, ed. C. M. Hladik, H. Pagezy, O. F. Linares, A. Hladik, A. Semple & M. Hadley, pp. 147–160. Paris: UNESCO & Parthenon.

Hladik, C. M. (1985). Discussion. In *L'Environnement des Hominidés au Pleistocène*, ed. Y. Coppens, pp. 447–452. Paris: Masson.

Hladik, C. M. (1990a). Gustatory perception and food taste. In *Food and Nutrition in the African Rain Forest*, ed. C. M. Hladik, S. Bahuchet & I. de Garine, pp. 67–68. Paris: UNESCO/MAB-CNRS.

Hladik, C. M. (1990b). Les stratégies alimentaires des primates. In *Primates, recherches actuelles*, ed. J. J. Roeder & J. R. Anderson, pp. 35–52. Paris: Masson.

Hladik, C. M. (1993). Fruits of the rain forest and taste perception as a result of evolutionary interactions. In *Tropical Forests, People and Food: Biocultural Interactions and Applications to Development*, Man and the Biosphere series, Vol. 13, ed. C. M. Hladik, H. Pagezy, O. F. Linares, A. Hladik, A. Semple & M. Hadley, pp. 65–74. Paris: UNESCO & Parthenon.

Hladik, C. M., Charles-Dominique, P. & Petter, J. J. (1980). Feeding strategies of five nocturnal prosimians in the dry forest of the West coast of Madagascar. In *Nocturnal Malagasy Primates, Ecology, Physiology and Behaviour*, ed. P. Charles-Dominique, H. M. Cooper, A. Hladik *et al.*, 41–73, New York: Academic Press.

Hladik, C. M. & Chivers, D. J. (1978). Ecological factors and specific behavioural patterns determining primate diet. In *Recent Advances in Primatology*, ed. D. J. Chivers & J. Herbert, pp. 433–444. London: Academic Press.

Hladik, C. M. & Gueguen, I. (1974). Géophagie et nutrition minérale chez les primates sauvages. *Comptes Rendus des Academie de Séances*, Paris, **279**, 1393–1396.

Steiner, E. J. & Glaser, D. (1984). Differential behavioral responses to taste stimuli in non human primates. *Journal of Human Evolution*, **13**, 709–723.

Waterman, P. G., Ross, J. A. M., Bennett, E. L. & Davies, A. G. (1988). A comparison of the floristics and leaf chemistry of the tree flora in two Malaysian rain forests and the influence of leaf chemistry on populations of colobine monkeys in the Old World. *Biological Journal of the Linnean Society*, **34**, 1–16.

6

Classification of foods for comparative analysis of the gastro-intestinal tract

PETER LANGER and DAVID J. CHIVERS

When classifying food, the following questions are often asked:

1. What are the fields of study that need food classification?
2. What criteria have to be considered in this classification in different fields of study?
3. What is the discrimination (or resolution) of classifications in different fields: is it always similar, or is it necessary to recognise coarser- or finer-grain classifications in different fields of study?

Food classifications are needed and applied in animal and human nutrition science, comparative physiology and ecology; they discriminate food characters with as much detail as necessary and/or possible. When classifying food materials, one should keep in mind that categorizing is always dangerous because it can restrict thoughts and ignore the origins of data used and required for the categories (Winkler, 1984).

The resolution (or discrimination) used in the classification of food can be based on physical properties, as well as chemical composition and food availability and shows the following tendencies in different areas of study.

Ecology Modern ecological studies emphasize, among other criteria, the availability or abundance of food in relation to area and time (e.g. Rattray, 1960; Osbourn, 1980; van Dyne *et al.*, 1980; Owen-Smith, 1982, 1988; Duncan, 1991). Chemical and physical properties, and food availability, have to be given in as much detail as possible. Detailed data are often not available, however, and the ecologist has to work with low resolution (coarse grain) data. Perrin (1988) of the University of Natal, Pietermaritzburg supplied a highly differentiated trophic classification for the mammals living in the Eastern Cape Coast area of South Africa. Table 6.1 is a modified compilation

Table 6.1. *Trophic classification of mammals – the ecologist's view*

Predators: mammals which prey upon other animals
Insectivorous mammals: feeding mainly on insects
Carnivorous mammals: feeding mainly on the flesh of vertebrates

Herbivores: mammals which subsist primarily on plants or plant material
Large herbivorous mammals
- Grazers: feed mainly on grass foliage
- General herbivores: feed on herbs, grasses or shrubs and trees
- Browsers: feed on leaves of herbs and shoots of shrubs and trees
- Omnivores: feed mainly on roots and bulbs, grasses, seeds and fruit, but also take carrion, reptiles, birds, eggs and insects

Medium-sized herbivorous mammals
- Grazers: feed mainly on grass foliage
- Browsers: feed on leaves of herbs and shoots of shrubs and trees
- Frugivores/folivores: feed on fruit and leaves of trees and shrubs

Small herbivorous mammals: mammals which all gather their food by gnawing
- Commensal omnivores: mammals which live together sharing food which may be of any kind, mainly of plant origin
- Non-commensal omnivores: mammals which do not live together, but do take all kinds of food, mainly of plant origin
- Frugivores: fruit-eating mammals
- Granivores: mammals living on grain and seeds
- Arboreal granivores/insectivores: inhabitants of trees which live on grain, seed and insects
- Grazers: feed mainly on grass foliage
- Fossorial herbivores: mammals which live in burrows and feed on plants

From Perrin (1988).

from his publication. It shows that ecological investigations can also deal with a fine-grain classification of food.

Palaeo-ecology Palaeo-ecologists cannot usually deal with the chemical composition of food, but studies of dentition and tooth wear give reliable ideas about the physical properties of the food available to fossil animal species (e.g. Lucas, Ch. 13; Kay and Covert, 1984; Fortelius, 1985). The high resolution of such data is sometimes impressive, but low resolution data often have to be used as well. The study of body size of fossil species and of palaeo-climatology indicates features of food availability and abundance during geological periods.

Nutrition Very detailed (high resolution) data are needed by nutritionists. Tables of nutrient requirements for domestic animals are issued by different authorities worldwide, in terms of energy content, physical properties, chemical composition and pasture quality (carrying capacity, or food availability either per area or per time-unit) (e.g. Rattray, 1960; National Academy of Sciences, 1973; DLG-Futterwerttabellen, 1982, 1984a,b; Meyer and Heckötter, 1986; Duncan, 1991).

Comparative functional anatomy The comparative functional anatomist uses criteria to discriminate between the physical properties of food (e.g. Chivers and Hladik, 1980; Clemens, 1980; Eisenberg, 1981; Langer, 1984; Stevens, 1988). It is not necessary to know the precise chemical composition of food; it is much more important to define the relative proportions of animal or plant matter, soluble or structural carbohydrates and nutrient or bulk (ballast) contents. In relation to food availability, a high resolution is established because it is important to determine whether large or small volumes are ingested over specific time periods, whether certain food constituents are selected or avoided or whether voided faeces are reingested (coprophagy and caecotrophy).

Comparative physiology High resolution (fine-grain) data are also needed by comparative physiologists (e.g. Schmidt-Nielsen *et al.*, 1980; Hume, 1982; Stevens, 1988). Comparisons of large intestine fermenters with forestomach fermenters, or of eutherian and marsupial herbivores, and, of course, the comparison between animals eating food of animal or plant origin, need precise data on physical and chemical properties (especially important functionally) and on abundance.

Evolutionary biology Evolutionary biologists studying the form and function of the digestive system apply a more generalized view, with high resolution when classifying the physical properties, chemical composition and availabilities of foods (e.g. Stevens, 1988; Langer, 1991; Langer and Snipes, 1991).

When studying the ruminant stomach (Hofmann, 1969, 1973) or the mammalian herbivore stomach (Langer, 1988), the categorization just refers to food consisting of plant material but a discussion of a wide range of mammals can only make sense with consideration of a wide range of food materials, both of animal and of plant origin. This is, at least, the case for classifying the chemical and physical properties and the chemical composition of food. In a very recent study Taylor (1992) deals with *Bettongia gaimardi*, a mycophagous marsupial but in discussing the range of food of this species in Tasmania, he has high discrimination by considering about 20 items of plant and fungus origin and at least seven items from invertebrates.

In the following account, physical properties and chemical composition, as well as the availability of food, will not be discussed separately, because the diversity of terms used in the literature is too great. A general overview of the different possibilities in categorizing food materials is given, but food classes for different fields of investigation or for a variety of food characteristics, such as physical, chemical or quantitative properties will not be specified. Each investigation and each investigator emphasizes different aspects and, therefore, they need different sets of food categories according to special requirements. Here, however, the danger exists that the basis of the data used and required for the categories (Winkler, 1984) might be forgotten.

Types of food classification

To obtain an overview, excerpts from the relevant literature (56 references) were made (Table 6.2). A combination of food classes with categories of behaviour, as given by Jarman (1974), McNab (1988) and Sedlag (1988) or with categories of habitat (Eisenberg, 1981; Freeland *et al.*, 1988; McNab, 1978, 1988) is not discussed here. Groups are formed according to five fields of investigation, namely, ecology, nutrition, comparative functional anatomy, comparative physiology and evolutionary biology. Langer attributed the publications to one or more fields of investigation and there is probably some bias in the listing of those references that deal with a wide range of studies. Within the fields of investigation, six mammalian groups ('taxa'), as they were discussed in the references, are grouped in alphabetical order: 'herbivores', Mammalia, Marsupialia, Primates, Ruminantia, 'ungulates'. It can be seen that the number of categories varies from only 2 to 32 and that a few papers offer more than one type of categorization. The discrimination or resolution is also tabulated. High resolution is achieved when a wide range of food materials from animal to plant origin is categorized. When, however, only a small range of food types is considered, when the mammal is a frugivore or a grazer, this presents only low discrimination.

In all fields of investigation, considered classifications with high discrimination of food classes are more frequent than those with low discrimination. It is interesting that almost equal numbers of the ecological studies considered show either high or low discrimination in food classification, whereas in studies of nutrition, comparative functional anatomy, comparative physiology and evolutionary biology high resolution (discrimination) predominates. When phytophagous groups (herbivores, ungulates, or Ruminantia) are considered, only small discrimination is applied. However, in Primates, Marsupialia and Mammalia in general, the resolution of food classification is

Table 6.2. *Compilation of data from the literature on the categorization of food*

Fields of investigation	Taxa or group	Number of categories	Resolution of categorizations	Authors
Ecology	Herbivores	2	l	Hörnicke (1982)
	Herbivores	5	l	Eisenberg (1978)
	Mammalia	3	l	Case (1979)
	Mammalia	16	h	Eisenberg (1981)
	Mammalia	3	l	Southwood (1985)
	Mammalia	3	h	Evans and Miller (1968)
	Mammalia	4	l	Lindroth (1989)
	Mammalia	5	h	Thompson (1987)
	Mammalia	7	h	Brambell (1972)
	Mammalia	7	h	Sedlag (1988)
	Mammalia	8	h	Arita *et al.* (1990)
	Mammalia	3	l	Kerley (1989)
	Mammalia	32	h	Perrin (1988)
	Marsupialia	3	h	Freeland *et al.* (1988)
	Primates	3	h	Waterman (1984)
	Primates	5	l	Richard (1978)
	Primates	5	l	Schlichte (1978)
	Primates	5	h	Coe (1984)
	Primates	5	h	van Roosmalen (1984)
	Primates	6	h	Rodman (1978)
	Primates	7	l	Struhsaker (1978)
	Primates	7	h	Ripley (1984)
	Primates	7	h	van Roosmalen (1984)
	Primates	8	l	Oftedal (1991)
	Primates	12	l	Rudran (1978)

	Primates	2	h	Andrews and Aiello (1984)
	Primates	5	h	Andrews and Aiello (1984)
	Ruminantia	3	l	Hofmann and Stewart (1972)
	Ruminantia	5	l	Jarman (1974)
	Ungulates	3	l	Gwynne and Bell (1968)
Nutrition	Herbivores	2	l	Hörnicke (1982)
	Herbivores	4	l	van Soest (1982)
	Herbivores	8	l	Demment and van Soest (1985)
	Mammalia	3	l	Englyst (1989)
	Mammalia	3	h	Evans and Miller (1968)
	Mammalia	5	h	Thompson (1987)
	Mammalia	6	h	Karasov and Diamond (1988)
	Mammalia	7	h	Brambell (1972)
	Mammalia	7	h	Sedlag (1988)
	Primates	2	h	Andrews and Aiello (1984)
	Primates	3	h	Waterman (1984)
	Primates	5	h	Andrews and Aiello (1984)
	Ungulates	5	l	Bodmer (1990)
Comparative functional anatomy	Herbivores	6	l	Langer (1986)
	Mammalia	15	h	Langer (1991)
	Mammalia	6	l	Langer (1988)
	Mammalia	7	l	Langer (1987c)
	Mammalia	8	h	Langer (1987b)
	Mammalia	7	h	Langer and Snipes (1991)
	Marsupialia	5	h	Langer (1979)
	Primates	3	h	Chivers and Hladik (1980)
	Primates	5	h	Kay and Covert (1984)
	Ruminantia	3	l	Hoffman and Stewart (1972)
	Ungulates	3	l	Langer (1987a)

Table 6.2 Continued

Fields of investigation	Taxa or group	Number of categories	Resolution of categorizations	Authors
Comparative physiology	Mammalia	21	h	McNab (1988)
	Mammalia	3	l	Englyst (1989)
	Mammalia	6	h	Karasov and Diamond (1988)
	Mammalia	6	h	McNab (1980)
	Mammalia	7	h	Nagy (1987)
	Mammalia	7	h	Sedlag (1988)
	Marsupialia	5	h	McNab (1978)
	Primates	2	h	Andrews and Aiello (1984)
	Primates	5	h	Andrews and Aiello (1984)
Evolutionary biology	Mammalia	16	h	Eisenberg (1981)
	Marsupialia	9	h	Lee and Cockburn (1985)

h, a wide range of food materials (from animal to plant) is categorized; l, a small range of food types is considered.

Table 6.3. *Terms that can be used to classify food*

- Generalist feeder; **polyphage**; omnivore
- Specialist feeder; **oligophage**
  - Animal matter feeder; **zoophage**; faunivore: these can be sub-divided into various types of food
    - Invertebrate (microfaunivore)
      - Arthropod
        - Ant (myrmecophage)
        - Insect (insectivore)
        - Crustacean (crustacivore)
      - Mollusc (molluscivore)
        - Squid
        - Clam
        - Mussel
    - Zoo-plankton (planktonivore)
    - Vertebrate (macrofaunivore)
      - Vertebrate other than mammal: fish (piscivore)
      - mammal (carnivore)
    - Blood (sanguinivore)
  - Plant matter feeder; **phytophage**; florivore; these can be divided into three types of feeders:
    - Bulk and roughage eaters
    - Concentrate selectors
    - Intermediate feeders
  - Plant matter food includes
    - Fungus (mycophage)
    - Lichen (lichenophage)
    - Moss (bryophytophage)
    - Wood
    - Bark
    - Branch
    - Root
    - Underground storage organ
      - Bulb
      - Corm
      - Tuber
    - Diaspores
      - Fruit (frugivore)
      - Nut
      - Cereal grain (granivore): seed, hull
    - Gall
    - Exudates
      - Saps
      - Gums
      - Resins
    - Blossoms/flowers
    - Pollen
    - Nectar (nectarivore)
    - Browse (browser)
      - Legume
      - Non-leguminous angiosperm
        - Twig
        - Herb (herbivore)
          - Forb: shoot; leaf (folivore); leaf stalk or petiole; bud
          - Grass (grazer, **poëphage**, graminivore): stem; sheath; leaf

more often high than low. This means that both food of animal and of plant origin is included in the categorization. It can be concluded that the majority of the publications considered categorize the food of mammals over a wide range of theoretically available food materials.

Conclusion

The categories from 56 references, as they are listed in Table 6.2, are partly redundant and an effort was made to categorize the types of food with as little tautology as possible (Table 6.3). In some cases, terms of Greek (zoophage) and of Latin (faunivore) origin are considered synonymous.

The grasses play an important role amongst phytophagous mammals. About a quarter of the world's land area is covered with grasslands (Shantz, 1954) and approximately 71% of the metabolizable energy consumed by British livestock is derived from grazing (Holmes, 1980). To emphasize the relative importance of grass-eaters or graminivores, it was thought appropriate to apply a term that is formally similar to polyphage, zoophage, and phytophage, which are all of Greek origin. Plant taxonomists call the grass family Poaceae (Hamby and Zimmer, 1988) and because of this **poëphage** was introduced for mammals that live mainly on grass (Werner, 1961).

In phytophagous mammals, the different types of plant matter cannot be classified unambiguously and, because of different stages of maturity, desiccation or wilting, it is not appropriate in a wide-range classification of food materials to categorize them according to the three groups differentiated by Hofmann and Stewart (1972). These authors divided the phytophages (florivores) into bulk and roughage eaters, concentrate selectors and intermediate feeders. Therefore, we need to categorize foods as precisely as possible. We cannot insist that other authors follow our classification, but we can offer unambiguous terms to researchers working on the comparative analysis of the gastro-intestinal tract. This will hopefully lead to a nomenclatural discussion and then to the establishment of a widely-applicable terminology.

References

Andrews, P. & Aiello, L. (1984). An evolutionary model for feeding and positional behaviour. In *Food Acquisition and Processing in Primates*, ed. D. J. Chivers, B. A. Wood & A. Bilsborough, pp. 429–466. New York: Plenum Press.

Arita, H. T., Robinson, J. G. & Redford, K. H. (1990). Rarity in neotropical forest mammals and its ecological correlates. *Conservation Biology*, **4**, 181–192.

Bodmer, R. E. (1990). Ungulate frugivores and the browser-grazer continuum. *Oikos*, **57**, 319–325.

Brambell, M. R. (1972). Mammals: Their nutrition and habitat. In *Biology of Nutrition*, ed. R. N. T. W. Fiennes, pp. 613–648. London: Pergamon Press.
Case, T. J. (1979). Optimal body size and an animal's diet. *Acta Biotheoretica*, **28**, 54–69.
Chivers, D. J. & Hladik, C. M. (1980). Morphology of the gastro-intestinal tract in primates: comparisons with other mammals in relation to diet. *Journal of Morphology*, **166**, 337–386.
Clemens, E. T. (1980). The digestive tract: Insectivore, prosimian and advanced primate. In *Comparative Physiology: Primitive Mammals*, ed. K. Schmidt-Nielsen, L. Bolis & C. R. Taylor, pp. 88–99. Cambridge: Cambridge University Press.
Coe, M. (1984). Primates: Their niche structure and habitats. In *Food Acquisition and Processing in Primates*, ed. D. J. Chivers, B. A. Wood & A. Bilsborough, pp. 1–32. New York: Plenum Press.
Demment, M. W. & van Soest, P. J. (1985). A nutritional explanation for body-size patterns of ruminant and nonruminant herbivores. *American Naturalist*, **125**, 641–672.
DLG-Futterwerttabellen (1982). *DLG-Futterwerttabellen für Wiederkäuer*, 5 Aufl. Frankfurt: DLG-Verlag.
DLG-Futterwerttabellen (1984a). *DLG-Futterwerttabellen für Pferde*, 2 Aufl. Frankfurt: DLG-Verlag.
DLG-Futterwerttabellen (1984b). *DLG-Futterwerttabellen für Schweine*, 5 Aufl. Frankfurt: DLG-Verlag.
Duncan, P. (1991). *Horses and Grasses*. New York: Springer-Verlag.
Eisenberg, J. F. (1978). The evolution of arboreal herbivores in the class Mammalia. In *The Ecology of Arboreal Folivores*, ed. G. G. Montgomery, pp. 135–152, Washington, DC: Smithsonian Institution Press.
Eisenberg, J. F. (1981). *Mammalian Radiations*. Chicago: University of Chicago Press.
Englyst, H. (1989). Classification and measurement of plant polysaccharides. *Animal Feed Science and Technology*, **23**, 27–42.
Evans, E. & Miller, D. S. (1968). Comparative nutrition, growth and longevity. *Proceedings of the Nutrition Society*, **27**, 121–129.
Fortelius, M. (1985). The functional significance of wear-induced change in the occlusal morphology of herbivore cheek teeth, exemplified by *Dicerorhinus etruscus* upper molars. *Acta Zoologica Fennica*, **170**, 157–158.
Freeland, W. J., Winter, J. W. & Raskin, S. (1988). Australian rock-mammals: A phenomenon of seasonally dry tropics. *Biotropica*, **20**, 70–79.
Gwynne, M. D. & Bell, R. H. V. (1968). Selection of vegetation components by grazing ungulates in the Serengeti National Park. *Nature*, **220**, 390–393.
Hamby, R. K. & Zimmer, E. A. (1988). Ribosomal RNA sequence for inferring phylogeny within the grass family (Poaceae). *Plant Systematics and Evolution*, **160**, 29–37.
Hofmann, R. R. (1969). *Zur Topographie und Morphologie des Wiederkäuermagens im Hinblick auf seine Funktion*. Beiheft 10 zum Zentralblatt für Veterinärmedizin, Berlin: Verlag Paul Parey.
Hofmann, R. R. (1973). *The Ruminant Stomach*. Nairobi: East African Literature Bureau.
Hofmann, R. R. & Stewart, D. R. M. (1972). Grazer or browser: A classification based on the stomach structure and feeding habits of East African ruminants. *Mammalia*, **36**, 226–240.

Holmes, W. (1980). *Grass, its Production and Utilization.* Oxford: Blackwell Scientific.
Hörnicke, H. (1982). Comparative aspects of digestion in nunruminant herbivores. In *Exogenous and Endogenous Influences on Metabolic and Neural Control.* ed. A. D. F. Addink & N. Spronk, pp. 57–68. Oxford: Pergamon Press.
Hume, I. D. (1982). *Digestive Physiology and Nutrition of Marsupials.* Cambridge: Cambridge University Press.
Jarman, P. J. (1974). The social organisation of antelope in relation to their ecology. *Behaviour*, **48**, 215–267.
Karasov, W. H. & Diamond, J. M. (1988). Interplay between physiology and ecology in digestion. *BioScience*, **38**, 602–611.
Kay, R. F. & Covert, H. H. (1984). Anatomy and behaviour of extinct primates. In *Food Acquisition and Processing in Primates*, ed. D. J. Chivers, B. A. Wood & A. Bilsborough, pp. 467–508. New York: Plenum Press.
Kerley, G. I. H. (1989). Diet of small mammals from the Karoo, South Africa. *South African Journal of Wildlife Research*, **19**, 67–72.
Langer, P. (1979) Phylogenetic adaptations of the stomach of the Macropodidae Owen, 1839, to food. *Zeitschrift für Säugetierkunde*, **44**, 321–333.
Langer P. (1984). Anatomical and nutritional adaptations in wild herbivores. In *Herbivore Nutrition in the Subtropics and Tropics*, ed. F. M. C. Gilchrist & R. I. Mackie, pp. 185–203. Craighall, South Africa: Science Press.
Langer, P. (1986). Large mammalian herbivores in tropical forests with either hindgut- or forestomach-fermentation. *Zeitschrift für Säugetierkunde*, **51**, 173–187.
Langer, P. (1987a). Evolutionary patterns of Perissodactyla and Artiodactyla (Mammalia) with different types of digestion. *Zeitschrift für zoologische Systematik und Evolutionsforschung*, **25**, 212–236.
Langer, P. (1987b). Formenmannigfaltigkeit mehrkammeriger Mägen bei Säugetieren. *Natur und Museum*, **117**, 47–60.
Langer, P. (1987c). Der Verdauungstrakt bei pflanzenfressenden Säugetieren. *Biologie in unserer Zeit*, **17**, 9–14.
Langer, P. (1988). *The mammalian herbivore stomach. Comparative anatomy, function and evolution.* Stuttgart: Gustav Fischer Verlag.
Langer, P. (1991). Evolution of the digestive tract in mammals. *Verhandlungen der Deutschen Zoologischen Gesellschaft*, **84**, 169–193.
Langer, P. & Snipes, R. L. (1991). Adaptation of gut structure to function in herbivores. In *Physiological Aspects of Digestion and Metabolism in Ruminants*, ed. T. Tsuda, Y. Sasaki & R. Kawashima, pp. 349–384. San Diego: Academic Press.
Lee, A. K. & Cockburn, A. (1985). *Evolution Ecology of Marsupials.* Cambridge: Cambridge University Press.
Lindroth, R. L. (1989). Mammalian herbivore–plant interactions. In *Plant–Animal Interactions.* ed. W. G. Abrahamson, pp. 163–206, New York: McGraw Hill.
McNab, B. K. (1978). The comparative energetics of neotropical marsupials. *Journal of Comparative Physiology*, **125**, 115–128.
McNab, B. K. (1980). Food habits, energetics, and the population biology of mammals. *American Naturalist*, **116**, 106–124.
McNab, B. K. (1988). Complications inherent in scaling the basal rate of metabolism in mammals. *Quarterly Review of Biology*, **63**, 25–54.
Meyer, H. & Heckötter, E. (1986). *Futterwerttabellen für Hunde und Katzen*, 2 Aufl. Hannover: Schlütersche Verlagsanstalt.

Nagy, K. A. (1987). Field metabolic rate and food requirement scaling in mammals and birds. *Ecological Monographs*, **57**, 111–128.

National Academy of Sciences (1973). *Nutrient Requirements of Swine*, 7th edn. Washington, DC: National Academy of Sciences.

Oftedal, O. T. (1991). The nutritional consequences of foraging in primates: the relationship of nutrient intakes to nutrient requirement. *Philosophical Transactions of the Royal Society of London*, **B334**, 161–170.

Osbourn, D. F. (1980). The feeding value of grass and grass products. In *Grass, its Production and Utilization*, ed. W. Holmes, pp. 70–124. Oxford: Blackwell Scientific.

Owen-Smith, N. (1982). Factors influencing the consumption of plant products by large herbivores. In *Ecology of Tropical Savannas*, ed. B. J. Huntley & B. H. Walker, pp. 359–404. Berlin: Springer-Verlag.

Owen-Smith, N. (1988). *Megaherbivores. The Influence of Very Large Body Size on Ecology*. Cambridge: Cambridge University Press.

Perrin, M. R. (1988). Terrestrial mammals. In *A field guide to the Eastern Cape Coast*, eds. R. Lubke, F. Gess & M. Bruton, pp. 289–313. Grahamstown: Centre of the Wildlife Society of Southern Africa.

Rattray, J. M. (1960). *The Grass Cover of Africa*. Rome: FAO.

Richard, A. F. (1978). Variability in the feeding behavior of a Malagasy prosimian, *Propithecus verreauxi*: Lemuriformes. In *The ecology of arboreal folivores*, ed. G. G. Montgomery, pp. 519–533, Washington, DC: Smithsonian Institution Press.

Ripley, S. (1984). Environmental grain, niche diversification and feeding behaviour in primates. In *Food Acquisition and Processing in Primates*, ed. D. J. Chivers, B. A. Wood & A. Bilsborough, pp. 33–72. New York: Plenum Press.

Rodman, P. S. (1978). Diets, densities, and distributions of Bornean primates. In *The Ecology of Arboreal Folivores*, ed. G. G. Montgomery, pp. 465–478, Washington, DC: Smithsonian Institution Press.

Rudran, R. (1978). Intergroup dietary comparisons and folivorous tendencies of two groups of blue monkeys (*Cercopithecus mitis stuhlmanni*). In *The Ecology of Arboreal Folivores*, ed. G. G. Montgomery, pp. 483–503, Washington, DC: Smithsonian Institution Press.

Schlichte, H.-J. (1978). The ecology of two groups of blue monkeys, *Cercopithecus mitis stuhlmanni*, in an isolated habitat of poor vegetation. In *The Ecology of Arboreal Folivores*, ed. G. G. Montgomery, pp. 505–517. Washington, DC: Smithsonian Institution Press.

Schmidt-Nielsen, K., Bolis, L. & Taylor, C. R. (1980). *Comparative Physiology: Primitive Mammals*. Cambridge: Cambridge University Press.

Shantz, H. L. (1954). The place of grasslands in the earth's cover of vegetation. *Evolution*, **35**, 143–145.

Sedlag, U. (1988). *Wie leben Säugetiere*? Frankfurt: Verlag Harri Deutsch.

Southwood, T. R. E. (1985). Interactions of plants and animals: patterns and processes. *Oikos*, **44**, 5–11.

Stevens, C. E. (1988). *Comparative Physiology of the Vertebrate Digestive System*. Cambridge: Cambridge University Press.

Struhsaker, T. T. (1978). Interrelations of red colobus monkeys and rain-forest treees in the Kibale Forest, Uganda. In *The Ecology of Arboreal Folivores*, ed. G. G. Montgomery, pp. 397–422, Washington, DC: Smithsonian Institution Press.

Taylor, R. J. (1992). Seasonal changes in the diet of the Tasmanian bettong

(*Bettongia gaimardi*), a mycophagous marsupial. *Journal of Mammalogy*, **73**, 408–414.

Thompson, S. D. (1987). Body size, duration of parental care, and the intrinsic rate of natural increase in Eutherian and metatherian mammals. *Oecologia*, **71**, 201–209.

van Dyne, G. M., Brockington, N. R., Scocs, Z., Duek, J. and Ribic, C. A. (1980). Large herbivore subsystem. In *Grasslands, System Analysis and Man*, ed. A. I. Breymeyer & G. M. van Dyne, pp. 269–537. Cambridge: Cambridge University Press.

van Roosmalen, M. G. M. (1984). Subcategorizing foods in primates. In *Food Acquisition and Processing in Primates*, ed. D. J. Chivers, B. A. Wood & A. Bilsborough, pp. 167–175. New York: Plenum Press.

van Soest, P. J. (1982). *Nutritional Ecology of the Ruminant*. Corvallis, OR: O & B Books.

Waterman, P. G. (1984). Food aquisition and processing as a function of plant chemistry. In *Food Acquisition and Processing in Primates*, ed. D. J. Chivers, B. A. Wood & A. Bilsborough, pp. 177–211. New York: Plenum Press.

Werner, C. F. (1961). *Wortelemente lateinisch-griechischer Fachausdrücke in den biologischen Wissenschaften*, 2 Aufl. Leipzig: Akademische Verlagsgesellschaft Geest & Portig.

Winkler, P. (1984). The adaptive capacities of the Hanuman langur and the categorizing of diet. In *Food Acquisition and Processing in Primates*, ed. D. J. Chivers, B. A. Wood & A. Bilsborough, pp. 161–166. New York: Plenum Press.

7

The 'carnivorous' herbivores

R.J. MOIR

Few mammalian neonates have the capacity to use the food of their herbivore parents readily; even precocial young are unable to cope with the quality of the components of the adult diet or to harvest them precisely (Oftedal, 1980). Altricial young have no chance at all. The maternal pre-processing of food and presentation of nutrients as milk simplifies the processes of digestion and masticatory development for the young and is advantageous to the parent in permitting her to harvest wide food resources to produce milk. The burden of milk production is frequently very heavy, particularly where seasonal conditions result in inadequate quality and quantity of food. The milk-fed young continue to receive the same quality, but not necessarily quantity, of diet but it is done at the expense of the dam's own tissues. Many mammals under such conditions rear a reduced litter and do not survive to breed again; a few animals, such as the polar bear, normally survive terminal pregnancy and lactation, without eating, solely from body reserves (Schmidt-Nielsen, 1986).

McCance and Widdowson (1964) stress the capacity of the newborn mammal to incorporate a very high proportion of milk protein into tissue despite the immaturity of tissues and systems particularly in the gut, the kidney, and in thermal regulation. They point out that in both the newborn pig and puppy, 92–94% of dietary protein is incorporated into tissue; the same is true, but to a lesser extent, for the relatively slow-growing human infant. 'The whole principle of feeding in the newborn period, and indeed later, should be to saturate the growth requirements of the animal in question . . . but at the same time to provide the excretory organs with the minimum amount of work.' Clearly, poor quality proteins or excess salts or water will stress the 'internal milieu' and the excretory organs, jeopardizing the neonate's survival.

In mammals generally, the newborn is equipped with a suitable suite of enzymes to digest the food components of milk including a lactase for the milk lactose, a sugar rarely found elsewhere in nature, and a casein-clotting enzyme, chymosin (rennin), and proteinases. The meconium at birth shows strong proteolytic activity clearly of endogenous origin as it is sterile. Milk coagulation, under the influence of chymosin and acid, produces two meals or fractions with quite different passage characteristics. Ternouth *et al.* (1974, 1975) have shown that when 80–90% of whey fluids have passed from the abomasum only 25–30% of the total milk protein has done so. The blood sugar and fat levels are fully in keeping with these findings, as are the proportions of amino acids in the abomasum (Yvon *et al.*, 1985). The multiple protein nature of milk enables some manipulation of the amino acid supply through variation of the several whey proteins as occurs in the wallaby where specific demands, such as the heightened sulphur-containing amino acid requirements of the embryo for hair and nail growth, can be met (Renfree *et al.*, 1981).

The undeveloped newborn frequently has poorly developed thermal regulatory mechanisms and they may be 'almost poikilotherms' (Brody, 1945), with anomalous low maintenance energy demands that further enhance the efficiency of growth under some conditions. This follows only for equal intakes where the lower poikilothermic maintenance-demand rate enables a higher partial growth rate. However, ingestion rate normally is geared to metabolic rate (Kleiber, 1961; Peters, 1983); based on Farlow's data (1976) adult endothermic carnivores and herbivores both follow the relationship y (ingestion rate in watts) $= 10.7\ W^{0.7}$ whereas the ingestion rate of the carnivorous poikilothermic tetrapods, y (watts) $= 0.78\ W^{0.82}$. According to Pullar and Webster (1977) the costs of depositing 1 g of protein or fat are almost identical, which they confidently stated as 53 kJ ME/g (where ME is metabolizable energy). Deposition 'costs' should remain constant irrespective of thermal state providing intake can be maintained and metabolized.

Lactivory is an adapted form of carnivory in that the food source, milk, is derived directly from animal cells (Moir, 1991). It follows that no young mammal is initially herbivorous (White, 1985); all are carnivores for some period. While this is a general rule for mammals, there is ample evidence that it applies to a large range of invertebrates as well as diverse vertebrates; the initial development of the young borne of eggs, whether oviparous or viviparous, is explicitly dependent on succour from the female in the form of yolk or nutrient transfer.

Breeding in birds is usually confined because of light cycles and seasonal influences on the harvestable supply of the high quality dietary components – usually insects, fruit and unripe grain. This is true of both precocious and altricious neonates. The precocious hatchling has yolk sac reserve for a short while but must rapidly learn shapes and sizes of nutritive objects such as larvae and seeds. They may be totally independent of their parents as in the Megapodes, follow parents but gather their own food (ducks), be shown food by parents or follow parents and be fed by them. The diet of some altricious forms has been well recorded as, in most circumstances, it dominates the behavioural pattern of the reproductive cycle.

Darwin's finches on Espanola fed hatchlings on a mixture of seeds and insects. The larger species, *Geospiza conirostris*, fed less seed (17%) and up to 95% by volume of insects to nestlings, whereas the smaller *G. fuliginosa* fed more seed (35%) and consequently less insects (Downhower, 1976). The practice of feeding a mixture of seeds and insects to hatchlings is common to all finches, but adults continue to feed themselves entirely on seeds. The very rapid growth and maturation of the *Quelea quelea* in Nigeria (Ward, 1965) depends on a high input of insects during the first 5–6 days together with the unripe seeds of *Echinochloa pyramidales*. Rowley (1990) showed that galahs also used unripe seeds, in this case Erodium seeds, along with insect larvae to feed hatchlings.

The several specific references to unripe seeds being fed to young hatchlings suggests possible nutritional advantages in the diet. The developing seed has initially a high protein and non-protein-nitrogen (NPN) content and, although the total amount does not diminish, its proportion as a ratio of total dry matter falls with development. Jennings and Morton (1963) have shown a further important point: the amino acid composition of the NPN and the protein change with time. At 8 days, the whole endosperm contained 9% lysine in the total protein or NPN but the lysine fell quite rapidly until it was only 3.3% after 33 days. The variation in the diet needed to meet specific growth requirements of the hatchlings can be met by varying the prey. Royama (1970) found that spiders were fed to great tit nestlings up to day 10; it has been suggested that this may meet the sulphur-containing amino acid needs for feathering.

A number of birds produce a cellular crop 'milk' developed under the stimulus of prolactin in both parents. The cells lining the crop grow prolifically and almost fill that organ at the time of hatching. This mass of cells is diluted with fluid when fed to the chicks. The dry matter of pigeon crop milk contains about 33% fat and 60% protein, but no carbohydrate (Davies, 1939). This author commented that the high true protein content, high fat and lecithin, rich potassium and phosphorus, medium calcium and low sodium and chloride gave a food similar to that provided by the egg. Grain is added after several days. It has been claimed that 5 g per day of the pigeon crop milk results in greatly improved gains to grain-fed chickens, suggesting that the crop milk contains a growth stimulatory substance (Pace *et al.*, 1952). The crop milk of the emperor penguin is essentially similar to that of the pigeon but, unlike that of the pigeon, is reported to contain almost 8% carbohydrate (Prevost and Vilter, 1962, cited by Schmidt-Nielsen, 1983). The male penguin may have to supply the chick for much longer than a few days.

Among the reptiles, carnivory is always a prelude to herbivory. Pough (1973) examined the diet and energetics of a wide range of lizards. In the families Agamidae, Gerrhosauridae, Iguanidae and Scincidae, juveniles of large herbivorous species are carnivorous until they reach body weights, of 50–300 g; species that weigh more than 300 g are almost all herbivores whereas those weighing less than 50–100 g are carnivores. Pough concludes that for an unspecialized lizard evolution of a large body size both requires and permits herbivorous diet.

In discussion, Pough puts forward evidence that the energy content of flesh and insects is substantially higher than than that of plant material (21.7–22.6 kJ/g versus 15–17.6 kJ/g) and the efficiencies of assimilation were of the order of 85% compared with 50–55% for general plant material. In support, Mayhew (1963) found that juvenile *Sauromalus obesus* increased their rate of growth from 2 mm and 2 g per month to 9 mm and 20 g per month when switched from a herbivorous to an insectivorous diet. Carnivorous populations of painted turtles, similarly, grew more rapidly than either omnivorous (intermediate) or herbivorous populations (Gibbons, 1967).

Rimmer (1986) has shown that the herbivorous reef-fish *Kyphosus cornelii*, particularly those below 50 mm in length, ingested considerable animal material, largely crustaceans, together with fleshy red algae; as the fish grew, animal material in the diet diminished and green and brown algae increased. Papers cited by Rimmer indicated that this progression was common among reef fish.

Koch (1967), in discussing symbiosis, stated that previous work by Buchner had established that a distinct relationship exists between endosymbiosis and the type of nourishment of insects. Koch suggested that the force of Buchner's ideas was strong enough to be a law of nature. When considered in relation to insects, all plant sap suckers bear symbionts and cannot reach sexual maturity without them; all blood suckers – as long as they ingest blood for the whole of their lives – are also symbiont bearers but those that have a bacteria-rich nutrition during their developmental stage and then turn to blood-sucking as adults do not. Predators do not have, nor do they require, symbionts. However, the larvae of the Lepidoptera, purely phytophagous throughout their lives, have so far not been recorded to bear symbionts. White (1985), in asking the question 'when is a herbivore not a herbivore?' considers that if these larvae are true herbivores they are only just so and that their survival is dependent upon the level of nitrogen in the leaves they eat exceeding a critical minimum (White, 1984). He states that it seems that very young herbivores have near, perhaps total, dependence on access to animal or microbial protein for survival and growth. That is, a young herbivore is not a herbivore. The phytophagous larvae may be an exception.

The phytophagous-larval exceptions to the general rule are important in drawing attention to and developing the nutritional base for the growth and development of juvenile animals. Exceptions prove the rule not when they are seen to be exceptions but when they have been shown to be either outside, or reconcilable with, the principles they seem to contradict (Fowler, 1965). Carnivory, insectivory, cannabalism, rich microbial (prokariotic) diets and the herbivory of the lepidopteran larvae are all exceptions to the rule developed by Koch and therefore share the reasons for the exception; it follows that the products of the symbionts share the same reasons. The exceptions are compatible with the principle that most single plant foods are inadequate for the growth of juvenile animals and their development to sexual maturity. There are two possible major reasons. First, the amino acid structure of many single protein foods is either imbalanced or more frequently deficient in specific amino acids, mainly lysine, methionine or threonine; alternatively, some amino acids, particularly lysine, are 'unavailable' although present and therefore are effectively deficient in that material. The beneficial effects of supplementary histidine, isoleucine and methionine on symbiont-deprived aphids clearly indicates some of the amino acid dietary contribution of the symbionts (Mittler, 1971). Second, other growth factors, such as vitamins, are absent or inadequate. In the latter case, while it is generally assumed that the cobalt-containing vitamin B_{12} or cobalamin is essential to all animals, it is not present in plants and it is produced only by bacteria or unicells. Its presence (and need?) in some insects is still entirely equivocal. Wakayama *et al.* (1984) have shown recently that vitamin B_{12} was not detected in some insects, e.g. the house fly, whereas others such as termites had large amounts present; it was present in some larvae but not in their adults. While possibly associated with dietary intake or the presence of vitamin B_{12}-producing bacteria in the gut, the pea aphid, a symbiont bearer, did not contain detectable amounts. Ten non-insect invertebrates, from centipedes to shrimps, all had relatively large amounts of the vitamin and, although more primitive than insects, used the same B_{12} propionate catabolic pathway as vertebrates (Halarnkar *et al.*, 1987). In reviewing the topic, Halarnkar and Blomquist (1989) indicate that in insects, as in plants and many microorganisms, propionate is metabolized through hydroxypropionate to acetate but, where vitamin

B_{12} is plentiful, some insects may use the B_{12}-dependent pathway common to all other animals. Whether selection of prey for B_{12}-rich sources occurs is not known.

If we consider the extraordinary contribution to basic science and the commercial value of the work on metabolic body size concepts by Brody (see Brody, 1945) and by Kleiber (1932, 1961), it is encouraging to see, after a 50-year gestation, the current interest and developments in 'scaling' and its application to biological thinking. In Peters' book *The Ecological Implications of Body Size* (1983), some hundreds of allometric relationships are given in the 50 pages of appendices, mainly developed since 1975.

Hemmingsen's work (1956–60) entitled *Energy Metabolism as Related to Body Size and Respiratory Surfaces and its Evolution* is a particularly important contribution as it extends the $W^{0.75}$ metabolic body size concept to the poikilotherms and unicells covering a body mass range from 10^{-12} to 10^5 g. Each group has the same slope but differs in intercept thus confirming the uniqueness of the several life forms but the ubiquity of the energy concepts. Most developed relationships are for adults, but growth introduces its own demands. Taylor (1985) developed a concept of metabolic age based on two formal genetic scaling rules where metabolic age ($A^{0.27}$) gave a scale for the development of metabolic body size to its mature body weight $A^{0.73}$. The scaled relationship of heat production in MJ of NE/day per $A(\text{kg})^{0.73}$ (NE is net energy) with metabolic age shows the enormous demand of growth above that of maintenance on energy resources in mammals.

Taylor (1985) states that the equilibrium maintenance requirement is the food energy required to maintain an unchanged body weight and composition by a fully mature animal or by an animal held in equilibrium at an immature body weight. The metabolizable energy for maintenance per day per unit of mature metabolic weight (ME) $m/A^{0.73} = C\mu$, where μ is the degree of maturity and C is the mean value for daily maintenance requirement per unit of mature metabolic weight in a fully mature animal ($\mu = 1$). This value approximates 0.6 MJ of metabolizable energy and applies to all degrees of maturity. Based on equations for ten species, the mean curve for the standardized heat production, q, of a normally growing mammal at different degrees of maturity is given as $q = 0.3\ \mu\ (1 - 0.54 \ln \mu)$ where the units for q are megajoules of basal energy per day per unit of mature metabolic weight. Allden's (1969) relationship between digestible energy (DE), intake (I) and maintenance plus growth in relation to body mass (W) is a useful tool as it is derived from grazing animals at various levels of growth: I (Intake kcal DE) $= 176.4\ W\ (\text{kg})^{0.75}\ (1 + 4.02\ G)$, where G is daily growth in kg, substantiates this growth demand. Figure 7.1 represents the energy maintenance

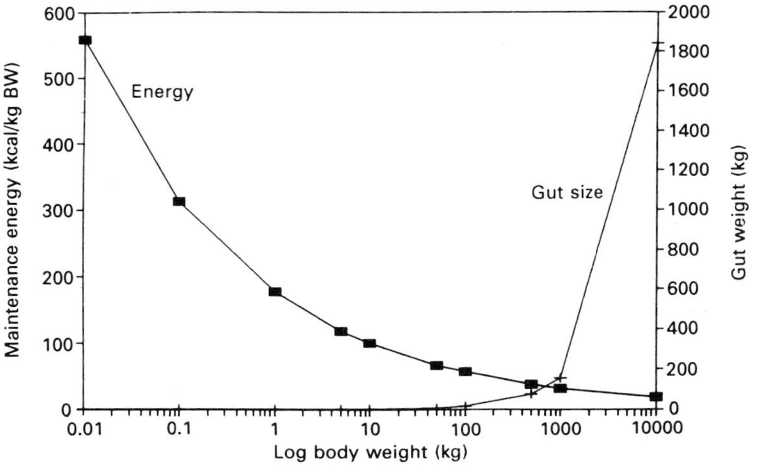

Fig. 7.1. Maintenance energy and gut weight in relation to body weight. The left-hand curve displays arithmetically the rising maintenance energy requirement as body weight (log base) diminishes, based on the relationship I (kcal DE) = 176.4 W(kg)$^{0.75}$(1 + 4.02 G) (Allden, 1969). The right-hand curve shows the relationship of gut weight (arithmetic values) to body weight derived from Parra's (1978) work: log Y = 1.0763 log X − 1.0289.

demand in arithmetic terms against body weight and illustrates the increasing call for energy above maintenance energy in small growing animals. The gut to accommodate and process the food is, on the other hand, minimal. The second curve to the right in Fig. 7.1, also in arithmetic terms, indicates mature gut size in relation to log body weight. This curve is based on the relationship developed by Parra (1978) for the log of the weight of gastro-intestinal contents (Y) and body weight (X): log Y = 1.0768 log X − 1.0289. The values in Fig. 7.1 are recapitulated in Fig. 7.2. Demment and van Soest (1982) obtained an essentially similar regression for a wide range of African herbivores. Brody (1945) found the slope of the gut mass curve for mammals was 0.941 ± 0.044 and for birds 0.985 ± 0.047. Calder (1974) gives a value of 1.003.

Pearson (1948) suggests that the critical size for endotherms (mammals) is of the order of 2.5 g. Below this, no mammal would be able to gather enough food to support its infinitely rapid metabolism and would have to have a lowered body temperature or some other fundamental method of conserving energy. He was, in this case, referring to the metabolic rate of the shrews, some of which are only 4.5 g. He states that: 'The young of many

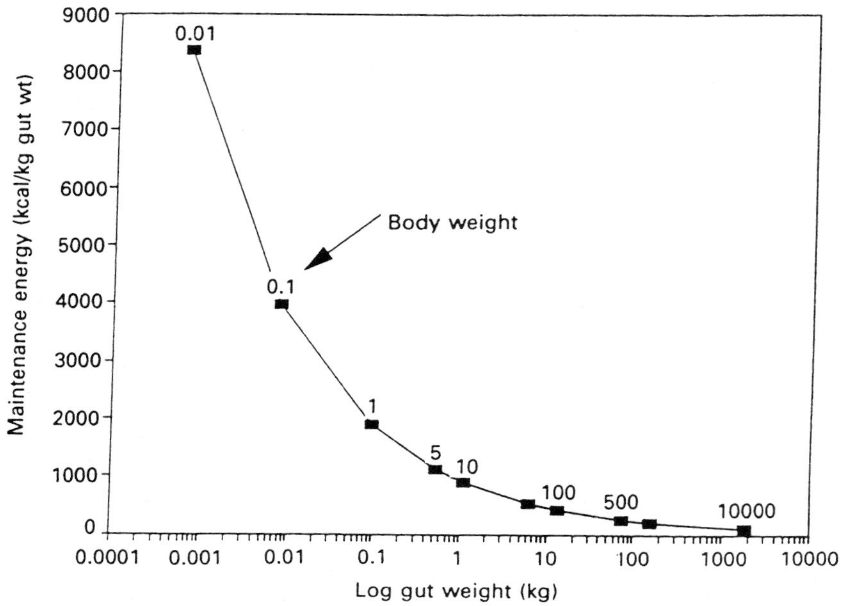

Fig. 7.2. The relationship of whole animal maintenance energy per unit of gut weight against log gut weight. Body weights are given on the curve.

mammals and birds are, of course, much smaller than this critical size, but the low body temperature of these tiny offspring keeps their rate of metabolism slow enough so that their parents can satisfy their food requirements'. Hopson (1972) suggests that the parents only provide their offspring with shelter and heat to provide near endothermic levels; thermo-regulation only develops as they approach mature size so that, prior to that time, a relatively large proportion of the food provided is converted to protoplasm because of this avoidance of the costs of endothermy. If this is the case, then does ingestion control differ from that in the adult, or does the realized heat production, i.e. metabolic plus growth heat production, represent controlling parameters as occurs in the gestating, lactating or egg-laying female? The efficiency of gain in the newly-born or hatched individual is dependent upon the quality of the food available to it, the accessibility of the food and the metabolic power of the developing animal. The initial caloric efficiency in doubling birth weight (independently of body weight) is quoted by Brody (1945: pp. 47–53 and 264–71) and Kleiber (1961; p. 330) from many sources to closely approximate 35%. This value falls rapidly as the animal increases in size. The constancy of the efficiency is related to metabolic power or capacity to metabolize the nutrients, but in poikilotherms this is only partly temperature-related. Brody (1945) fitted van't Hoff–Arrhenius equations to data on feeding and growth of grasshoppers. The Q_{10} increase for intake was 3.0 and that for growth was 2.8. Clearly, activity and metabolism both increased at the same rate but the senescence Q_{10} was 1.4. Else and Hulbert (1981) compared the 'standard metabolism' and body composition of a reptile *Amphibolurus nuchalis* and a mammal *Mus musculus*. The lizard and the mouse were of similar body weight (34.3 ± 4.9 g and 32.1 ± 1.4 g). Body temperatures were 37.0 ± 0.1°C. The O_2 consumption (ml/g per h) was 0.20 ± 0.03 for the reptile and 1.62 ± 0.16 for the mouse. Although these are resting rates, maximum metabolism of mice has been shown to be up to seven times greater than standard metabolism while that of the lizards may be five times. This suggests that while the gain in accrued energy from a given input could be similar for both groups, it would take longer for the poikilotherm to accomplish. As a consequence the food required to satisfy the high energy and high growth demands of a juvenile, as well as the low excretory products, must be of a much higher quality with respect to energy, protein and other nutrients; its ingestibility and digestibility must also be of the highest order. Carnivory (including insectivory) satisfies these requirements, as does milk, prokaryotic cells and most metabolizing eukaryotic unicells. As in animals, large or small, cell sizes are all within an order of magnitude of 10 μm (Teissier, 1939) and have essentially similar amino acid composition of the cell wall and machinery proteins, all animal tissues tending towards a common composition.

Beach *et al.* (1943) and many others subsequently have shown that the amino acid composition of muscle in mammals, birds, amphibians, fish and crustaceans was essentially similar. The compositions of other tissues except hair and wool were also similar and led Mitchell (1950) to ask whether 'the amino acid requirements of the growing animal are determined in the last analysis by the amino acid composition of the tissue proteins formed during growth'. As a principle, this is readily demonstrated by reference to the work of McCance and Widdowson (1964), who showed that over 90% of the protein of milk fed to pigs or puppies was incorporated into cells.

Figure 7.3 shows the amino acid profiles of basic animal nutriments including those from rumen bacterial, protozoal and yeast unicell protein (Fig. 7.3*e*) and single cell algae, *Chlorella, Scenedesmus*, and the blue-green *Anabaena* (Fig. 7.3*f*). All the amino acid profiles are very similar indeed, i.e. all metabolizing animal, unicell and prokaryote cell proteins fit a similar pattern. Although digestibility of the nitrogen varies to some extent, the biological value of the protein is high, largely 80% and up to 94% for egg. In these diagrammatic profiles of the ten major essential amino acids the most

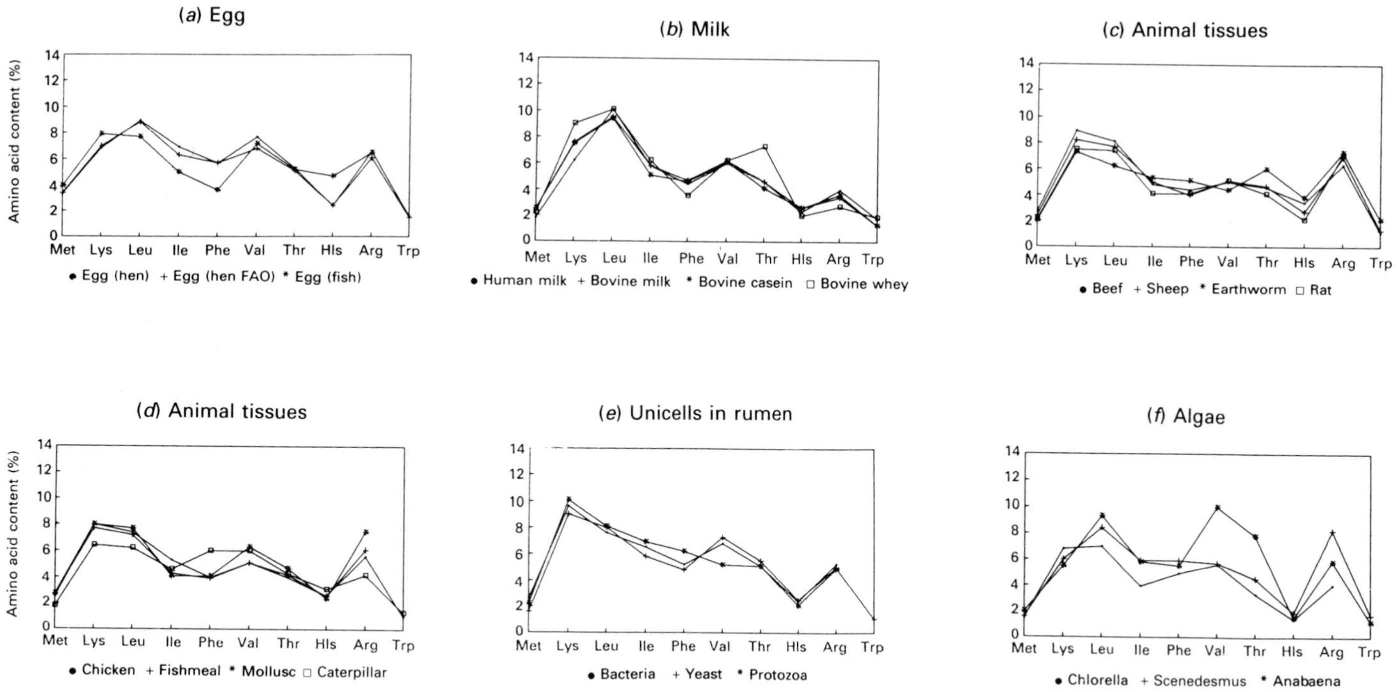

(a) Egg
Amino acid content (%)
14
12
10
8
6
4
2
0
Met Lys Leu Ile Phe Val Thr HIs Arg Trp
• Egg (hen) + Egg (hen FAO) * Egg (fish)
(b) Milk
• Human milk + Bovine milk * Bovine casein □ Bovine whey
(c) Animal tissues
• Beef + Sheep * Earthworm □ Rat
(d) Animal tissues
• Chicken + Fishmeal * Mollusc □ Caterpillar
(e) Unicells in rumen
• Bacteria + Yeast * Protozoa
(f) Algae
• Chlorella + Scenedesmus * Anabaena

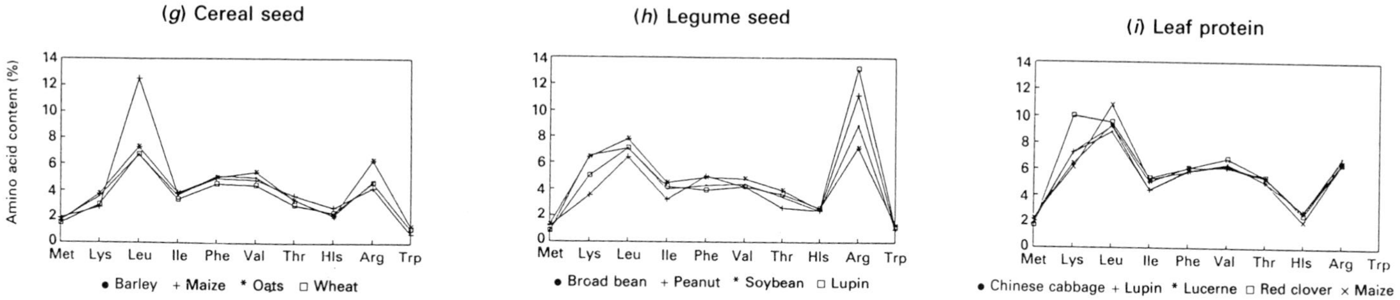

Fig. 7.3. Essential amino acid profiles of food protein groups. Methionine and lysine, the two amino acids most likely to be deficient, are given prominence on the left so that poor quality proteins may be readily discerned. Amino acid content is given as g/100 g protein (16 g N).

(*a*) Egg proteins are exceptional for their methionine level but all the amino acid, methyl and single carbon compounds must be supplied from this source. The two hen egg analyses (Mitchell, 1950 and FAO, 1970–72) were some years apart; fish egg protein differs quite substantially. (*b*) The amino acid profiles of milk and constituent whey proteins are similar for all species and compare well with those of animal tissues in (*c*) and (*d*) and the unicells in (*e*). With the exception of caterpillar tissue (*d*), all the proteins display higher than 2% methionine and 6% lysine. The proteins of the unicellular algae (*f*) tend to be marginal. The seeds of legumes (*h*) are low in methionine and tend to be low in lysine; cereal seeds (*g*) are frankly deficient in both but these are storage proteins and not immediately involved in metabolism. The leaf proteins (*i*) profiled here are substantially fraction 1, or chloroplast proteins, principally ribulose 1:5 bisphosphate carboxylase/oxygenase (rubisco), the CO_2-fixing enzyme which is also a major storage protein. The similarity of all the other metabolizable proteins and milk protein will be recognised easily.

Sources of amino acid compositions: hen egg (Mitchell, 1950; FAO, 1970–72); milks (Heine *et al.*, 1991); earthworm (Taboga, 1980); bacteria and protozoa (Purser and Buechler, 1966); *Anabaena cylindrica* (Fowden, 1962); Chinese cabbage and lupin (Byers, 1971); Lucerne (alfalfa) (Fafunso and Byers, 1977). All other values are from FAO (1970–72).

commonly deficient acids, methionine and lysine, are placed first to underscore their primacy; less than 2% or 6% of these amino acids, respectively, indicates proteins that are inadequate to support growth.

Most of the plant storage organs fail to satisfy the juvenile animal needs. The seeds of all graminae (Fig. 7.3*g*) and many legumes (Fig. 7.3*h*) are high in energy and protein and are readily digestible but are very deficient in lysine and sulphur-containing amino acids. Legume seeds are sometimes of high quality chemically, e.g. soybean, but many contain antimetabolites which reduce their value or even prevent their use. Further, the lysine of many plant materials is unavailable and, as a result, the remainder of the amino acids are degraded and not utilized. Carpenter and Woodham (1974) showed that there was a very strong relationship between the gross protein value of various proteins for chickens and the FDNB-reactive, i.e. available, lysine (Carpenter, 1960) in those proteins. Similarly, recent work by Batterham (1987) feeding pigs with cottonseed meal, in which the available lysine was only 5 g/kg (out of 8 g/kg total lysine) resulted in a gain of 410 g daily; when the diet was supplemented with lysine to give 7 g/kg available lysine, the daily gain was 480 g, the same as that seen for a soya meal with 7 g available of 8 g/kg lysine. Kobayashi *et al.* (1990) found that of 12 common feedstuffs no less than nine were deficient for pigs in lysine and two in methionine. Fishmeal was adequate in both amino acids, although it was low in arginine for chickens; chickens also showed wider needs of methionine, tryptophan and threonine. The world production of 70 000 tons of L-lysine and 250 000 tons of DL-methionine to supplement the storage (seed) proteins largely fed to domestic animals (Araki, 1990) underlines the difficulties experienced by the animal in nature.

Probably the most plentiful protein on earth (Ellis, 1979) is the chloroplast enzyme ribulose bisphosphate carboxylase/oxygenase, known as fraction 1 protein or 'rubisco'. It is found wherever photo or chemical energy is used to fix CO_2 into organic substances. According to Akazawa (1979), it occurs in the Bacteriophyta, the Cyanobacteria, the Euglenophyta, Chlorophyta and Tracheophyta. The presence of rubisco in the sulphur bacteria (organelles of this substance occur in some purple sulphur bacteria) indicates its antiquity. It is both an enzyme and a storage protein and Friedrich and Huffaker (1980) have shown that it may constitute as much as 81% of the total protein in barley leaves. While rubisco is known to be a substantial portion of most leaf proteins, its distribution in C_4 grasses (Ku *et al.*, 1979) and the lower digestibility of those grasses affect its availability in tropical plant species.

Rubisco is a large molecule, usually eight large peptides (about 75% of its molecular weight) and eight smaller ones. The large peptides are produced by the genome of the prokaryotic chloroplast organelle; the smaller units are produced by the eukaryotic host cell and combined in the chloroplast. The development of the large units has been extremely conservative and unicell proteins are immunologically indistinguishable from spinach and tobacco rubisco. The smaller peptide units of rubisco do vary significantly in their amino acid composition without very greatly influencing the total amino acid profile (Kawashima and Wildman, 1970).

Figure 7.3*i* illustrates the amino acid profile of extracted leaf proteins, mainly fraction 1 proteins which are close to pure rubisco. It is clear that the amino acid profile of this substantially prokaryotic protein has remained unaltered through time and is remarkably similar to the machinery proteins in animals. As it is a prokaryotic protein then the phytophagous lepidopteral larvae are no longer exceptions to Buchner's rule as proposed by Koch (1967).

The apparent carnivory of juvenile herbivores is a necessary tactic to satisfy the requirement for available amino acids, particularly the lysine and methionine, required in a highly concentrated and digestible form that is constantly and universally available. While plant proteins, particularly rubisco, have the required amino acid structure, digestibility, biological value and concentration required, the levels and availability fluctuate because of seasonal and structural changes (Minson and Wilson, 1980) and other interfering substances (Carpenter, 1960), except in very specific niches. Therefore, the only way for young herbivores to realize their growth potential and survive to reach reproductive maturity is to behave nutritionally as 'carnivores'.

References

Akazawa, T. (1979). Ribulose 1, 5-biphosphate carboxylase. In *Photosynthesis 11*, Ch. 17, ed. M. Gibbs & E. Latyko, pp. 208–229. Berlin: Springer-Verlag.

Allden, W. G. (1969). The summer nutrition of Weaner sheep: the voluntary feed intake, body weight change, and wool production of sheep grazing the mature herbage of sown pasture in relation to the intake of dietary energy under a supplementary feeding regime. *Australian Journal of Agricultural Research*, **20**, 499–512.

Araki, K. (1990). Production of amino acids. In *Nutrition: Protein and Amino Acids*. ed. A. Yoshida, H. Naito, Y. Niiyama & T. Suzuki, pp. 303–322. Berlin: Springer-Verlag.

Batterham, E. S. (1987). Availability of lysine for grower pigs. In *Manipulating Pig Production*. ed. J. L. Barnett *et al.*, pp. 121–123. Werribee, Australia: Australasian Pig Science Association.

Beach, E. F., Munks, B. & Robinson, A. (1943). The amino acid composition of animal tissue protein. *Journal of Biological Chemistry*, **148**, 431–439.

Brody, S. (1945). *Bioenergetics and Growth, with Special Reference to the Efficiency Complex in Domestic Animals*. New York: Reinhold.

Byers, M. (1971). Amino acid composition and *in vitro* digestibility of some protein fractions from three species of leaves of various ages. *Journal of Science & Food Agriculture*, **22**, 242–251.
Calder, W. A. (1974). Consequences of body size for avian energetics. In *Avian Energetics*. ed. R. A. Paynter, pp. 86–151. Cambridge, MA: Harvard University Press.
Carpenter, K. J. (1960). The estimation of the available lysine in animal protein foods. *Biochemical Journal*, **77**, 604–610.
Carpenter, K. J. & Woodham, A. A. (1974). Protein quality of feeding-stuffs. 6. Comparisons of the results of collaborative biological assays for amino acids with those of other methods. *British Journal of Nutrition*, **32**, 647–660.
Davies, W. L. (1939). The composition of the crop milk of pigeons. *Biochemical Journal* **33**, 898–901.
Demment, M. W. & van Soest, P. J. (1982). *Body Size, Digestive Capacity and Feeding Strategies of Herbivores*. Petit Jean Mountain, Morrilton AR: Winrock International Livestock Research and Training Center.
Downhower, J. T. (1976). Darwin's finches and the evolution of sexual dimorphism. *Nature*, **263**, 558–563.
Ellis, R. J. (1979). The most abundant protein in the world. *Trends in Biological Sciences*, **4**, 241–244.
Else, P. L. & Hulbert, A. J. (1981). Comparison of the 'mammal machine' and the 'reptile machine': energy production. *American Journal of Physiology*, **240**, R3–R9.
Fafunso, M. & Byers, M. (1977). Effect of prepress treatments of vegetation on the quality of the extracted leaf protein. *Journal of Science and Food Agriculture*, **28**, 375–380.
FAO (1970–1972). *Amino-acid Content of Foods and Biological Data on Proteins*. Rome: Food Policy and Food Science Service, Nutrition Division, FAO.
Farlow, J. O. (1976). A consideration of the trophic dynamics of a late cretaceous large-dinosaur community (Oldman Formation). *Ecology*, **57**, 841–857.
Fowden, L. (1962). Amino acids and proteins. In *Physiology and Biochemistry of Algae*, ed. R. A. Lewin, pp. 189–209. New York: Academic Press.
Fowler, H. W. (1965). *A Dictionary of Modern English Usage*, 2nd edn. (Revised Sir Ernest Gowers) New York: Oxford University Press.
Friedrich, J. W. & Huffaker, R. C. (1980). Photosynthesis, leaf resistances, and ribulose-1,5-biphosphate carboxylase degradation in senescing barley leaves. *Plant Physiology*, **65**, 1103–1107.
Gibbons, J. W. (1967). Variation in growth rates in three populations of the painted turtle, *Chrysemys picta*. *Herpetologica*, **23**, 292–303.
Halarnkar, P. P. & Blomquist, G. J. (1989). Comparative aspects of propionate metabolism. *Comparative Biochemistry & Physiology*, **92B**, 227–231.
Halarnkar, P. P., Chambers, J. D., Wakayama, E. J. & Blomquist, G. J. (1987). Vitamin B12 levels and propionate metabolism in selected non-insect arthropods and other invertebrates. *Comparative Biochemistry & Physiology*, **88B**, 869–873.
Heine, W. E., Klein, D. P. & Reeds, P. J. (1991). The importance of α-lact-albumin in animal nutrition. *Journal of Nutrition*, **121**, 277–283.
Hemmingsen, A. (1956–1960). Energy metabolism as related to body size and respiratory surfaces and its evolution. *Copenhagen Steno Memorial Hospital Reports* 6–9.
Hopson, J. A. (1972). Endothermy, small size and the origin of mammalian reproduction. *American Naturalist*, **106**, 446–452.

Jennings, A. C. & Morton, R. K. (1963). Changes in carbohydrate, protein and non-protein nitrogenous compounds of developing wheat grain. *Australian Journal of Biological Sciences*, **16**, 318–331.
Kawashima, N. & Wildman (1970). Fraction 1 protein. *Annual Review of Plant Physiology*, **21**, 325–358.
Kleiber, M. (1932). Body size and metabolism. *Hilgardia*, **6**, 315–353.
Kleiber, M. (1961). *The Fire of Life, an Introduction to Animal Energetics*. New York: Wiley.
Kobayashi, T., Doi, Y. & Takami, T. (1990). Application of amino acids in foods and feeds. In *Nutrition: Protein and Amino Acids*. ed. A. Yoshida, H. Naito, Y. Niiyama & T. Suyaki, pp. 285–299. Tokyo: Japan Science Society Press, Berlin: Springer-Verlag.
Koch, A. (1967). Insects and their endosymbionts. In *Symbiosis*, Vol. 11, ed. S. M. Henry, Ch. 1, pp. 1–106. New York: Academic Press.
Ku, M. S. B., Schmitt, M. R. & Edwards, G. E. (1979). Quantitative determination of RuBP carboxylase-oxygenase protein in leaves of several C3 and C4 plants. *Journal of Exponential Botany*, **30**, 89–98.
Mayhew, W. W. (1963). Some food preference of captive *Sauromalus obesus. Herpetologica*, **19**, 10–16.
McCance, R. A. & Widdowson, E. (1964). Protein requirements in the newborn. In *Mammalian Protein Metabolism*, Vol. 11, ed. H. N. Munro & J. B. Allison, pp. 225–245. New York: Academic Press.
Minson, D. J. & Wilson, J. R. (1980). Comparative digestibility of tropical and temperate forage – a contrast between grasses and legumes. *Journal of the Australian Institute of Agricultural Science,* **46**, 247–249.
Mitchell, H. H. (1950). Some species and age differences in amino acid requirements. In *Protein and Amino Acid Requirements of Mammals*, ed. A. A. Albanese, pp. 1–32. New York: Academic Press.
Mittler, T. E. (1971). Dietary amino-acid requirements of the aphid *Myzus persical* affected by antibiotic uptake. *Journal of Nutrition*, **101**, 1023–1028.
Moir, R. J. (1991). The role of microbes in digestion. In *Microbiology of Animals and Animal Products*. ed. J. B. Woolcock Ch. 3, pp. 29–59. Amsterdam: Elsevier.
Oftedal, O. T. (1980). Milk and mammalian evolution. In *Comparative Physiology: Primitive Mammals*. ed. K. Schmidt-Nielsen, L. Bolis & C. R. Taylor, Ch. 3, pp. 31–42. Cambridge: Cambridge University Press.
Pace, D.M., Landott, P. A. & Mussehl, F. E. (1952). The effect of pigeon crop-milk on growth in chickens. *Growth*, **16**, 279–285.
Parra, P. (1978). Comparison of foregut and hindgut fermentation in herbivores. In *The Ecology of Arboreal Folivores*. ed. G. Montgomery, pp. 205–229. Washington, DC: Smithsonian Institute Press.
Pearson, O. P. (1948). Metabolism of small mammals, with remarks on the lower limit of mammalian size. *Science*, **108**, 44.
Peters, R. H. (1983). *The Ecological Implications of Body Size*. Cambridge: Cambridge University Press.
Pough, F. H. (1973). Lizard energetics and diet. *Ecology*, **54**, 837–844.
Prevost, J. & Vilter, V. (1962). Histologie de la sécrétion oesophagienne du Manchot Empereur. In *Proceedings of 13th International Ornithological Congress*, Vol. 2. American Ornithologists Union. pp. 1085–1094. Cited by K. Schmidt-Nielsen (1983), p. 140.
Pullar, J. D. & Webster, A. J. F. (1977). The energy cost of fat and protein deposition in the rat. *British Journal of Nutrition*, **37**, 353–363.

Purser, D. B. & Buechler, S. M. (1966). Amino acid composition of rumen organisms. *Journal of Dairy Science*, **49**, 81–84.

Renfree, M. B., Meier, P., Fang, C. & Battaglia, F. C. (1981). Relationship between amino acid intake and accretion in a marsupial *Macropus eugenii*. 1. Total amino acid composition of the milk throughout pouch life. *Biology of the Neonate*, **40**, 29–37.

Rimmer, D. W. (1986). Changes in diet and the development of microbial digestion in juvenile buffalo bream, *Kyphosus cornelii*. *Marine Biology*, **92**, 443–448.

Rowley, I. (1990). *Behavioural Ecology of the Galah, Eolaphus roseicapillus in the Wheat Belt of Western Australia*. CSIRO/Royal Australian Ornithologists Union.

Royama, T. (1970). Factors affecting the hunting behaviour and selection of food by the Great Tit (*Parus major*). *Journal of Animal Ecology*, **39**, 619–668.

Schmidt-Nielsen, K. (1983). *Animal Physiology Adaptation and Environment*, 3rd edn. Cambridge: Cambridge University Press.

Schmidt-Nielsen, K. (1986). Why milk? *News in Physiological Science*, **1**, 140–142.

Taboga, L. (1980). The nutritional value of earthworms for chickens. *British Poultry Science*, **21**, 405–410.

Taylor, St. C. S. (1985). Use of genetic size scaling in evaluation of animal growth. *Journal of Animal Science*, **61** (Suppl. 2), 118–143.

Teissier, G. (1939). Biometrie de la cellule. *Tabulae Biologicae*, **19** (1), 1–64.

Ternouth, J. H., Roy, J. H. B. & Siddons, R. C. (1974). The effects of addition of fat to skim milk and of 'severe' preheating treatment of spray-dried skim-milk powder. *British Journal of Nutrition*, **31**, 13–26.

Ternouth, J. H., Roy, J. H. B., Thompson, S. Y., Toothill, J., Gillies, C. M. & Edwards, J. D. (1975). Further studies on the addition of fat to skim-milk and the use of non-milk proteins in milk-substitute diets. *British Journal of Nutrition*, **33**, 181–196.

Wakayama, E. J., Dilwith, J. W., Howards, R. W. & Blomquist, G. J. (1984). Vitamin B_{12} levels in selected insects. *Insect Biochemistry*, **14**, 175–179.

Ward, P. (1965). Feeding ecology of the black-faced dioch *Quelea quelea* in Nigeria. *Ibis*, **107**, 173–197.

White, T. C. R. (1984). The abundance of invertebrate herbivores in relation to the availability of nitrogens in stressed food plants. *Oecologia*, **63**, 90–105.

White, T. C. R. (1985). When is a herbivore not a herbivore? *Oecologia*, **67**, 596–597.

Yvon, M., Pelissier, J-P., Guilloteau, P. & Toullec, R. (1985). *In vivo* milk digestion in the calf abomasum. III. Amino acid compositions of the digesta leaving the abomasum. *Reproduction and Nutritional Development*, **25**, 495–504.

8

Nutritional ecology of fruit-eating and flower-visiting birds and bats

CARLOS MARTINEZ DEL RIO

Volant vertebrates include two extant groups: birds and bats. Both of these taxa are commonly used as model organisms for studies of community and population ecology (Kunz, 1982; Wiens, 1983) and foraging behaviour (Morrison *et al.*, 1988). Despite this, the nutrition and digestive function of birds and bats are too often considered inscrutable black boxes, or worse, ignored by ecologists and behaviourists (Wheelwright, 1991). There are several reasons why the processes by which these animals digest and absorb nutrients should be of interest to animal ecologists and behaviourists. The design and adaptability of the gut may determine diet diversity and hence niche width (Martínez del Rio *et al.*, 1988). Digestive processes may set limits on metabolizable energy and nutrient intake and thus determine rates of growth and reproduction (Kenward and Sibly, 1977; Karasov *et al.*, 1986). In addition, nutrient assimilation is one of the factors that mediates the interaction of animals with their environment (Martínez del Rio and Restrepo, 1992).

Out of neglect, the contributions of nutrition and digestive physiology to ecology and behaviour have been minor. We know little about how digestion influences feeding choices (Martínez del Rio and Karasov, 1990), constrains meal sizes and frequencies (Carpenter *et al.*, 1991), determines rates of growth and fat accumulation (Konarzewski *et al.*, 1989), or affects interspecies interactions. With the notable exception of several studies on pollinator energetics (summarized in Heinrich, 1975), comparative avian/chiropteran physiologists have also overlooked the potential of their field as a tool to understand interactions among species. Ecological physiologists have devoted most effort to documenting the physiological responses of animals to abiotic challenges (McNab, 1982; Feder *et al.*, 1987). Here I advocate that physiological traits reflect and shape characteristics of other organisms, most notably prey and food-plants (Martínez del Rio and Restrepo, 1992).

I will focus on fruit-eating and flower-visiting birds and bats to illustrate how knowledge of digestive and nutritional physiology can contribute to our understanding of animal behaviour and species interactions. I emphasize pollination and seed dispersal systems where plant characteristics can be influenced by the physiological processes of animal mutualists. Specifically, I discuss the impact of digestive and metabolic traits on nutrition and feeding choices. Finally, I speculate about the consequences of nutritional constraints on the evolution of nectar and fruit pulp composition.

Nutrition has been an implicit but important ingredient in the development of 'paradigms' about bird–plant interactions (Snow, 1981; McKey, 1975). The physiological foundations of these paradigms are based on scanty empirical evidence (Martínez del Rio and Restrepo, 1992). Here I attempt to clarify these physiological assumptions and to critically examine them using available data. I suggest that the nutritional approach traditionally favoured in bird/bat–plant interaction studies ('proximal nutrient analysis', defined below) is inappropriate.

I restrict this essay to the consideration of the three principal organic compounds contained in nectar and fruit pulp: carbohydrates, lipids and proteins. In addition to these compounds, fruit and nectar contain secondary compounds, vitamins, and minerals (Goldstein and Swain, 1963; Hiebert and Calder, 1983). These substances influence animal–plant interactions and, presumably, mediate interactions between animals, plants and the organisms that parasitically use the rewards that plants offer (arthropods, fungi, etc.; Janzen, 1977). Although these compounds are undoubtedly nutritionally important, too few data on their effects in pollinators and seed dispersers are available (Buchholz and Levey, 1990) for me to discuss them here.

Bird/bat preferences and the chemical composition of plant rewards

Plants secrete floral nectar and surround seeds with pulp that attract pollinators and seed dispersers. Hence fruit and nectar chemistry should correlate with animal preferences for different nutrients (Baker, 1975; Calder, 1979). The fruit and nectar preferences of both birds and bats have received considerable attention (Moermond and Denslow, 1985; Stashko and Dinerstein, 1988; Thomas, 1988). Nectar-feeding birds and the plants they visit show fairly clear preference patterns (see Stiles, 1976 and below). In contrast, the vast majority of field and laboratory experiments that have attempted to relate bird and bat preferences with fruit composition have failed to reveal strong patterns (see references in Foster, 1990 for birds and Bonaccorso and Gush, 1987 for bats). There are two potential reasons for the absence of clear correlations between fruit chemistry and frugivore preferences: (a) fruit-eating birds do not use nutrient composition in fruit selection or (b) researchers measure nutrient composition in ways that are irrelevant for frugivores.

Nutrient composition in fruit is commonly measured using **proximate nutrient analyses** that emphasize a dichotomy between relatively undigestible fibres and readily assimilable cell contents (van Soest, 1983). Information derived from proximate nutrient analysis has been impressively successful in the prediction of the food assimilation efficiency of herbivores (see Grajal *et al.*, 1989) but may be inadequate for studying fruit- and nectar-eating animals. Karasov (1990) attempted to apply a modification of 'the summative' equation of van Soest (1983) to fruit-eating birds and found large discrepancies between the assimilation efficiencies predicted from proximal analyses of fruit and the observed assimilation measurements.

Because most bird/bat frugivores do not depend on fermenting microorganisms for nutrient assimilation (Klite, 1965), they must rely on a plurality of enzymatic and transport pathways to assimilate specific fats, carbohydrates and proteins (Alpers, 1987). The microorganisms associated with fermenting herbivores turn most nutrients into volatile fatty acids and microorganism biomass. They turn nutrients into common denominators (volatile fatty acids and bacterial protein) that are assimilated by a relatively small set of pathways (Stevens, 1990). The synthetic information derived from proximal nutrient analyses fails to convey the complexity of the nutritional interactions of birds/bats with fruit. A set of recent studies on sugar and fat digestion by birds emphasizes the relevance of detailed nutritional analyses for preference studies in nectar- and fruit-eating animals.

Sugars in nectar and fruit

Floral nectar is probably the simplest avian/chiropteran food on earth. Nectar is a dilute solution of different sugars (mainly sucrose, glucose and fructose) containing small amounts of amino acids, vitamins, lipids, alkaloids, and electrolytes (Baker and Baker, 1986). The simplicity of nectar facilitated detailed comparative studies of its chemical composition (Percival, 1961; Baker and Baker, 1983). Herbert and Irene Baker collected and analysed nectar from more than 200 species of bird-pollinated plants (Baker and Baker, 1983) and discovered a peculiar pattern in sugar composition. Plants pollinated by hummingbirds secrete nectar that contains large amounts of sucrose ($> 50\%$ total sugars) whereas plants pollinated by passerine birds secrete nectars containing mainly glucose and fructose (in a roughly 1 : 1 ratio) and with extremely low sucrose contents ($< 10\%$ total sugars; Baker and Baker, 1982). The pattern is consistent. In plant genera known to have both passerine- and hummingbird-pollinated species (*Erythrina*, *Puya*, *Campsis*, *Fuchsia* and *Fritillaria*), hummingbird-pollinated plants secrete high sucrose content nectar whereas passerine pollinated plants secrete high hexose content nectars (Martínez del Rio *et al.*, 1992). In *Erythrina*, a plant genus for which a detailed phylogeny is available, changes from hummingbird- to passerine-pollination (and vice versa) are accompanied by concomitant shifts in nectar composition (A. Bruneau, personal communication). A recent survey of fruit pulps (Baker *et al.*, 1992) uncovered another remarkable pattern. The pulp of most bird-dispersed fruits is rich in glucose and fructose but contains only very small amounts of sucrose. In contrast, cultivated fruits used for human consumption and mammal-dispersed fruits contain higher sucrose proportions. Bat-pollinated plants exhibit an analogous dichotomy in nectar composition. Old World flowers pollinated by megachiropteran bats (family Pteropodidae) seem to have nectars richer in sucrose than those New World flowers pollinated by microchiropteran bats (family Phyllostomidae; Freeman *et al.*, 1991; Baker *et al.*, 1992).

Martínez del Rio *et al.* (1992) summarized the relationship between fruit and nectar sugar composition and bird preferences. Given a choice, hummingbirds prefer sucrose over glucose and fructose, whereas many nectar- and fruit-eating passerines prefer glucose and fructose over sucrose. Thus most birds prefer the sugar composition of the plants that they feed on. That birds show sugar preferences at all is puzzling: sugars that are chemically very similar and that contain the same energy content per unit gram? Neither energetic models of foraging (Stephens and Krebs, 1986), which ignore the intricacies of digestion (Speakman, 1987), nor standard proximal nutrient nutritional ecology, which ignores the details of nutrient assimilation, can provide an explanation.

The physiological details of sugar assimilation provide a hypothesis to explain differential sugar preferences in birds (Martínez del Rio and Stevens, 1989). Sucrose is a disaccharide that is hydrolysed into its monosaccharide constituents, glucose and fructose, before assimilation (Alpers, 1987). Glucose and fructose, in contrast, are absorbed directly in the intestine (Semenza and Corcelli, 1986). Martínez del Rio *et al.* (1988) hypothesized that the preference for glucose and fructose over sucrose in nectar- and fruit-eating passerines reflects a physiological inability to assimilate sucrose efficiently. Physiological measurements support this hypothesis: preference for hexoses in passerines seems to be associated with poor sucrose assimilation (Martínez del Rio *et al.*, 1992). Two physiological mechanisms seem to account for inefficient sucrose digestion in passerines: lack of intestinal sucrase activity (Martínez del Rio and Stevens, 1989) and extremely fast passage rates

(Martínez del Rio *et al.*, 1989). For sucrase-deficient birds, sucrose is a useless energy source that can cause osmotic diarrhoea (Brugger and Nelms, 1991). A whole phylogenetic lineage of frugivorous and nectarivorous species (the sturnid–muscicapid lineage, *sensu* Sibley and Ahlquist, 1984) appears to be sucrase deficient (Martínez del Rio, 1990a). Fast passage rates appear to be characteristic of small frugivores (Levey and Grajal, 1991) and appear not to provide the processing time required to assimilate substrates, such as sucrose, that must be first hydrolysed to be absorbed (Martínez del Rio *et al.*, 1989). On the other hand, hummingbirds have digestive characteristics that make them extremely well suited to digest and absorb sucrose (Martínez del Rio, 1990b). They have the highest rates of carrier-mediated intestinal glucose transport and exceedingly high rates of intestinal sucrose hydrolysis (Karasov *et al.*, 1986; Martínez del Rio, 1990a). These traits allow them to use sucrose as efficiently (but not more efficiently) as 1 : 1 mixtures of glucose and fructose (Martínez del Rio, 1990b). It is not yet clear, nevertheless, why hummingbirds show such strong and persistent preference for sucrose over glucose and fructose (Martínez del Rio *et al.*, 1992).

The dichotomy in sugar composition between the nectars of Old and New World plants visited by bats suggests contrasting sugar preferences and digestive abilities between pteropodid and phyllostomid bats. Unfortunately the relationship between bat preferences and physiology and the sugar composition of nectar and fruit has not been investigated. Hernández and Martínez del Rio (1992) examined intestinal sucrase in several species of fruit- and nectar-feeding phyllostomid bats. Sucrase activities, standardized by intestinal area, were 33–50% lower in two species of nectar-feeding bats (*Glossophaga soricina* and *Leptonycteris curasaoe*) than in hummingbirds. Because total sucrose hydrolysis results from the integration of intestinal surface area and area-specific rates of hydrolysis, it is difficult to assess if differences in area-specific sucrase activity between hummingbirds and bats translate into functional differences in the ability to assimilate sucrose and in preference differences. It is also not known if mega- and microchiropteran flower- and fruit-eating bats differ in digestive carbohydrase profiles as would be expected from the sugar composition of their respective food plants. Ogunbiyi and Okon (1976) reported the presence of sucrase and maltase in the small intestine of the African pteropodid *Eidolon helvum*. Unfortunately, methodological differences in enzyme assays preclude quantitative comparison with the data collected by Hernández and Martínez del Rio (1992).

Lipids in fruit and the specialized–generalized plant–frugivore dichotomy

Lipid-rich fruits such as those produced by the Lauraceae, Palmae, Areaceae and Myricaceae have played a significant, albeit confusing, role in the development of plant–frugivore coevolutionary paradigms. Howe and Smallwood (1982) synthesized the ideas of Snow (1971), McKey (1975) and Howe and Estabrook (1977) into a 'unified field theory' of plant–frugivore interactions. This unified paradigm emphasizes a dichotomy between specialized and generalized plants and their frugivores where degrees of specialization between plants and frugivores are correlated (i.e. generalized plants produce fruits dispersed by generalized frugivores). Howe (1992) provides a favourable view and a detailed narrative of the development of this paradigm.

The specialized–generalized dichotomy of coincidence is based on imprecise definitions of specialization and generalization (see Sherry (1990) for an illuminating discussion of these terms in the context of avian foraging), and, not surprisingly, it has not fared well in 15 years of empirical research (Herrera, 1986; Martínez del Rio and Restrepo, 1992). Although currently the specialized–generalized paradigm has little more than historical value, one of its elements continues to plague research on the nutritional ecology of avian frugivores (e.g. Stiles, 1992). The specialized–generalized model assumes that specialized plants produce lipid-rich fruits whereas generalized plants produce juicy, carbohydrate-rich fruits. It follows that birds feeding preferentially on fat-rich fruits should be more specialized. This peculiar lipid chauvinism has been perpetuated by the wide use of Herrera's fruit profitability measure (Herrera, 1981) which only considers lipid and protein contents and disregards carbohydrates in its assessment of a fruit's nutritional quality. The carbohydrate-rich fruits consumed by presumed generalists simply have negligible profitability and nutritional value if one uses Herrera's index!

Perhaps the best counter-example to the specialized (fatty)–generalized (sugary) dichotomy is the dichotomy found in Neotropical mistletoes. Two New World mistletoe families produce fruits with contrasting chemical compositions: Restrepo (1987) reports that, in Colombia, species in the Viscaceae produce watery, carbohydrate-rich fruits (less than 10% lipids) that are dispersed by euphonias and tanagers, whereas species in the Loranthaceae produce lipid-rich fruits (more than 20% lipids) that are consumed by more or less generalist flycatchers (Davidar, 1983; Restrepo, 1987). In euphonias, the intestine is an uninterrupted tube and the gizzard is reduced to a narrow zone between the proventriculus and the duodenum (Wetmore, 1914). Euphonias are very dependent on mistletoe fruits and it can be hypothesized that this arrangement is efficient at processing the sticky mucilage-covered fruits of mistletoes. In the Old World too, mistletoe specialists (Dicaedidae; Docters van Leeuwen, 1954) appear to have guts that are convergent with the apparently specialized and highly-derived guts of euphonias (Desselberger, 1931; Richardson and Wooler, 1988a).

Although the specialist–generalist view probably wrongly associates chemical composition with degree of specialization in both birds and plants, it rightly recognizes that some fruit pulps contain primarily lipids whereas others contain primarily carbohydrates (Stiles, 1992). Are lipid-rich (or carbohydrate-rich) fruits preferred by some types of frugivores and are the lipid versus carbohydrate preferences of birds correlated with bird metabolic and digestive abilities? Few studies address these questions directly. Stiles (1992) demonstrated that some frugivorous birds (*Dumetella carolinensis* and *Turdus migratorius*) prefer artificial high-lipid fruits over low-lipid fruits. Because he kept carbohydrate and protein levels constant in his artificial fruits while varying lipid content, his experiments may only demonstrate that birds can differentiate among fruits with different caloric contents and preferentially ingest those with higher energy.

Wax digestion in avian frugivores

Fruit nutritional analyses have traditionally relied on non-polar solvent extraction to estimate lipid contents (Herbst, 1988). These extractions are summarized as dry weight percentage of lipids and are made up of a complex mixture of acylglycerols, waxes, phospholipids, volatile oils and pigments (Nawar, 1985). The processes by which vertebrates assimilate different lipids are quite specific (Carey *et al.*, 1983). For animals differing in the expression of specific lipases and esterases, long-chain fatty alcohols, waxes and many triacylglycerides are not nutritionally equivalent (Jackson and Place, 1990). To assess the nutritional value of the lipid fraction in a fruit it is necessary to know two things: (a) the specific identity of its constituents and (b) the digestive abilities of the consuming frugivore.

Place and Stiles (1992) have provided a fine example of a fruit–frugivore interaction in which the specific identity of a fruit's lipids influences its seed disperser assemblage. The fruits of the North American plant genera *Myrica* and *Toxicodendron* contain a waxy coating of mono- and diglycerides of myristic, palmitic and stearic fatty acids (Place and Stiles, 1992). These waxy acylglycerides are notoriously difficult to assimilate by most terrestrial vertebrates (Renner and Hill, 1961). Among birds, both chickens and quail show impaired digestion when fed on either waxy lipids or the fruits of *Myrica* (Martin *et al.*, 1951). In paired tests, American robins (*T. migratorius*) preferred all non-waxy fruit species offered (eight species) over the waxy fruits of *Myrica* and *Rhus* (Stiles, 1992).

Waxy fruits are consumed by a variety of woodpeckers and quite surprisingly by a warbler (*Dendroica coronata*) and a swallow (*Tachycineta bicolor*). The last two species form large flocks that feed very intensively on *Myrica* spp. fruits during migration. Place and Stiles (1992) have demonstrated that both *D. coronata* and *T. bicolor* assimilate the waxy lipids in *Myrica* (assimilation efficiency > 80%). They have also identified some of the physiological traits associated with these relatively high assimilation efficiencies: retrograde reflux of intestinal contents into the gizzard, slow gastro-intestinal transit of dietary lipids, an elevated bile salt concen-tration in gall bladder and intestine and a bile composition dominated by taurine salt conjugates that prevent bile salt conjugation in the acid environment of the gizzard (Place and Stiles, 1992). In support of Place and Stiles' (1992) characterization, Bosque and de Parra (1992) found very long food retention in the gut of oilbirds (*Steatornis capensis*) which feed on oily fruits.

Frugivores that ingest primarily carbohydrate-rich fruits appear to show contrasting digestive traits (Karasov and Levey, 1990). In cedar waxwings (*Bombycilla cedrorum*), for example, food transit times are short (Martínez del Rio *et al.*, 1989), retrograde reflux of small intestinal contents seems to be rare or non-existent (Levey and Duke, 1992), and waxwings often (but not always) lack a detectable gall bladder (M. Witmer and C. Martínez del Rio, unpublished observation).

Greenberg (1981) found that the fruits of *Lindackeria laurina* (Flacourtiaceae), a plant with an aril with 'a distinctly waxy texture and odor', were consumed and presumably dispersed in Panama almost exclusively by two North American migrant warblers (*D. pennsylvanica* and *D. castanea*). It is likely that these two species share the traits that enable *D. coronata* to assimilate the waxy coating of *Myrica*. Although both *D. coronata* and *T. bicolor* feed on the lipid-rich fruits of a 'specialized' plant species (i.e. one with a very small diversity of consumers), it is noteworthy that they are far from being 'specialized' frugivores. Both species are primarily insectivorous (Martin *et al.*, 1951) and *D. coronata* feeds on a variety of fruits (Bent, 1953) and even on floral nectar in its wintering grounds (Gryj *et al.*, 1990). The specialized–generalized paradigm certainly does not apply to *Myrica* and its associated frugivores!

The contrast between the digestive traits of *D. coronata* and *B. cedrorum* suggest potential trade-offs in the assimilation of lipid- and carbohydrate-rich fruits. Vertebrates digest lipids by a complex process that involves bile secretion, lipid emulsification, hydrolysis of fatty acid esters and the absorption of the hydrolysed products in the enterocytes. Several steps in this process are potentially hampered by the digestive traits of sugar-eating frugivores. Short retention times may hinder effective fat emulsification and may prevent the reabsorption of valuable bile constituents in the small intestine (Lindsay and March, 1967). The gall bladder appears to be absent in hummingbirds (quintessential sugar feeders) and its presence is variable among frugivores (both within and among species, Gorham and Ivy, 1938; H. Witmer and C. Martínez del Rio, unpublished observation). The absence of a gall bladder may result in the inability to concentrate bile which may prevent the assimilation of a lipid-rich diet. The absence of a gall bladder may also indicate low production of bile and, thus, low faecal losses in frugivores with short food retention times.

Protein in nectar

An unanswered question permeates the literature on the nutritional composition of the rewards that plants offer to pollinators and seed dispersers: do nectar and fruit pulp contain sufficient protein to sustain the growth of young and to keep adults in nitrogen balance (Morton, 1973; Foster, 1978; Baker and Baker, 1986)? For nectar, the answer is almost certainly no. The protein content of floral nectars consumed by birds and bats is low (less than 0.01% weight/volume and less than 0.04% of dry weight; calculated from values in Baker and Baker, 1982) and no known nectarivorous bird feeds its young exclusively on nectar. Hummingbirds feed young on insects and later in development on a mixture of nectar and insects (Carpenter and Castronova, 1980). It is also unlikely that the amino acids contained in nectar can satisfy the equilibrium protein requirements of adult nectar-feeding birds and bats. Bats (*Leptonycteris curasoae*) fed exclusively on nectar lose weight and die within 4–10 days, in spite of ingesting enough energy to satisfy metabolic requirements (Howell, 1974).

Brice and Grau (1991) estimated minimal protein requirements for the hummingbird *C. costae* (*ca* 3.7 g) to be roughly 30 mg of protein per day. Average hummingbird nectars contain *ca* 0.07 mg amino acids/ml (Baker, 1977). The amount of nectar needed to satisfy energy requirements (10 ml) contains only about 2% of the minimal protein requirements of a *C. costae* adult (recalculated from Brice and Grau, 1991). In contrast, *C. costae* individuals can satisfy their minimal protein requirements with roughly 30 arthropods (weighing 7.5 mg each), which can be captured in less than 10 min of active arthropod catching (Brice and Grau, 1991; see also Stiles, 1971).

Although hummingbirds and honeyeaters have extremely low protein requirements (Paton, 1982; Brice and Grau, 1991), they cannot satisfy them solely on a nectar diet.

Howell (1974) estimated that the minimal requirement of the nectar-feeding bat *L. curasoae* (15–20 g) was 84–106 mg protein/day. Thus, protein requirements of nectar-feeding bats (*L. curasoae*) and hummingbirds (*C. costae*) standardized by metabolic mass ($mass^{0.75}$), are almost identical (approximately 11 mg of protein/$g^{0.75}$). Hummingbirds and bats, however, seem to respond very differently to protein-free diets. After 10 days on a diet containing exclusively sucrose, hummingbirds (*C. costae*) behave normally in spite of 16% weight loss (Brice, 1992). Bats (*L. curasoae*) on a protein-free diet, in contrast, die within 10 days (Howell, 1974). The physiological mechanisms that allow hummingbirds to subsist for extended periods without ingesting proteins are unknown.

Nectar-feeding birds obtain most of their energy requirement from nectar and presumably most of their protein from insects (Gass and Montgomerie, 1981). Brice (1992) provided *C. anna* hummingbirds with three diets: a sucrose solution and live fruit flies, a sucrose solution only, and fruit flies and water only. She found that birds maintained weight on the sucrose and flies diet, lost weight very slowly on the sucrose diet (less than 16% of initial mass in 10 days), and refused to eat fruit flies if no sucrose solution was offered. Brice's results contrast with Hainsworth's (1977) suggestion that hummingbirds should be able to obtain all their energy needs from arthropods (see also Wolf and Hainsworth, 1971; Montgomerie and Redsell, 1980).

The processes by which arthropod protein and the sugars contained in nectar are assimilated differ in both enzymatic and transport pathways and in the digestive organs in which digestion takes place: all sugar assimilation occurs in the small intestine, sugars are hydrolysed by membrane-bound intestinal disaccharidases and then transported by intestinal brush border carriers (Alpers, 1987). To assimilate proteins contained in arthropods, birds have to mechanically break the chitinous exoskeleton in their muscular stomach and to use pepsin proteolytic hydrolysis which requires low pH and is restricted to the stomach (Castro, 1984). In many nectar-feeding species the proventricular and pyloric openings lie next to each other and are positioned in the same axis, allowing nectar to flow into the small intestine moving directly from the crop into the duodenum and bypassing the gizzard that can be processing arthropods concurrently (Desselberger, 1932). In honeyeaters, both the muscular (gizzard) and glandular stomach are extremely reduced (Richardson and Wooler, 1986, 1988b). Richardson and Wooler (1988b) have hypothesized that the reduced muscularity of nectar-feeding meliphagids is correlated with the consumption of soft-bodied arthropods and may limit the ability of these birds to utilize hard-bodied arthropods. Although quantitative comparative evidence is lacking, it is likely that the stomach is also reduced in hummingbirds (see Desselberger, 1932). The energy content of arthropods is high enough to sustain the metabolic needs of nectar-feeding birds if, and only if, these birds have the digestive abilities to process arthropods in large amounts. It may be that digestive traits that permit efficient assimilation of nectar (and the use of small amounts of arthropods) are inefficient at processing large numbers of arthropods. It is unfortunate that there are no data available on arthropod assimilation efficiency by nectar-feeding birds. I predict that arthropod assimilation will be low.

Protein in fruit

Discussions on the amount of protein in fruit have centred around the inadequacy of fruit as a protein source for growing birds (Morton, 1973; Ricklefs, 1974; Foster, 1978). Morton (1973) and Ricklefs (1974) suggested that fruit protein contents are so low that young fed exclusively on fruit develop more slowly than those fed on insects. Morton (1973) concludes that, because increased nesting length increases predation, birds that feed young only fruit should be rare, and total frugivory should be restricted to birds nesting in areas with reduced predation. Foster (1978) used fruit protein content values to estimate quantities of fruit needed to meet protein requirements in growing birds. She concluded that in most instances, because protein/energy ratios in fruit are low, fruit ingestion required to meet protein needs supplied energy far in excess of that required by the young. Thus, nestling birds fed exclusively on fruit appear to be protein limited. Nestlings confined to a fruit diet probably accumulate excess assimilated energy as fat and grow slowly (O'Connor, 1984). The maintenance protein requirements of adult fruit-eating birds are almost certainly lower than those of growing birds and in most frugivorous birds the proportion of dietary fruit increases with age.

The value of a fruit as a protein source for a frugivore depends on three factors: (a) fruit protein content and composition, (b) the frugivore's digestive ability and (c) the frugivore's protein requirements (e.g. the protein requirements of lactating female bats are considerably higher than those of non-reproductive individuals; Herbst, 1986). Current data on these three factors are remarkably inadequate. More than 20 years of research on the interaction between birds and fruit have produced a long list of fruit species for which proximal nutrient analyses have been done (Stiles, 1980; Johnson *et al.*, 1985; Herrera, 1987 among others). Unfortunately, fruit protein content estimates derived from these proximal nutrient analyses rely on the false assumptions that usable nitrogenous compounds occur in a constant ratio to unusable compounds, that no nitrogenous compounds are toxic and that 6.25 is an accurate factor to transform nitrogen into available protein (proximal nutrient analyses rely on Kjeldahl total nitrogen determination for protein estimates; Herbst, 1988). The few available estimates of total protein in wild fruit, estimated either directly or from amino acid analysis after hydrolysis, suggest that proximal analyses overestimate available protein by 30–50% (recalculated from Foster, 1978 and Herbst, 1986). Herbst (1986) suggests that accurate conversion factors must be determined for each fruit type (see also Milton and Dintzis, 1981).

Because fruits are considered a relatively poor source of dietary protein (Mattson, 1980; Milton and Dintzis, 1981), it is often assumed that frugivorous birds and bats have unusually low nitrogen requirements and/or high nitrogen extraction efficiencies (Thomas, 1984; Levey and Karasov, 1989). Surprisingly, there is little experimental evidence to support these hypotheses. In favor of these hypotheses, Bosque and de Parra (1992) found that developing *S. capensis* chicks have nitrogen requirements that are about 0.3 times lower than the nitrogen equilibrium predicted by Robbins (1981). Against these hypotheses, however, are results of several studies that have reported nitrogen intake and output for birds and bats fed fruit diets (Walsberg, 1975; Worthington, 1989; Herbst, 1986; Studier *et al.*, 1988). Exclusively frugivorous diets that provide adequate energy intake also provide nitrogen in amounts equal to or in excess of those predicted for nitrogen equilibrium in birds (0.43 g $N/kg^{0.75}$ per day, Robbins, 1981). These results should be interpreted cautiously, however. First, all these studies relied on estimates from Kjeldahl analysis which yields both usable (proteins, nucleic acids and amino sugars) and non-usable (some alkaloids, inorganic nitrogen) nitrogen. Crude nitrogen intake cannot be equated with intake of utilizable nitrogen. Second, estimated nitrogen requirements in birds depend on the method of measurement. Commonly, more dietary protein is needed to maintain body weight than to maintain nitrogen balance (Levielle and Fisher, 1958; Martin, 1968; Brice and Grau, 1991). Nitrogen requirements as determined by the nitrogen balance method seem to underestimate nitrogen required for body mass maintenance. In summary, we have little evidence for (or against) the presumed low protein requirements of fruit-eating birds and for the commonly-held assumption that fruits are poor protein sources for adult birds.

Thomas (1984) suggested that Old World and New World fruit-eating bats differed in their means of obtaining protein and speculated that this difference resulted in differences in intake regulation and function as seed dispersers. He argued that New World phyllostomid bats rely on protein-poor fruits supplemented with insects. Although Old World pteropodids also rely on protein-poor fruits, they apparently do not supplement their diet with insects. To satisfy their protein requirements, Thomas (1984) argued that pteropodids ingest and assimilate energy above their requirements for daily maintenance and storage. The general validity of Thomas' hypothesis is questionable on two counts: (a) bat-dispersed fruits probably show significant interspecific variation in protein content and (b) the importance of insects in the diet of New World fruit-eating phyllostomids varies both seasonally and among species (Howell and Burch, 1974; Gardner, 1977; Fleming, 1988). Thomas (1984) founded his argument on nutritional analyses of a small and taxonomically biased fruit species sample (four species, three of which were figs, *Ficus* spp.). Old World fruit-eating bats feed on fruits in at least 145 plant genera in 50 families (Marshall, 1985). Protein content within this diverse sample is likely to vary considerably. For example, protein content varies more than three-fold in fruits consumed by New World fruit-eating bats (Fleming, 1988). Additionally, all the fruit species analysed by Herbst (1986) had protein contents that satisfied the maintenance requirements of non-reproductive, though not those of lactating, *Carollia perspicillata* (Carollinae; Phyllostomidae). Only one of the fruit species, however, provided a balanced mixture of amino acids. Herbst (1986) suggested that by selectively mixing fruit species, some New World bats may not need to supplement their diet with insects.

Fruit pulp is accompanied by a relatively large seed mass (30–40% of total dry fruit mass) that is non-assimilable (Moermond and Denslow, 1985). Seeds dilute the concentration of nutrients in fruits and make them bulky. Animals in negative energy balance on a fruit diet have high nitrogen excretion rates and hence negative protein balances as a result of endogenous nitrogen loss resulting from catabolism of body protein (Levey and Karasov, 1989). The failure of many non-specialized frugivores to maintain body mass and positive nitrogen balances on fruit diets may be a consequence of the inability of these animals to process the large amounts of fruit required to maintain a positive energy balance (Levey and Karasov, 1989; M. Witmer, personal communication). Hypotheses on the adequacy of fruit as a protein source for birds and bats were posed 20 years ago. We still lack adequate data on both bird and bat protein requirements and useable protein in fruit to properly evaluate these hypotheses.

Pollen as food

Flower-visiting animals often ingest pollen in addition to nectar. Pollen is a potentially rich food source and seems to be an important nitrogen and energy source for some animals such as bats and some small marsupials (Howell, 1974; Richardson *et al.*, 1986). The nutritional role of pollen for flower-visiting birds is unclear, however. Wooler *et al.* (1988) found 23–48% assimilation of pollen grains in New Holland honeyeaters (*P. novahollandiae*) and purple-crowned lorikeets (*Glossopsitta porphyrocephala*). Paton (1981) and Brice *et al.* (1989), in contrast, found less than 10% assimilation of pollen in adult hummingbirds (*C. anna* and *C. costae*), honeyeaters (*P. novahollandiae*), lorikeets (*T. haemadotus*) and cockatiels (*Nimphycus hollandicus*). Brice and Gau (1989) concluded that none of the pollens used in their study (*Zauschsneria* spp., *Callistemon* spp. and *Eucalyptus* spp.) could furnish significant amounts of protein or energy to birds. The nutritional value of pollen probably depends on the consuming bird's digestive traits (Wooler *et al.*, 1988) and the chemical and physical characteristics of the consumed pollen. Different plant species produce pollen with extremely disparate chemical compositions (Howell, 1974; Baker and Baker, 1979) and variation in the digestibility of different pollens fed to the same species of bird has been reported (Brice *et al.*, 1989). More comparative research on the physiological and ecological aspects of the assimilation of pollen by birds is required to settle the role of pollen in the nutrition of flower-visiting birds.

Although some glossophagine bats (Phyllostomidae), and presumably also some macroglossines (Pteropodidae), eat insects throughout much of the year (Howell and Burch, 1974; Start and Marshall, 1976), all are almost exclusive flower-feeders during at least some months (Howell and Hodgkin, 1976). During these months, pollen is probably the only reliable source of protein. Although the data documenting the efficiency with which bats digest pollen are scanty, it is almost certain that they do digest it. *Syconicteris australis* digests approximately 50% of the pollen grains of *Banksia* spp. and *Callistemon* spp. that it ingests (Law, 1992a). Howell (1974) and Law (1992b) demonstrated that *L. curasoae* and *S. australis* (Pteropidae) maintain positive nitrogen balances with pollen as a sole source of protein. The exine outer coat of pollen grains is composed of cellulose and sporollenin and is very resistant to degradation (Stanley and Linskens, 1974). The mechanisms by which bats break this barrier to digestion have not been established. Howell (1974) suggested that pollen digestion occurred in the stomach of *L. curasoae* and Rasweiler (1977) argued that pollen digestion occurs in the enlarged fundic caecum characteristic of the stomachs of glossophagine and macroglossine bats (Forman, 1990). Law (1992a), however, found that the fraction of empty ('digested') pollen grains in *S. australis* increased from 25% to 59.9% from the stomach to the distal section of the intestine, suggesting that most pollen digestion occurs in the intestine. In the pollinivorous marsupial *Tarsipes rostratum* (Tarsipedidae: Marsupialia) most of the pollen digestion also seems to take place in the small intestine (Richardson *et al.*, 1986). Pollen is clearly an important component of the nutrition of flower-visiting bats, but we know extremely little about the mechanisms by which it is assimilated.

Howell (1974) hypothesized that the chemical characteristics of pollen produced by bat-pollinated plants were moulded by the preferences and nutritional needs of bats. She demonstrated that the pollen of bat-pollinated Cactaceae and Agavaceae contained higher protein contents than bee-pollinated congeners. Amino acid profiles of bat-pollinated Agavaceae and Cactaceae are also apparently rich in proline and tyrosine, which Howell (1974) hypothesized are important for the bat's nutritional balance. The generality of this pattern remains to be established. Voss *et al.* (1980) found that the pollen of *Markea neurantha* (Solanaceae) does not provide a full complement of amino acids. In common with many plant products, *M. neurantha* pollen is deficient in methionine. In a broad survey of pollen amino acids, I. Baker and H. G. Baker (unpublished data) found pollen deficient in methionine.

Pollen digestion probably occurs through the germination pores rather than from mechanical breakdown of the exine. Consequently, a digested pollen grain consists solely of an empty exine. Pollen provides both a marker for transit time (the exine) and a simple way to estimate digestive efficiency (the fraction of empty grains) (Turner, 1984). Research on pollen digestion by birds and bats is eased by these two characteristics. More research on pollen assimilation by vertebrates can yield significant insights into the nutrition of anthophylous animals and on their reciprocal evolution with plants.

Nutrients in nectar and the evolution of bird–plant interactions

Determining the role of coevolution (*sensu stricto*, Janzen, 1980) in bird–plant and bat–plant interactions has been one of the implicit goals of many studies (Heithaus, 1982; Wheelwright, 1988a). Although in recent years, preoccupation with coevolution has abated (Wheelwright, 1991), many questions about the evolutionary origin and maintenance of bird/bat–plant interactions remain unanswered. Can information on the physiology of these animals illuminate the evolution of their interaction with plants? Physiological and behavioural data emphasize current performance and can therefore provide extremely valuable clues about processes responsible for maintaining a given pattern of association between fruit/nectar chemistry and birds. Ecological patterns, however, are partially the result of history and the origin and historical maintenance of such patterns is rarely explored (Losos, 1990). An approach that emphasizes exclusively behaviour and physiology, unfortunately, cannot provide insights into the evolution of the mechanisms and processes that established the present enormous variation in nectar and fruit pulp composition and the avian responses to them.

Martínez del Rio *et al.* (1992) used a four-step procedure in their attempt to unravel the evolution of sugar constituents in plants pollinated and dispersed by birds: (a) they identified an ecological pattern (the apparent convergence in sugar composition in nectar and fruit of plants with common groups of pollinators and dispersers); (b) they postulated and tested a process that can account for the maintenance of this pattern (variation in the sugar preferences of birds); (c) they investigated the physiological and ethological mechanisms responsible for this process; and, finally, (d) they integrated this pattern/process/mechanism information in a phylogenetic and biogeographical context. Although the results of their analysis were far from definitive, they concluded that the approach was useful. They assert that physiological processes and their effect on bird behaviour provide necessary (maybe essential) but not sufficient elements for the evolutionary explanation of the chemical diversity of plant rewards. Phylogenetic and biogeographical information is needed as well.

The increased availability of phylogenetic information for both birds and bats (Hood and Smith, 1982; Sibley and Ahlquist, 1990) and plants (Donoghue, 1989) and the development of methodologies to include historical information in comparative studies (Maddison, 1990; Harvey and Pagel, 1991; Janson, 1992) will contribute in elucidating the evolutionary causes and consequences of bird–plant associations. McDade (1992) and Stein (1992) provide excellent recent examples of how phylogenetic and biogeographical information can be used to elucidate the evolution of floral and hummingbird morphology.

Quo vadis nutritional ecology of nectar- and fruit-eating birds?

In spite of the mostly recent research reviewed here, the nutritional ecology of nectar- and fruit-eating birds and bats remains a collection of tenuous patterns and unexplored mechanisms. We need more and better data on the nutritional characteristics of fruit and nectar. The intricacies of digestion in nectar- and fruit-eating birds suggest that data from proximal nutrient analysis are too coarse-grained to be useful tools in nutritional and coevolutionary (*sensu lato*) studies of these animals (Martínez del Rio and Restrepo, 1992). We also need more basic data on the digestive and metabolic abilities of nectar- and fruit-eating birds and bats in a variety of taxa. Artificial fruits and nectar of controlled nutrient composition can be used to estimate nutritional requirements (Brice and Grau, 1989) and to determine preferences for specific nutrients (Martínez del Rio *et al*., 1989; Stiles, 1992). We also need to complement our aviary studies with field studies where many additional variables influence foraging and metabolism. The methods used by Studier *et al*. (1988) to study the nutritional budget of free-flying cedar waxwings (*B. cedrorum*) hold considerable promise for nutritional field studies. Cedar waxwings are 'good' frugivores (i.e. dispersers) that do not digest ingested seeds. Studier *et al*. (1988) analysed nutrients in fruit and faecal samples of free-living birds and used seeds as natural feed markers to determine nutritional budgets.

Although I have emphasized how the physiological traits of pollinators and seed dispersers can reflect and shape their ecological interaction with plants, nectar- and fruit-eating birds are interesting subjects of physiological studies in their own right. They can contribute to the advance of physiological ecology and basic physiology in at least two ways. First, fruit- and nectar-eating birds seem to illustrate clearly the existence of trade-offs in physiological traits (e.g. the physiological traits associated with high assimilation efficiencies of lipids appear to be different from those associated with the assimilation of carbohydrates; see p. 109).

Second, many frugivores experience dramatic ontogenetic and seasonal shifts in diet (Wheelwright, 1988b; Walsberg and Thompson, 1990 and references therein) and may provide useful models for phenotypic adaptations associated with dietary changes (Levey and Karasov, 1989). The dramatic temporal changes in diet and the strong and consistent interspecific differences in the chemical composition of the food that fruit- and nectar-eating animals ingest make them a unique system to explore the phenotypic and evolutionary plasticity of digestive functions.

Acknowledgements

F. Bozinovic, N. Conklin, D. Emlen, A. Masters and D. Phillips read and criticized several preliminary versions of this chapter. H. G. Baker provided unpublished information, comments and encouragement. The ideas presented here are the result of the pioneering work of Irene Baker on the chemical composition of nectar, pollen and fruit. This chapter is dedicated to her memory.

References

Alpers, D. H. (1987). Digestion and absorption of carbohydrates and proteins. In *Physiology of the Gastrointestinal Tract*, Vol. 2, ed. L. R. Johnson, pp. 1469–1486. New York: Raven Press.

Baker, H. G. (1975). Sugar concentrations in nectars from hummingbird flowers. *Biotropica*, **7**, 37–41.

Baker, H. G. (1977). Non-sugar constituents of nectar. *Apidologie*, **8**, 349–356.

Baker, H. G. & Baker, I. (1979. Starch in angiosperm pollen grains and its evolutionary significance. *American Journal of Botany*, **66**, 591–600.

Baker, H. G. & Baker, I. (1982). Chemical constituents of nectar in relation to pollination mechanisms and phylogeny. In *Biochemical Aspects of Evolutionary Biology*, ed. H. M. Nitecki, pp. 131–171. Chicago: Chicago University Press.

Baker, H. G. & Baker, I. (1983). Floral nectar sugar constituents in relation to pollinator type. In *Handbook of Experimental Pollination Ecology*, ed. C. E. Jones & R. J. Little, pp. 131–171. New York: Scientific and Academic Editions.

Baker, H. G. & Baker, I. (1986). The occurrence and significance of amino acids in floral nectar. *Plant Systemș in Evolution*, **151**, 175–186.

Baker, I., Baker, H. G., & Hodges, S. A. (1992). Patterns in the sugar composition of nectar and fruit juices taken by Microchiroptera and Megachiroptera. *Biotropica*, in press.

Bent, A. C. (1953). Life histories of North American wood warblers. *US National Museums Bulletin*, **20**, 239–258.

Bonaccorso, F. J. & Gush, T. J. (1987) Feeding behavior and foraging strategies of captive phyllostomid fruit bats: an experimental study. *Journal of Animal Ecology*, **56**, 907–920.

Bosque, C. & de Parra, O. (1992). Digestive efficiency and rate of food passage in oilbird nestlings. *Condor*, **94**, 557–571.

Brice, A. T. (1992). The essentiality of nectar and arthropods in the diet of the anna's hummingbird (*Calypte anna*). *Comparative Biochemistry and Physiology*, **101A**, 151–152.

Brice, A. T. & Grau, C. R. (1989). Hummingbird nutrition: a purified diet for long-term maintenance. *Zoology and Biology*, **8**, 233–237.

Brice, A. T. & Grau, C. R. (1991). Protein requirements for maintenance of the adult Costa's hummingbird (*Calypte costa*). *Physiological Zoology*, **64**, 611–626.

Brice, A. T., Dahl, K. H. & Grau, C. R. (1989). Pollen digestibility by hummingbirds and psittacines. *Condor*, **91**, 681–688.

Brugger, K. E. & Nelms, C. O. (1991). Sucrose avoidance by American robins (*Turdus migratorius*): implications for control of bird damage in fruit crops. *Crop Protection*, **10**, 455–460.

Buchholz, R. & Levey, D. J. (1990). The evolutionary triad of microbes, fruits, and seed dispersers: an experiment in fruit choice by cedar waxwings, *Bombycilla cedrorum*. *Oikos*, **59**, 202–204.

Calder, W. A. (1979). On the temperature dependency of optimal nectar concentration for birds. *Journal of Theoretical Biology*, **78**, 185–196.

Carey, M. C., Small, D. M. & Bliss, C. M. (1983). Lipid digestion and absorption. *Annual Review of Physiology*, **45**, 651–677.
Carpenter, F. L. & Castronova, J. L. (1980). Maternal diet selectivity in *Calypte anna*. *American Midland Naturalist*, **103**, 175–179.
Carpenter, F. L., Hixon, M., Hunt, A. & Russel, R. W. (1991). Why hummingbirds have such large crops. *Evolutionary Ecology*, **5**, 405–414.
Castro, G. A. (1984). Digestion and absorption. In *Gastrointestinal Physiology*, ed. L. R. Johnson, pp. 105–128. St Louis, MO: C. V. Mosby.
Davidar, P. (1983). Birds and neotropical mistletoes: effects on seedling recruitment. *Oecologia*, **60**, 271–273.
Desselberger, H. (1931). Der Verdauungskanal der Dicaediden nach Gestalt und Funktion. *Journal für Ornithologie*, **79**, 353–370.
Desselberger, H. (1932). Ueber der Verdauungskanal nektarfressender Vögel. *Journal für Ornithologie*, **80**, 309–318.
Docters van Leeuwen, W. M. (1954). On the biology of some Loranthaceae and the role birds play in their life-history. *Beaufortia*, **4**, 105–208.
Donoghue, M. J. (1989) Phylogenies and the analysis of evolutionary sequences, with examples from seed plants. *Evolution*, **43**, 1137–1156.
Feder, M. E., Nennet, A. F., Burggren, W. W. & Huey, R. B. (1987). *New Directions in Ecological Physiology*. New York: Cambridge University Press.
Fleming, T. H. (1988). *The short-tailed fruit bat*. Chicago: University of Chicago Press.
Forman, G. L. (1990). Comparative macro- and micro-anatomy of stomachs of macroglossine bats (Megachiroptera: pteropodidae). *Journal of Mammalogy*, **71**, 555–565.
Foster, M. S. (1978). Total frugivory in tropical passerines: a reappraisal. *Tropical Ecology*, **19**, 131–154.
Foster, M. S. (1990). Factors influencing bird foraging preferences among conspecific fruit trees. *Condor*, **92**, 844–854.
Freeman, C. E., Worthington, R. D. & Jackson, M. S. (1991). Floral sugar composition of some south and southeast Asian species. *Biotropica*, **23**, 568–574.
Gardner, A. L. (1977). Feeding habits. In *Biology of Bats of the New World Phyllostomatidae*, Part II, ed. R. J. Baker, J. K. Jones & D. C. Carter, pp. 293–350. Special Publications No. 13. Austin, TA: The Museum, Texas Technical University.
Gass, C. L. & Montgomerie, R. D. (1981). Hummingbird foraging behavior: decision making and energy regulation. In *Foraging Behavior: Ecological, Ethological, and Psychological Approaches*, ed. A. C. Kamil & T. D. Sargent, pp. 159–199. New York: Garland STPM.
Goldstein, J. L. & Swain, T. (1963). Changes in tannins in ripening fruit. *Phytochemistry*, **2**, 371–383.
Gorham, F. W. & Ivy, A. C. (1938). General function of the gall bladder from the evolutionary standpoint. *Bulletin of the Field Museum of Natural History*, **XXII**, 159–213.
Grajal, A., Strahl, S. D., Parra, R., Dominguez, M. G. & Neher, A. (1989). Foregut fermentation in the Hoatzin, a neotropical leaf-eating bird. *Science*, **245**, 1236–1238.
Greenberg, R. (1981). Frugivory in some migrant tropical forest wood warblers. *Biotropica*, **13**, 215–223.
Gryj, E., Martínez del Rio, C. & Baker, I. (1990). Avian pollination and nectar use in *Combretum fruticosum* (Loefl.). *Biotropica*, **22**, 266–271.

Hainsworth, F. R. (1977). Foraging efficiency and parental care in *Colibri coruscans*. *Condor*, **79**, 69–75.

Harvey, P. H. & Pagel, M. D. (1991). *The Comparative Method in Evolutionary Biology*. Oxford: Oxford University Press.

Heinrich, B. (1975). Energetics of pollination. *Annual Review of Ecology Systems*, **6**, 139–170.

Heithaus, E. R. (1982). Coevolution between bats and plants. In *Ecology of Bats*, ed. T. H. Kunz, pp. 327–368. New York: Plenum Press.

Herbst, L. H. (1986). The role of nitrogen from fruit pulp in the nutrition of the frugivorous bat *Carollia perspicillata*. *Biotropica*, **18**, 39–44.

Herbst, L. H. (1988). Methods of nutritional ecology of plant-visiting bats. In *Ecological and Behavioural Methods for the Study of Bats*, ed. T. H. Kunz, pp. 233–246. Washington, DC: Smithsonian Institution Press.

Hernández, A. & Martínez del Rio, C. (1992). Intestinal disaccharidases in five species of phyllostomoid bats. *Comparative Biochemistry and Physiology*, **103B**, 105–111.

Herrera, C. M. (1981) Are tropical fruits more rewarding than temperate ones? *American Naturalist*, **118**, 896–907.

Herrera, C. M. (1986). Vertebrate-dispersed plants: why don't they behave as they should? In *Frugivores and Seed Dispersal*, ed. A. Estrada & T. H. Fleming, pp. 5–18. Dordrecht, Netherlands: W. Junk.

Herrera, C. M. (1987). Vertebrate-dispersed plants of the Iberian peninsula: a study of fruit characteristics. *Ecological Monographs*, **57**, 305–331.

Hiebert, S. M. & Calder, W. A. (1983). Sodium, potassium, and chloride in floral nectars: energy-free contributions to refractive index and salt balance. *Ecology*, **64**, 399–402.

Hood, C. S. & Smith, J. D. (1982). Cladistic analysis of female reproductive histomorphology in phyllostomid bats. *Systematic Zoology*, **31**, 241–251.

Howe, H. F. (1992). Specialized and generalized dispersal systems: where does 'the paradigm' stand? *Vegetatio*, **107/108**, 3–13.

Howe, H. F. & Estabrook, G. F. (1977). On intraspecific competition for avian dispersers in tropical trees. *American Naturalist*, **111**, 817–832.

Howe, H. F. & Smallwood, J. (1982). Ecology of seed dispersal. *Annual Review of Ecology and Systematics*, **13**, 201–208.

Howell, D. J. (1974). Bats and pollen: physiological aspects of the syndrome of chiropterophily. *Comparative Biochemistry and Physiology*, **48A**, 263–276.

Howell, D. H. & Burch, P. (1974). Food habits of some Costa Rican bats. *Reviews in Biology of the Tropics*, **21**, 281–294.

Howell, D. H. & Hodgkin, N. (1976). Feeding adaptations in the hairs and tongues of nectar-feeding bats. *Journal of Morphology*, **148**, 329–336.

Jackson, S. & Place, A. R. (1990). Gastrointestinal transit and lipid assimilation efficiencies in three species of high latitude seabirds. *Journal of Experimental Zoology*, **255**, 141–154.

Janson, C. H. (1992). Measuring evolutionary constraints: a Markov model for phylogenetic transitions among seed dispersal syndromes. *Evolution*, **46**, 136–158.

Janzen, D. H. (1977). Why fruit rot, seeds mold and meat spoils. *American Naturalist*, **111**, 691–713.

Janzen, D. H. (1980). When is it coevolution? *Evolution*, **34**, 611–612.

Johnson, R. A., Wilson, M. F., Thompson, J. N. & Bertin, R. I. (1985). Nutritional values of wild fruits and consumption by migrant frugivorous birds. *Ecology*, **66**, 819–827.

Karasov, W. H. (1990). Digestion in birds: chemical and physiological determinants, and ecological implications. In *Avian Foraging: Theory, Methodology, and Applications*, ed. M. L. Morrison, C. J. Ralph, J. Verner & J. R. Jehl, Studies in Avian Biology No. 13. Lawrence, KA: Cooper Ornithological Society.

Karasov, W. H. & Levey, D. J. (1990). Digestive trade-offs and adaptations of frugivorous birds. *Physiological Zoology*, **63**, 1248–1270.

Karasov, W. H., Phan, D., Diamond, J. M. & Carpenter, F. L. (1986). Food passage and intestinal nutrient absorption in hummingbirds. *Auk*, **103**, 453–464.

Konarzewski, M., Kozlowski, J. & Ziólko, M. (1989). Optimal allocation of energy to growth of the alimentary tract of birds. *Functional Ecology*, **3**, 589–596.

Kenward, R. E. & Sibly, R. M. (1977). A woodpigeon (*Columba livia*) preference explained by a digestive bottleneck. *Journal of Applied Ecology*, **14**, 815–826.

Klite, P. D. (1965). Intestinal bacterial flora and transit time of three neotropical bat species. *Journal of Bacteriology*, **90**, 375–379.

Kunz, T. H. (1982). *Ecology of Bats*. New York: Plenum Press.

Law, B.S. (1992a) Physiological factors affecting pollen use by Queensland blossom bats (*Syconycteris australis*). *Functional Ecology*, in press.

Law, B. S. (1992b). The maintenance of nitrogen requirements of the Queensland blossom bat (*Syconycteris australis*). *Physiological Zoology*, **65**, 634–648.

Levey, D. J. & Duke, G. E. (1992). How do frugivores process fruit? Gastrointestinal transit and glucose absorption in cedar waxwings (*Bombycilla cedrorum*). *Auk*, **109**, 722–730.

Levey, D. J. & Grajal, A. (1991). Evolutionary implications of fruit processing and intake limitation in cedar waxwings. *American Naturalist*, **138**, 171–189.

Levey, D. J. & Karasov, W. H. (1989). Digestive responses to diet switching between fruits and insects in temperate birds. *Auk*, **106**, 675–686.

Levielle, G. A. & Fisher, H. (1958). Amino acid requirements for maintenance in the adult rooster. I. Nitrogen and energy requirements in normal and protein depleted animals receiving whole egg protein and amino acid diets. *Journal of Nutrition*, **66**, 441–453.

Lindsay, O. B. & March, B. E. (1967). Intestinal absorption of bile salts in the cockerel. *Poultry Science*, **46**, 164–171.

Losos, J. B. (1990). A phylogenetic analysis of character displacement in Caribbean *Anolis lizards*. *Evolution*, **44**, 558–569.

Maddison, W. P. (1990). A method for testing the correlated evolution of two binary characters: are gains and losses concentrated on certain branches of a phylogenetic tree. *Evolution*, **44**, 539–557.

Marshall, A. G. (1985). Old World phytophagous bats (Megachiroptera) and their food plants: a survey. *Zoological Journal of the Linnaean Society*, **83**, 351–369.

Martin, A. C., Zim, H. S. & Nelson, A. L. (1951). *American Wildlife and Plants*. New York: McGraw Hill.

Martin, E. W. (1968). The effects of dietary protein on the energy and nitrogen balance of the tree sparrow (*Spizella arborea arborea*). *Physiological Zoology*, **41**, 313–331.

Martínez del Rio, C. (1990a). Dietary and phylogenetic correlates of intestinal sucrase and maltase activity in birds. *Physiological Zoology*, **63**, 987–1011.

Martínez del Rio, C. (1990b). Sugar preferences in hummingbirds: the influence of subtle chemical differences on food choice. *Condor*, **92**, 1022–1030.

Martínez del Rio, C. & Karasov, W. H. (1990). Digestion strategies in nectar- and

fruit-eating birds and the composition of plant rewards. *American Naturalist*, **136**, 618–637.
Martínez del Rio, C. & Restrepo, C. (1992). Ecological and behavioral consequences of digestion in frugivorous animals. *Vegetatio*, **107/108**, 205–216.
Martínez del Rio, C. & Stevens, B. R. (1989). Physiological constraint on feeding behavior; intestinal membrane disaccharidases of the starling. *Science*, **243**, 794–796.
Martínez del Rio, C., Stevens, B. R., Daneke, D. & Andreadis, P. T. (1988). Physiological correlates of preference and aversion for sugars in three species of birds. *Physiological Zoology*, **61**, 222–229.
Martínez del Rio, C., Levey, D. J. & Karasov, W. H. (1989). Physiological basis and ecological consequences of sugar preferences in cedar waxwings. *Auk*, **106**, 64–71.
Martínez del Rio, C., Baker, H.G. & Baker, I. (1992). Ecological and evolutionary implications of digestive processes: bird preferences and the sugar constituents of floral nectar and fruit pulp. *Experientia*, **48**, 554–561.
Mattson, W. J. (1980). Herbivory in relation to plant nitrogen content. *Annual Review of Ecological Systems*, **11**: 119–161.
McDade, L. A. (1992). Pollinator relationships, biogeography, and phylogenetics. *BioScience*, **42**, 21–26.
McKey, D. (1975). The ecology of coevolved seed dispersal systems. In *Coevolution of Animals and Plants*, ed. L. E. Gilbert & P. Raven, pp. 159–191. Austin, TA: University of Texas Press.
McNab, B. K. (1982). Evolutionary alternatives in the physiological ecology of bats. In *Ecology of Bats*, ed. T. H. Kunz, pp. 151–199. New York: Plenum Press.
Milton, K. & Dintzis, F. R. (1981). Nitrogen to protein conversion factors for tropical plant samples. *Biotropica*, **13**, 177–181.
Moermond, T. C. & Denslow, J. S. (1985). Neotropical avian frugivores: patterns of behavior, morphology, and nutrition with consequences for food selection. In *Neotropical Ornithology*, ed. P. A. Buckley, M. S. Foster, E. S. Morton, R. S. Ridgely & N. G. Smith, pp. 865–897. Lawrence, KA: Ornithological Monographs 36.
Montgomerie, R. D. & Redsell, C. A. (1980). A nesting hummingbird feeding solely on arthropods. *Condor*, **82**, 463–464.
Morrison, M. L., Ralph, C. J., Verner, J. & Rehl, J. R. (1988). *Avian Foraging: Theory, Methodology, and Applications.* Studies in Avian Biology No. 13. Lawrence, KA: Cooper Ornithological Society.
Morton, E. S. (1973). On the evolutionary advantages and disadvantages of fruit-eating in tropical birds. *American Naturalist*, **107**, 8–22.
Nawar, W. W. (1985). Lipids. In *Food Chemistry*, ed. O.R. Fennema, pp. 139–244. Basel: Marcel Dekker.
O'Connor, R. J. (1984). *The Growth and Development of Birds*. Brisbane: John Wiley.
Ogunbiyi, O. A. & Okon, E. E. (1976). Studies on the digestive enzymes of the African bat *Eidolon helvum* (Kerr). *Comparative Biochemistry and Physiology*, **55A**, 359–361.
Paton, D. C. (1981). The significance of pollen in the diet of the New Holland honeyeater, *Phylidornis novahollandiae* (Aves: Meliphagidae). *Australian Journal of Zoology*, **29**, 217–224.

Paton, D. C. (1982). The diet of the New Holland honeyeater, *Phylidornis novahollandiae*. *Australian Journal of Ecology*, **7**, 279–298.
Percival, M. S. (1961). Types of nectar in angiosperms. *New Phytologist*, **60**, 235–281.
Place, A. R. & Stiles, E. W. (1992). Living off the wax of the land: bayberries and yellow-rumped warblers. *Auk*, **109**, 334–345.
Rasweiler, J. J. (1977). The care and management of bats as laboratory animals. In *Biology of Bats*, Vol. 3, ed. W. A. Wimsatt, pp. 519–617. New York: Academic Press.
Renner, R. & Hill, W. F. (1961). Utilization of fatty acids by chicken. *Journal of Nutrition*, **74**, 259–264.
Restrepo, C. (1987). Aspectos ecologicos de la diseminación de cinco especies de muerdagos por aves. *Humboldtia*, **1**, 65–116.
Rhoades, D. F. & Bergdahl, J. C. (1981). Adaptive significance of toxic nectar. *American Naturalist*, **117**, 798–803.
Richardson, K. C. & Wooler, R. D. (1986). The structures of the gastrointestinal tracts of honeyeaters and other small birds in relation to their diets. *Australian Journal of Zoology*, **34**, 119–124.
Richardson, K. C. & Wooler, R. D. (1988a). The alimentary tract of a specialist frugivore, the mistletoebird, *Dicaeum hirundinaceum*, in relation to its diet. *Australian Journal of Zoology*, **36**, 373–382.
Richardson, K. C. & Wooler, R. D. (1988b). Morphological relationships of passerine birds from Australia and New Guinea in relation to their diets. *Zoology Journal of the Linnaean Society*, **94**, 193–201.
Richardson, K. C., Wooler, R. D. & Collins, B. G. (1986). Adaptations to a diet of nectar and pollen in the marsupial *Tarsipes rostratum* (Marsupialia: Tarsipedidae). *Journal of Zoology* (Lond.) **A208**, 285–297.
Ricklefs, R. E. (1974). Energetics of reproduction in birds. In *Avian Energetics*, ed. R. E. Paynter, pp. 152–202. Publication of Nuttal Ornithology Club 15.
Robbins, C. T. (1981). Estimation of the relative protein cost of reproduction in birds. *Condor*, **83**, 177–179.
Semenza, G. & Corcelli, A. (1986). The absorption of sugars and amino acids across the small intestine. In *Molecular and Cellular Basis of Digestion*, ed. P. Desnuelle, H. Sjöström & A. Norén, pp. 381–412. New York: Elsevier.
Sherry, T. W. (1990). When are birds dietarily specialized? In *Distinguishing Ecological from Evolutionary Approaches*, ed. M. L. Morrison, C. J. Ralph, J. Verner & J. R. Jehl, pp. 337–352. Studies in Avian Biology No. 13, Lawrence, KA: Cooper Ornithological Society.
Sibley, C. G. & Ahlquist, J. E. (1984). The relationships of the starlings (Sturnidae: Sturnini) and the mockingbirds (Sturnidae: Mimini). *Auk*, **101**, 230–243.
Sibley, C. G. & Ahlquist, J. E. (1990). *Phylogeny and Classification of Birds: A Study in Molecular Evolution*. New Haven: Yale University Press.
Snow, D. W. (1971). Evolutionary aspects of fruit-eating by birds. *Ibis*, **113**, 194–202.
Snow, D. W. (1981). Coevolution of birds and plants. In *The Evolving Biosphere*, ed. P. L. Foley, pp. 169–178. Cambridge: Cambridge University Press.
Speakman, J. R. (1987). Apparent absorption efficiencies for Redshank (*Tringa totanus* L.) and oystercatcher (*Haematopus ostralegus* L.). Implications for the predictions of optimal foraging models. *American Naturalist*, **130**, 677–691.
Stanley, R. G. & Linskens, H. F. (1974). *Pollen: Biology, Biochemistry, Management*. New York: Springer-Verlag.
Start, A. N. & Marshall, A. G. (1976). Nectarivorous bats as pollinators of trees in

West Malaysia. In *Variation, Breeding and Conservation of Forest Trees*, ed. J. Burley & B. T. Stiles, pp. 141–150. London: Academic Press.
Stashko, E. R. & Dinerstein, E. (1988). Methods for estimating fruit availability to frugivorous bats. In *Ecological and Behavioral Methods for the Study of Bats*, ed. T. H. Kunz, pp. 221–232. Washington, DC: Smithsonian Institution Press.
Stein, B. A. (1992). Sicklebill hummingbirds, ants, and flowers. *BioScience*, **42**, 27–33.
Stephens, D. W. & Krebs, J. R. (1986). *Foraging Theory*. Princeton, NJ: Princeton University Press.
Stevens, C. E. (1990). *Comparative Physiology of the Vertebrate Digestive System*. New York: Cambridge University Press.
Stiles, E. W. (1980). Patterns of fruit presentation and seed dispersal in bird-disseminated woody plants in the eastern deciduous forest. *American Naturalist*, **116**, 670–688.
Stiles, E. W. (1992). The influence of lipids on fruit preference by birds. *Vegetatio*, **107/108**, 227–236.
Stiles, F. G. (1971). Time, energy, and territoriality of the Anna hummingbird (*Calypte anna*). *Science*, **173**, 818–821.
Stiles, F. G. (1976). Taste preferences, color preferences and flower choice in hummingbirds. *Condor*, **78**, 10–26.
Studier, E. H., Szuch, E. J., Tompkins, T. M. & Cope, V. W. (1988). Nutritional budgets in free flying birds: cedar waxwings (*Bombycilla cedrorum*) feeding on Washington hawthorn fruit (*Crataegus phaenopyrum*). *Comparative Biochemistry and Physiology*, **89A**, 471–474.
Thomas, D. W. (1984) Fruit and energy intake budgets of frugivorous bats. *Physiological Zoology*, **57**, 457–467.
Thomas, D. W. (1988). Analysis of diets of plant-visiting bats. In *Ecological and Behavioral Methods for the Study of Bats*, ed. T. H. Kunz, pp. 211–220. Washington, DC: Smithsonian Institution Press.
Turner, V. (1984). *Banksia* pollen as a source of protein in the diet of two Australian marsupials *Cercatecus nanus* and *Tarsipes rostratus*. *Oikos*, **43**, 53–61.
van Soest, P. J. (1983). *Nutritional ecology of the ruminant*. Corvallis, OR: O & B Books.
Voss, R., Turner, M., Inouye, M., Fisher, M. & Cort, R. (1980). Floral biology of *Markea neurantha* Hemsley (Solanaceae), a bat pollinated epiphyte. *American Midland Naturalist*, **103**, 262–268.
Walsberg, G. E. (1975). Digestive adaptations of *Phainopepla nitens* associated with the eating of mistletoe berries. *Condor*, **77**, 169–174.
Walsberg, G. E. & Thompson, C. W. (1990). Annual changes in gizzard size and function in a frugivorous bird. *Condor*, **92**, 794–795.
Wetmore, A. (1914). The development of the stomach in the Euphonias. *Auk*, **31**, 458–461.
Wheelwright, N. T. (1988a). Fruit-eating birds and bird-dispersed plants in the tropics and temperate zone. *Trends in Ecology and Evolution*, **10**, 270–274.
Wheelwright, N. T. (1988b). Seasonal changes in food preferences of American robins in captivity, *Auk*, **105**, 374–378.
Wheelwright, N. T. (1991). Frugivory and seed dispersal: 'la coevolución ha muerto. Viva la coevolución. *Trends in Ecology and Evolution*, **16**, 312–313.
Wiens, J. A. (1983). Avian community ecology: an iconoclastic view. In *Perspectives in Ornithology*, ed. A. H. Brush & G. A. Clark, pp. 355–403. New York: Cambridge University Press.

Wolf, L. L. & Hainsworth, F. R. (1971). Time and energy budgets of territorial hummingbirds. *Ecology*, **52**, 980–988.

Worthington, A. H. (1989). Adaptation for avian frugivory: assimilation efficiency and gut transit time of *Manacus vitellinus* and *Pipra mentalis*. *Oecologia*, **80**, 381–389.

Wooler, R. D., Richardson, K. C. & Pagendham, C. M. (1988). The digestion of pollen by some Australian birds. *Australian Journal of Zoology*, **36**, 357–362.

9

Herbivory and niche partitioning

MICHAEL R. PERRIN

Green plants are abundant in natural environments and represent the major food resource of mammals. Plant diversity, growth forms and anatomical structure contribute to the diversity of herbivores. Physical defences, such as thorns and silicia bodies, are important in determining food selection, but structural polysaccharides, particularly cellulose, determine digestive styles which include pregastric and caecum–colon fermentation. Plant toxins (secondary compounds), representing a third order of defence, have resulted in the most specific and specialised defences and counter-adaptations of herbivores. These anatomical, cellular, and molecular defences modify the availability and palatability of plant foods to herbivores.

Plant–herbivore associations are many and varied and the diversity of terrestrial mammalian herbivores is great. By definition, each species occupies a unique (independent, realised) niche. Niches are *n*-dimensional hypervolumes defining the role and position of a species within a natural environment (Hutchinson, 1957, 1959). Initially, niches were simply defined in terms of space or diet, but current definitions are holistic and all facets of a species' biology are recognised in niche designation. Each species in a herbivore guild or community occupies a different niche, but only a small number of species, and hence niches, will be abundant and many will be rare. This is demonstrated by rank–abundance curves.

Three theoretical community organisation processes have been proposed to account for these distributions (see Fig. 9.1). Each one demonstrated that species, and hence niches, do not occupy equal or equivalent space in an assemblage (May, 1975). Sugihara's (1980) sequential breakage hypothesis proposed a biological mechanism to account for the lognormal distribution, while Branch's (1985) competitive elimination hypothesis provided an explanation for the method of speciation that might operate.

Although fundamental niche overlap may occur between syntopic herbi-

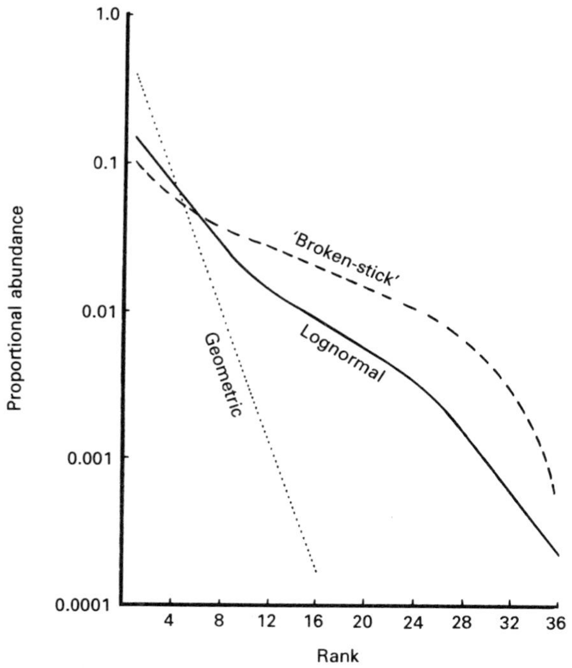

Fig. 9.1. Examples of rank–abundance diagrams.

vores, the realised niches, when resources are limiting, are distinct. Since evolution is a sequential process (whether by phyletic gradualism or punctuated equilibria) and since 'monogastric' digestive styles preceded 'polygastric' ones (e.g. in the ruminants (Hofmann, 1968; Langer, 1974), Perissodactyla (Janis, 1976) and Rodentia (Vorontsov, 1961, 1962)), herbivores exploited abundant and nutritious foods before uncommon and innutritious ones. Further, since adaptation is relative (Landry, 1970; Lewontin, 1980) and species track environmental change (van Valen's 1973 Red Queen hypothesis), herbivores became better adapted to exploit plant foods with cellular and molecular defences. Normally, generalists preceded specialists through evolutionary time, and *new* niches are often increasingly circumscribed. So when a new adaptive zone was colonised by the equids, and subsequently by the ruminants, speciation and niche differentiation were rapid. Herbivorous niches increased in diversity through time to produce a guild with a small number of common species and a larger number of rare species (Janis, 1976; Vrba, 1980, 1984).

To summarise, the niche concept is holistic and niches are species-specific;

niche spaces are very varied, unique and of no particular size. The niche concept has great theoretical value in community ecology and evolutionary biology because it can be regarded as a corollary of the unit of selection.

The pregastric fermenter's niche

Synopsis

The success of ruminants and pregastric fermenters (PGF) in general has been explained by their ability to digest cellulose cell walls, to access nutritious cell contents and to recycle urea for microbial protein synthesis. Kinnear and Main (1979), however, have explained the success of pregastric fermenters in terms of the nutritional niche concept.

In an attempt to marry aspects of ecological theory and animal nutrition, Kinnear and Main (1979) considered the pregastric microbial–mammal symbiosis in terms of the Hutchinsonian multidimensional niche and defined the fundamental nutritional niche as a set of axes representing nutrient variables (e.g. amino acids, carbon, phosphorus, lipids, vitamins, H_2O) essential for the persistence of a species. They illustrated the nutritional niche in two dimensions (Fig. 9.2), although the concept was generalised to higher dimensions. In the example, two essential nutrients were linearly ordered on rectangular coordinate axes. The lower limits, en_1' and en_2', denoted the minimum daily intake of two essential nutrients required by a species, and the upper limits (en_1'', en_2'') denoted the maximum amounts of the same essential nutrients that the species can tolerate. Thus a nutritional area was described in which a species could survive and persist.

Difficulties

I question the validity of a nutritional niche outside a physiological context, since it has no significance in ecology or evolution because only certain aspects of the animal's biology are defined. It has parallels with the trophic niche concept and has similar constraints because the concept is not holistic. To interpret community structure, or to identify interspecific competition, all aspects of the niche must be identified. For example, to say that two species have the same nutritional niche does not mean that they are competitors as they may be segregated spatially. Similarly, two species with the same nutritional niche may use different food plants to access the same nutrients. Although their nutritional niches are identical their trophic and Hutchinsonian niches are distinct.

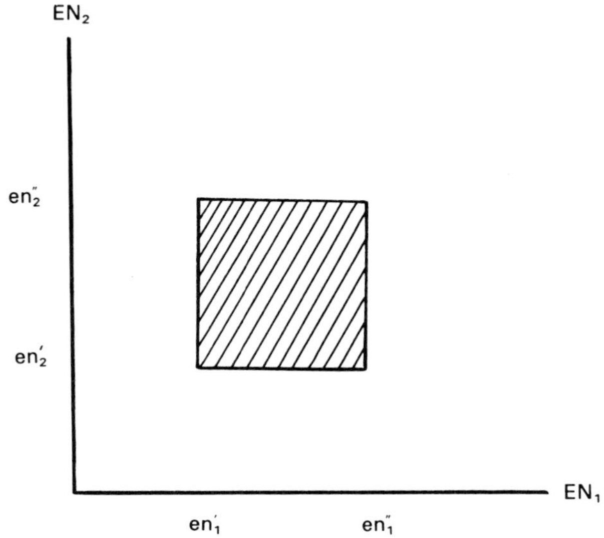

Fig. 9.2. The fundamental nutritional niche as represented by Kinnear and Main. Two essential nutrients (EN_1 and EN_2) are linearly ordered on rectangular coordinate axes with arbitrary limits denoted by en_1', en_2' etc. The lower limits (en_1' and en_2' on the respective axes) define the minimum amount of an essential nutrient required by a species, and the upper limits (en_1'', en_2'') denote the maximum amount of an essential nutrient the species can tolerate. Thus, an area (2-space) is described and a species can only survive and persist within the region bounded by the perimeter. Modified after Kinnear and Main (1979).

Second, herbivores select plant foods mainly on the basis of crude protein content and digestible energy availability (Owen-Smith and Novellie, 1982) rather than essential nutrients. Since nutrients covary in their availability, the critical niche parameter becomes the availability of the most limiting nutrient in relation to physiological demand. I ask, therefore, whether essential nutrients are meaningful niche parameters? Further, although herbivores have minimal requirements for essential nutrients, which may define the lower limit of a niche dimension, there may be no natural upper limit unless high concentrations of essential nutrients have negative influences. Similarly, plant toxins may have critical upper concentrations that contain nutritional niche space, but is there a meaningful lower limit? There is, but it is only marginally lower than the upper limit and is diffuse, particularly when many foods are eaten.

Kinnear and Main (1979) also suggest that symbiotic microbes with their unique biochemical and metabolic characteristics lower host requirements for certain essential nutrients (e.g. essential amino acids and vitamins) and extend tolerances to toxins and nutrient excesses or imbalances (Fig. 9.3). They equate these (adaptive) characteristics with (fundamental) nutritional niche expansion since the host can survive on an innutritious diet.

The hypothesis was illustrated by studies of a ruminant-like, potorine marsupial, the woylie, *Bettongia penicillata*. It has a mass of less than 2 kg and feeds predominantly on fungi, which are both deficient (in lysine) and imbalanced (excess of methionine) in amino acid composition (by known mammalian standards), and gum exudate, which is a high-energy carbohydrate source but contains no protein. It would seem that the fundamental nutritional niche has been expanded. However, the symbiotic bacteria (and protozoa) impose energetic and nutritional demands on the host (Reid, 1970) which reduce fundamental nutritional niche space. For example, carbohydrate yields 11 to 30% more energy when digested post-ruminally and microbial protein synthesis in the rumen is an inefficient use of extrinsic protein resources (Smith, 1975). Kinnear and Main (1979) identified these costs, but did not interpret them as fundamental nutritional niche reduction, and argued that PGF herbivores are at a competitive disadvantage against non-PGF herbivores.

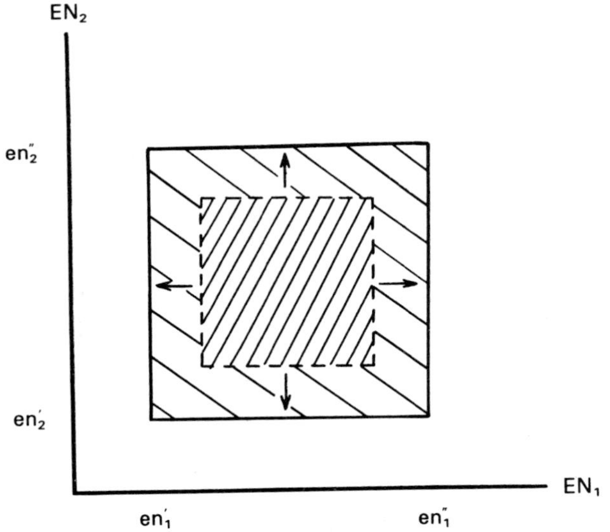

Fig. 9.3. Nutritional niche expansion resulting from the symbiotic interactions between the pregastric microbes and the host species as represented by Kinnear and Main. The microbes lower the requirements for certain essential nutrients and extend the tolerance to nutrient excesses or imbalances. Therefore, the nutritional niche space is expanded enabling the host species to survive and persist on innutritious diets. Modified after Kinnear and Main (1979).

Quite simply, microbial gastric symbionts have benefits (e.g. cellulases) and costs (e.g. energy demands) that alter the fundamental nutritional niche space of the host. Niche expansion occurs only if benefits exceed costs. Generally, the fundamental nutritional niche will be reconfigured or shifted by microbial symbiosis and not automatically expanded as stated by Kinnear and Main (1979).

I suggest that the *n*-dimensional niche of a herbivore is inclusive of its microflora (since it represents a corollary of the unit of selection). The microflora is part of the extended phenotype (*sensu* Dawkins, 1978) of the herbivore. The alternative option is to regard the host and the microflora as competitors for the same trophic resources, the host's digesta. However, the Hutchinsonian niches of herbivore and gastric symbionts are quite distinct and to adopt such a perspective is counter-productive. The gastro-intestinal microbe species compete amongst themselves for niches within their micro-environment, the gastro-intestinal tract.

To summarise, a major problem with the nutritional niche concept is the supposition that gastric symbionts always bring about niche *expansion*, rather than niche differentiation or diversification. Failure to recognise the constraints imposed by the microbes was a flaw that confounded theoretical development of the ideas.

Niche constraints

It is necessary and appropriate to identify some of the constraints that gastro-intestinal symbionts impose on the host and, incidentally, on its Hutchinsonian niche. The microflora requires a near constant supply of substrates necessitating regular foraging and feeding. If there is a shortage of carbohydrate or nitrogen in the diet, the activity, replication and species richness of the microflora is severely reduced (Williams-Smith, 1967; Janis, 1976), which may result in extinction of the flora and its host (Hungate, 1975). The necessity to maintain an adequate uptake of carbohydrate and protein forces the PGF host to select food from a variety of sources (Freeland and Jansen, 1974). The symbiotic microflora uses energy inefficiently (Phillipson and McNally, 1942) and generates large quantities of CO_2 and CH_4 (Dougherty, 1968), while the absorption of calcium, magnesium and phosphorus can be decreased (Reddy *et al.*, 1969). Digesta passage rate is slow to enable micro-organisms to perform their digestive functions effectively (Parra, 1978), particularly in PGFs. Dentition and many anatomical features of the digestive tract (Janis, 1976) are specialised to serve this function.

The restricted flow rate of digesta, necessitated by the microbes, prevents

PGF mammals from processing forage quickly, which could otherwise compensate for low-quality food or a dietary deficiency that cannot be corrected by forestomach microbes. The long time needed for microbial cellulolytic activity demands a large fermentation chamber and places a lower size limit on PGF mammals (Janis, 1976; Mellett, 1982). Homeothermy provides a stable thermal environment for the symbionts that favours stable digestive and metabolic functions. However, the gut microflora prevents torpor (diurnal or seasonal) as an adaptive stratagem and thereby precludes an avenue of energy conservation. Dissipation of microbially-generated heat can be problematic for large, tropical PGF mammals (McBee, 1971).

Comparative nutritional niche space of PGFs and Non-PGFs

Kinnear and Main (1979) applied their nutritional niche concept to the major dichotomy present among herbivores, that between PGFs and non-PGFs, so as to demonstrate relative niche spaces and potential for competitive interaction. Figure 9.4 shows their depiction of the nutritional niche of a PGF species. It was claimed that the symbiotic interaction between the PGF and its microflora resulted in nutritional niche expansion in the absence of non-PGF competitors. In fact, it is the extension of only two niche dimensions

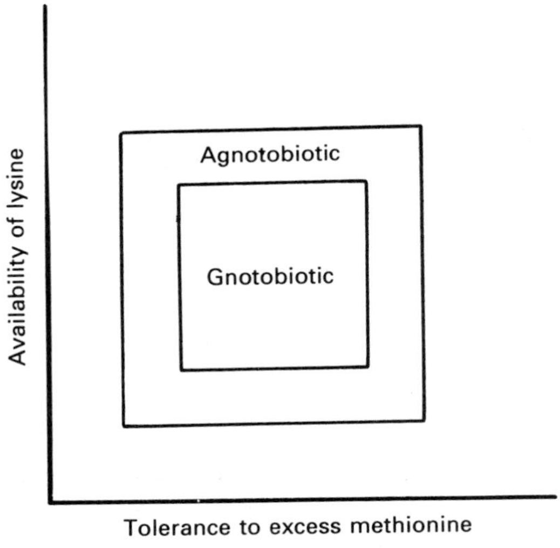

Fig. 9.4. The nutritional niche of a PGF species (e.g. *B. penicillata*). The inner area represents the fundamental niche space (i.e. in the absence of microbial symbionts) while the outer area represents realised niche space (i.e. in the presence of symbionts – which are regarded as negative competitors). Modified after Kinnear and Main (1979).

and not necessarily an increase in niche space. In their terminology the inner area shown in Figure 9.4 represents the fundamental nutritional niche of a herbivore species without gastro-intestinal symbionts. The outer area represents the realised niche of a herbivore species with its symbiotic microflora. Hutchinsonian theory (1957, 1959) defines the realised niche as a sub-space of fundamental niche space caused by niche space having been reduced by competitors. However, since Kinnear and Main (1979) interpret symbiosis as negative competition they had to 'reverse' the terminology, which is confusing. If the herbivore (with or without symbionts) is considered as the unit for niche definition (and selection), then it is the fundamental and not the realised niche that is increased by the beneficial characteristics of the symbionts (assuming that constraints do not have greater negative contributions).

Figure 9.5 shows the nutritional niche space of a PGF species under competition with a non-PGF species. PGF species are excluded from regions of their fundamental niche space because they are competitively inferior (Kinnear and Main, 1979) owing to the constraints of the microbial symbionts. One cannot accept that the figure represents the holistic niche spaces of PGF and non-PGF herbivores. Again, only two niche dimensions are identified and other niche dimensions can be chosen to present a converse argu-

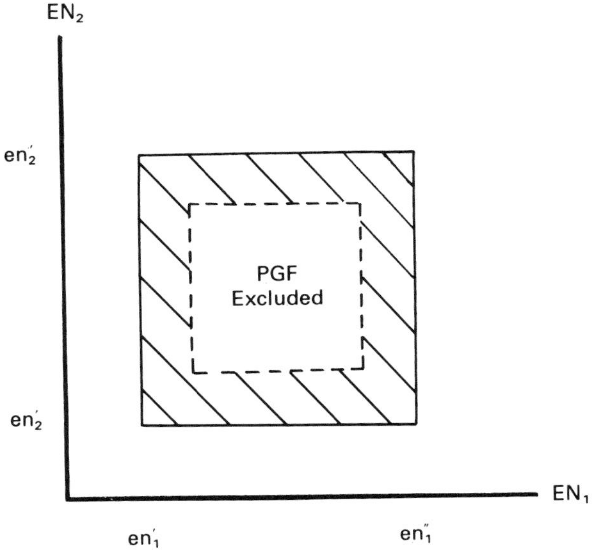

Fig. 9.5. The nutritional niche space of a PGF species under competition from a non-PGF species as represented by Kinnear and Main. PGFs are excluded from the inner niche space owing to competition from non-PGF species. Modified after Kinnear and Main (1979).

ment (Janis, 1976). Competition necessitates consideration of all niche parameters. The figure suggests that the realised nutritional niches of non-PGF herbivores are (a) contained within those of PGFs and (b) smaller than those of PGFs. For particular nutrient pairs this may be true, but the concept is unlikely to apply for all nutrients, and I suggest the generalisation is invalid. This is demonstrated below by modifying a niche parameter in one of the examples.

The woylie was used to illustrate the concept of nutritional niche expansion, since its diet is deficient in lysine but with an excess of methionine (Fig. 9.4). When a new parameter, such as the efficiency of energy use is considered, it becomes evident that the apparent niche expansion is realistically a niche shift (Fig. 9.6). Similarly, our data on the white-tailed rat (*Mystromys albicaudatus*) can be used to infer either nutritional niche expansion, in terms of increased amylase and lipase activity (Fig. 9.7); niche contraction, with reference to efficiency of energy utilisation and mineral uptake (Fig. 9.8); or a niche shift owing to increased lipase activity but reduced efficiency of energy utilisation (Fig. 9.9). The important conclusions to draw are (a) mammal–microbe symbioses allow for the colonisation of trophic niches unavailable to herbivores without a microflora, (b) niches become shifted or differentiated rather than expanded and (c) symbiosis facilitates coevolution and dietary specialisation. However, specialisation is not a necessary or implicit consequence of the symbiosis and many PGFs, like the

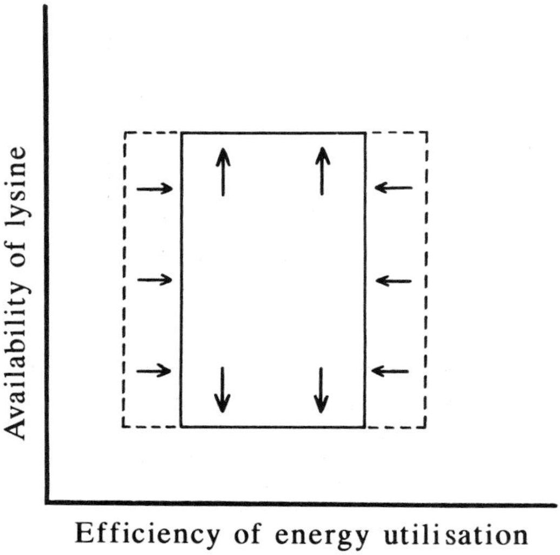

Fig. 9.6. Diagrammatic representation of a nutritional niche shift, owing to the selection of different niche parameters.

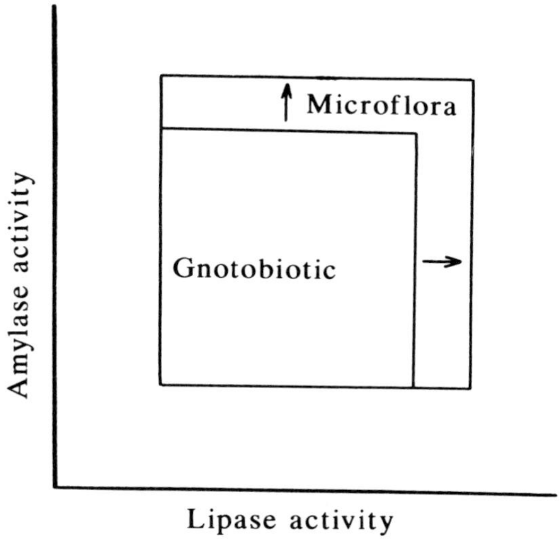

Fig. 9.7. Diagrammatic representation of (apparent) nutritional niche expansion in *M. albicaudatus* in terms of increased amylase and lipase activity.

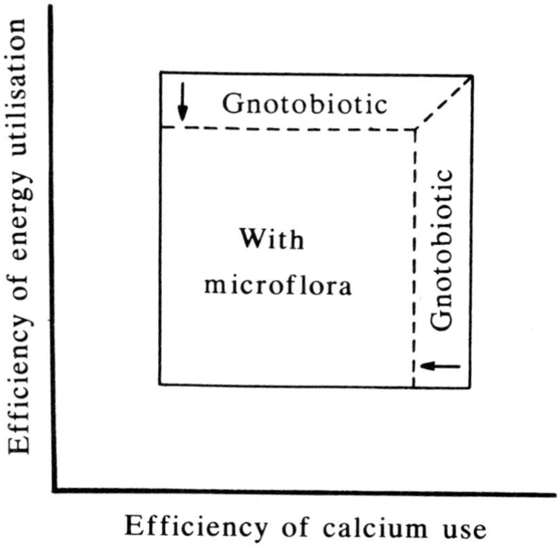

Fig. 9.8. Diagrammatic representation of (apparent) nutritional niche contraction in *M. albicaudatus* in terms of energy utilisation and mineral uptake.

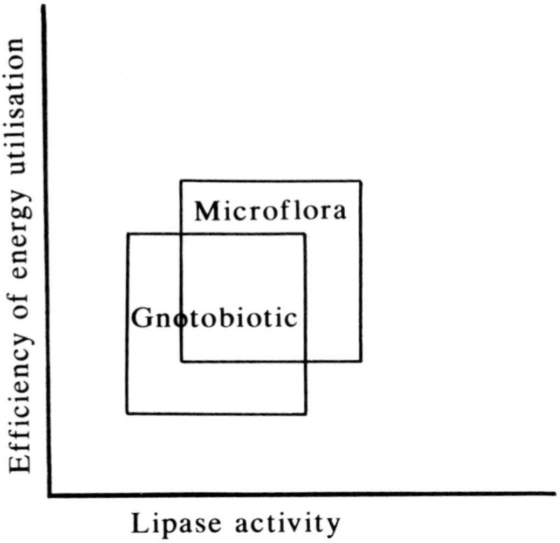

Fig. 9.9. Diagrammatic representation of a (apparent) nutritional niche shift in *M. albicaudatus* owing to increased lipase activity and reduced efficiency of energy utilisation.

impala (*Aepycerus melampus*), are generalist feeders. In fact, the microflora may facilitate changes in diet. The giant rat (*Cricetomys gambianus*) has a dynamic gastric microflora that rapidly shifts with diet composition and may facilitate use of diverse dietary items (Perrin, 1986).

Coexistence of congeners

Closely-related species-pairs of syntopic herbivores are likely to have very similar niches owing to their phylogenetic commonality and habitat equivalence. In such cases, differences between niches can often be demonstrated with reference to only a few parameters. Our research (Bruorton and Perrin, 1988, 1991; Bruorton *et al.*, 1991) on samango and vervet monkeys has shown that a major avenue for niche segregation relates to an increased capacity for folivory in the former species. This is demonstrated by increased caecal and colonic capacities and a greater abundance of microbial symbionts with cellulolytic activity (Fig. 9.10). Red duikers and blue duikers are also folivorous in the forests of Natal, but major niche parameters causing niche separation are quite different. Many aspects of the physiology and ecology of the two species are the same (Bowland, 1990; Perrin *et al.*, 1992; Faurie and Perrin, 1993) and ecological differentiation occurs through differences in body size and the differential use of a wide range of fallen leaves which

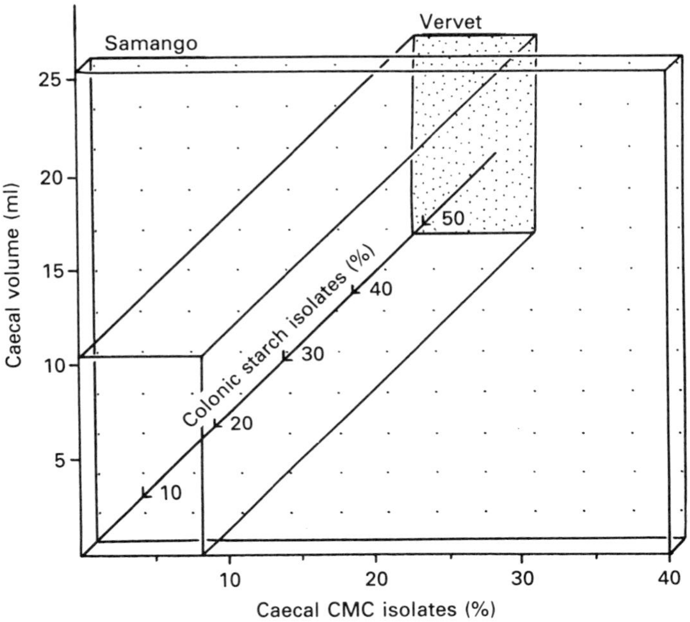

Fig. 9.10. Quantitative representation of three nutritional niche parameters known to be important in the ecological segregation of samango and vervet monkeys. Isolates (%) refers to the proportion of caecal and colonic samples exhibiting CMC (carboxymethylcellulose) activity.

comprise the diet. Both species use leaves containing condensed and hydrolysable tannins and of variable nutritional quality but of different species composition (Fig. 9.11). In these examples 'nutrients' (fibre and tannins) are important in niche definition but only when interpreted in relation to anatomy (hindgut capacity and body size). Nutrients alone are insufficient to separate the niches of congeneric species-pairs of syntopic herbivores.

Nutritional niche segregation: any need to invoke competition?

M. albicaudatus occurs at low densities (Dean, 1978) in the cold, semi-arid temperate grasslands of southern Africa (Brain, 1985). Although hard data are not available, it is unlikely that an omnivorous, generalist selector in a variable environment will be exposed to a great deal of interspecific competition. Further, the relic distribution of the only cricetine in Africa (Meester *et al.*, 1986) suggests unique characters, evolved in the past, may account

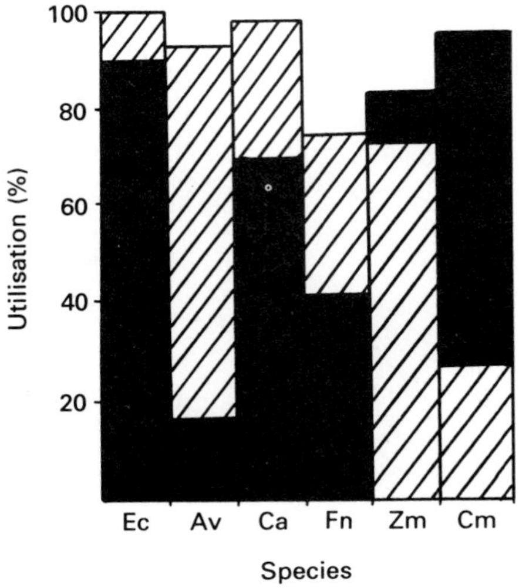

Fig. 9.11. Preference for six leaf species by blue duikers (■) and red duikers (□). Ec, *Ekebergia capensis*; Av, *Antidesma venosum*; Ca, *Carissa macroscarpa*; Fn, *Ficus natalensis*; Zm, *Ziziphus mucronata*; Cm, *Combretum molle*.

for its continued survival (the 'ghost of competition past', Connell, 1980). Here, I examine some of the specialised characteristics which may define the nutritional niche of *M. albicaudatus* but which do not invoke interspecific competition. *M. albicaudatus*, with a forestomach and caecal microflora, has a greater capacity to use soluble carbohydrates and fats than the related and ecologically similar pouched mouse *Saccostomus campestris* (Perrin and Kokkinn, 1986) that possesses only a caecal microflora (Fig. 9.12). This suggests that fibre fermentation predominates in the caecum. This is confirmed for *M. albicaudatus* by the higher concentrations of volatile fatty acids (VFAs) in the caecum than the forestomach (Fig. 9.13). The role of the microflora in the digestive processes of *M. albicaudatus* is shown in Fig. 9.14. The forestomach microbes aid amylosis and lipolysis while those of the caecum contribute to cellulolysis. (Since gnotobiotic and untreated *M. albicaudatus* maintain weight and condition on diets containing 3–6% hydrolysable tannin (tannic acid) or quercetin (a flavonoid) (Mahida, 1992), it indicates that it is not the gastric microbes, but an intrinsic microsomal enzyme system, that enables detoxification of phenolic compounds.)

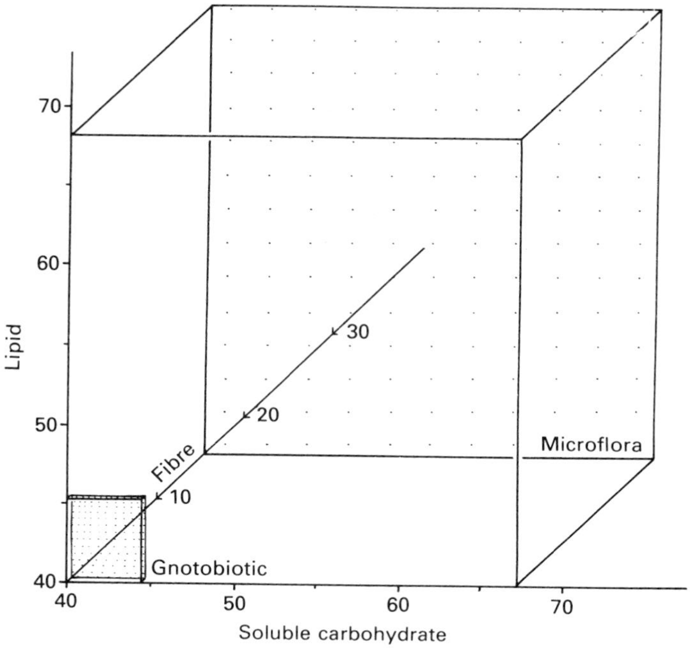

Fig. 9.12. Quantitative representation of three assimilation efficiencies (carbohydrate, lipid and fibre) for *M. albicaudatus*, with microflora and without its symbiotic microflora (gnotobiotic).

An ecologist's perspective

Kinnear and Main (1979) considered the niches of herbivores from a physiological viewpoint. What can be gained, if anything, from an ecological and behavioural approach? Owen-Smith (1985) used simulation modelling to explore the effects of vegetation structure in relation to animal phenotype (body mass, rate of movement, bite size, capacity of the fermentation chamber, size of the ostium, and rate of passage of digesta), on the feeding performance of African ungulates in terms of nutrient profits. He showed that a grazing non-ruminant, with fast digesta passage and low digestibility of cell walls, has a wider dietary breadth than a grazing ruminant of similar body mass (Fig. 9.15). This was due to its ability to use lower quality foods more profitably as a result of lessened digestive constraints. Although Owen-Smith's (1985) research did not aim to define nutritional or trophic niches, it was effective in explaining trophic resource partitioning and treated herbivore and microflora as a single unit of selection. His model went on to

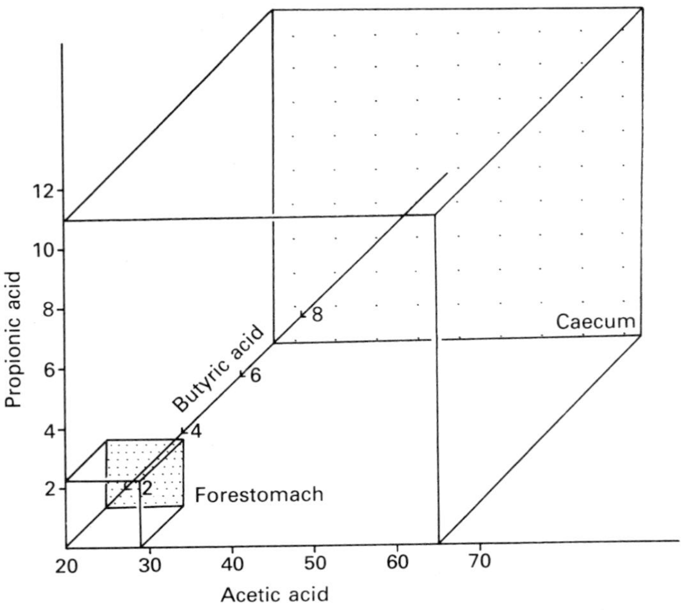

Fig. 9.13. Quantitative representation of three VFA concentrations present in the forestomach and the caecum of *M. albicaudatus*.

explain why there are fewer equids than bovids and why ruminants tend to be more specialist feeders than equids.

When nutrient profits were plotted in relation to dietary acceptance range (Fig. 9.16) for browsing bovids of varying mass, a 5 kg animal achieved maximum energy profit by eating only foods containing more than 11.8% protein, while a 450 kg animal needed to feed on foods down to 6% protein to maximise its energy profit. This relation is not due to differing metabolic rates per unit of body mass, which merely sets the baseline, but rather to the relation between food ingestion rate and digestion rate (Owen-Smith, 1985). The ascending (left-hand) portion of the plot for each species is controlled by the increase in food ingestion rate with expanding dietary range, while the descending (right-hand) portion reflects the digestive constraint, which becomes increasingly restrictive as dietary quality declines. The cut-off on the extreme right of the plots occurs when protein becomes more limiting than energy.

These plots represent one component of niche space (Owen-Smith, 1985), namely dietary breadth in relation to food quality. They suggest that, while

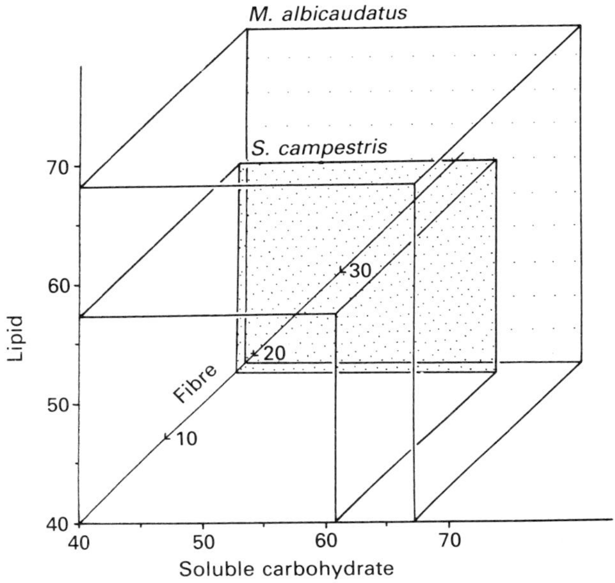

Fig. 9.14. Quantitative representation of three assimilation efficiencies for *M. albicaudatus* (with caecal and forestomach microflora) and *S. campestris* (with only caecal microflora).

dietary overlap exists, larger species achieve optimal performance by including food types of lower quality than those falling within the optimal set for smaller ungulates. Thus, an ecological or a foraging theory approach can be effective in identifying and delimiting the nutritional niches of herbivores.

Conclusion

The nutritional niche concept represented an attempt at generating a wider and more plausible explanation for mammalian herbivore diversity than the microbial fermentation of cellulose and the use of microbial protein. The approach was expansionist in that all essential nutrients and plant chemical defences were perceived as defining (nutritional) niche space, but reductionist in that, *inter alia*, allometric, spatial and temporal aspects of niche definition were excluded. A niche concept without holism loses much, if not all, of its theoretical significance and cannot be used effectively in community ecology or evolutionary theory. It may be meaningful, however, in categorising and defining nutrient requirement differences between similar (ecologically or

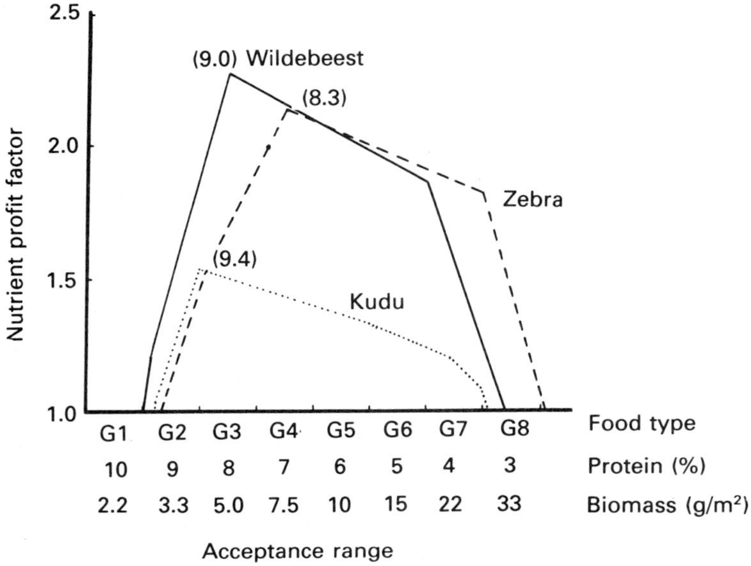

Fig. 9.15. Nutrient profitability spectrum. Nutrient profits in relation to dietary acceptance range predicted by the model for ungulates of the same body mass (180 kg) but differing in their digestive adaptations. Numbers in parentheses indicate the protein content (%) of the optimal dietary range. The figures are calculated for the late dry season with a maximal foraging time of 60% of the 24 h day. Maintenance energy requirement, 1.6 × basal metabolic rate; maintenance protein requirement, 2EUN + MFN where EUN is endogenous urinary nitrogen excretion and MFN is metabolic faecal nitrogen losses. G represents grasses. The numerical values for each group symbolise different proximate compositions. (From Owen-Smith, 1985.)

taxonomically) herbivores, which is important for animal production and herbivore conservation. Nutritional niche definition may facilitate multi-species management, particularly if used in association with data on foraging pattern (as defined by Owen-Smith, 1985).

I interpret the nutritional niche as a sub-space of the trophic niche, which is itself a sub-space of Hutchinson's *n*-dimensional (hypervolume) space. Historically the niche concept developed from a description of habitat of a species in the form of the spatial niche, through a definition of the species' role or 'profession' (food resource needs), as exemplified by the trophic niche, to the holistic multi-variate characterisation of a species in nature. I suggest that essential nutrients (essential nutrient axis) represent a significant but single component of niche space, paralleling Owen-Smith's (1985) niche dimension of dietary breadth in relation to food quality (foraging axis).

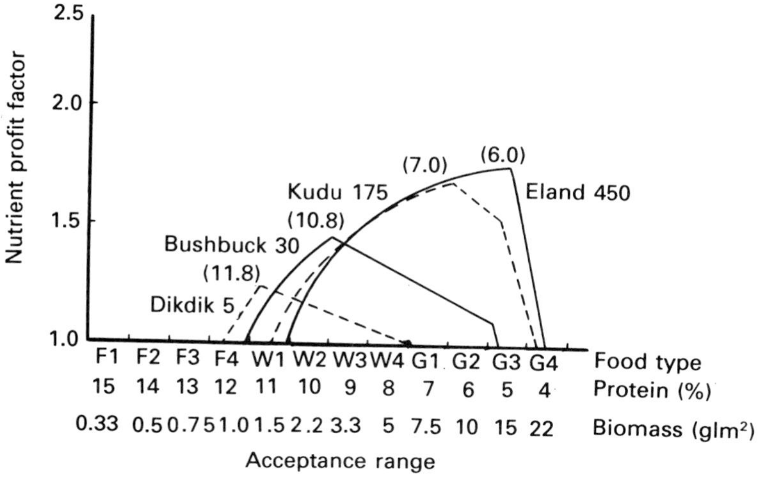

Fig. 9.16. Nutrient profitability spectrum. Nutrient profits in relation to dietary acceptance range predicted by the model for browsing bovids of varying body mass. Animal body masses in kg are eland, 450; kudu, 175; bushbuck, 30 and dikdik, 5. The protein content (%) of the optimal dietary range is given in parentheses. The nutrient profit factor is the ratio of nutrient gains to body maintenance requirements for energy or protein, whichever is more limiting. The figures are calculated for the late dry season with a maximal foraging time of 60% of the 24h day. Maintenance energy requirement, 1.6 × basal metabolic rate; maintenance protein requirement, 2EUN + MFN where EUN is endogenous urinary nitrogen excretion and MFN is metabolic faecal nitrogen losses. F, W and G represent forbs, woody dicots and grasses. The numerical values for each group symbolise different proximate compositions. (From Owen-Smith, 1985.)

Whereas the essential nutrient axis represents the ultimate set of nutrients required by the species, the foraging axis represents the proximate acquisition of those nutrients. Both of these variables are of paramount importance in defining trophic niche space and of considerable significance in defining Hutchinsonian (holistic) niche space; other dimensions might summarise anatomical, physiological, microbial and size attributes of the species.

Subsequent to the niche concept (Hutchinson, 1957), ecologists have derived the guild concept (Root, 1967). In many ecological studies, communities have been divided into guilds, that is, groupings of species that exploit the same kinds of resources in similar ways within a habitat. For example, in the arid zone Karoo of South Africa, gerbils, larks and ants comprise a granivorous guild (Kerley, 1989, 1991, 1992). Guild structure can be defined as the patterns of resource use among co-occurring species, with emphasis on the similarities and differences in how those species exploit resources.

The resource parameter invariably (but not implicitly) invoked in defining or structuring a guild is food. This is because food (quantity, quality) is often a critical limiting resource for animal populations (Lack, 1954, 1966; Cody, 1974). Therefore, the trophic and nutritional niche concepts are relevant to guild theory. I suggest that trophic niches and guilds are parallel concepts if food is the resource being competed for or partitioned. Unlike the fundamental niche that is genetically fixed, trophic niches and guilds are not so constrained and are dynamic. Guilds vary owing to historic and evolutionary differences that result in different species composition (e.g. the granivore guild in the Sonoran Desert comprises ants, birds and rodents, but the species are different from those of the Karoo).

It is possible, therefore, to comprehend and define complex niche parameters such as dietary breadth in relation to food quality (Owen-Smith, 1985) or essential nutrient requirements (derived from Kinnear and Main's nutritional niche concept). These have far greater intrinsic and independent value than a simple niche parameter (e.g. ability to digest cellulose) but they do not and cannot define an holistic niche. They form part of a trophic niche which may have meaning in terms of guild structure and in certain interspecies interactions. However, it is the *n*-dimensional hypervolume or Hutchinsonian niche that has real significance in community ecology and evolution.

The nutritional niche concept can be used to segregate syntopic congeners (e.g. forest primates and duikers); to identify the common adaptive strategem of species with parallel or convergent adaptations (e.g. 'polygastric' African rodents); and to define precise species-specific characteristics.

The nutritional niche concept has some theoretical merit when obvious flaws are eliminated and has application in single species management and conservation. To evaluate guild structure of herbivores, one must use the trophic niche, which *inter alia* incorporates essential nutrient and foraging behaviour parameters. For studies of community ecology and evolutionary biology, it is essential that Hutchinson's multi-dimensional niche concept be used.

References

Bowland, A. E. (1990). The ecology and conservation of blue duiker and red duiker in Natal. Ph.D. thesis. University of Natal, RSA.

Brain, C. K. (1985). Temperature induced environmental changes in Africa as evolutionary stimuli. In *Species and Speciation*, ed. E. S. Vrba, pp. 45–52. Transvaal Museum Monograph No.4. Pretoria: Transvaal Museum.

Branch, G. M. (1985). Competition: its role in ecology and evolution in intertidal communities. In *Species and Speciation*, ed. E. S. Vrba, pp. 97–104. Transvaal Museum. Monograph No. 4. Pretoria: Transvaal Museum.

Bruorton, M. R., Davis, C. & Perrin, M. R. (1991). The gut microflora of vervet and samango monkeys in relation to diet. *Journal of Applied and Environmental Microbiology*, **57**, 573–578.

Bruorton, M. R. & Perrin, M. R. (1988). The anatomy of the stomach and caecum of the samango monkey *Cercopithecus mitis erythrarchus* (Peters, 1852). *Zeitschrift für Säugetierkunde*, **53**, 210–224.

Bruorton, M. R. & Perrin, M. R. (1991). Comparative gut morphometrics of vervet (*Cercopithecus aethiops*) and samango (*C. mitis erythrarcus*) monkeys. *Zeitschrift für Säugetierkunde*, **56**, 65–71.

Cody, M. L. (1974). *Competition and the Structure of Bird Communities*. Princeton: Princeton University Press.

Connell, J. H. (1980). Diversity and the coevolution of competitors, or the ghost of competition past. *Oikos*, **35**, 131–138.

Dawkins, R. (1978). Replicator selection and the extended phenotype. *Zeitschrift für Tierpsychologie*, **47**, 61–76.

Dean, W. R. J. (1978). Conservation of the white-tailed rat in South Africa. *Biological Conservation*, (**3**)2, 133–140.

Dougherty, R. W. (1968). Physiology of eructation in ruminants. In *Handbook of Physiology*, section 6. Alimentary Canal, Vol. 5, ed., Heidel, pp. 2695–2698. Washington, DC: American Physiological Society.

Faurie, A. S. & Perrin, M. R. (1993). Preliminary water and energy kinetics of the blue duiker (*Philantomba monticola*). *Journal of African Zoology*, in press.

Freeland, W. J. & Jansen, D. H. (1974). Strategies in herbivory by mammals. *American Naturalist*, **109**, 269–289.

Hofmann, R. R. (1968). Comparison of rumen and omasum structure in East African game ruminants in relation to their feeding habits. In *Comparative Nutrition of Wild Animals*, ed. M. A. Crawford. Symposium of the Zoological Society of London, **21**, 179–194.

Hungate, R. E. (1975). The rumen microbial ecosystem. *Annual Review of Ecology and Systematics*, **6**, 39–66.

Hutchinson, G. E. (1957). Concluding remarks. *Cold Spring Harbour Symposium on Quantitative Biology*, **22**, 415–427.

Hutchinson, G. E. (1959). Homage to Santa Rosalia, or why are there so many kinds of animals? *American Naturalist*, **93**, 145–159.

Janis, C. (1976). The evolutionary strategy of the Equidae and the origins of rumen and caecal digestion. *Evolution*, **30**, 757–774.

Kerley, G. I. H. (1989). Diet of small mammals from the Karoo, South Africa. *South African Journal of Wildlife Research*, **19**, 67–72.

Kerley, G. I. H. (1991). Seed removal by rodents birds and ants in the semi-arid Karoo, South Africa. *Journal of Arid Environments,* **20**, 63–69.

Kerley, G. I. H. (1992). Trophic status of small mammals in the semi-arid Karoo, South Africa. *Journal of Zoology* (Lond.) **226**, 563–572.

Kinnear, J. E. & Main, A. R. (1979). Niche theory and macropodid nutrition. *Journal of the Royal Society of Western Australia*, **62**, 63–74.

Lack, D. (1954). *The Natural Regulation of Animal Numbers*. Oxford: Clarendon Press.

Lack, D. (1966). *Population Studies of Birds*. Oxford: Clarendon Press.

Landry, S. O. (1970). The Rodentia as omnivores. *Quarterly Review of Biology*, **45**, 351–372.

Langer, P. (1974). Stomach evolution in the Artiodactyla. *Mammalia,* **38**, 295–314.

Lewontin, R. C. (1980). Adaptation. In *Vertebrates: Adaptation*, ed. N. K. Wessells, pp. 114–125. San Francisco: W. H. Freeman.

Mahida, H. (1992). Aspects of the digestive processes of the white-tailed rat *Mystromys albicaudatus* (A. Smith, 1834). M.Sc. thesis. University of Natal, Pietermaritzburg, RSA.
May, R. M. (1975). Patterns of species abundance and diversity. In *Ecology and Evolution of Communities*, ed. M. L. Cody & J. M. Diamond, pp. 81–120. Cambridge, MA: Belknap.
McBee, R. (1971). Significance of intestinal microflora in herbivory. *Annual Review of Ecology and Systematics*, **2**, 165–176.
Meester, J. A. J., Rautenbach, I. L., Dippenaar, N. J. & Baker, C. M. (1986). *Classification of Southern African Mammals*. Transvaal Museum Monograph No. 5. Pretoria: Transvaal Museum.
Mellett, J. S. (1982). Body size, diet and scaling factors in large carnivores and herbivores. *Proceedings of the Third North American Palaeontological Conference*, **2**, 371–376.
Owen-Smith, N. (1985). Niche separation among African ungulates. In *Species and Speciation*, ed. E. S. Vrba, pp. 167–171. Transvaal Museum Monograph No. 4. Pretoria: Transvaal Museum.
Owen-Smith, N. & Novellie, P. (1982) What should a clever ungulate eat? *American Naturalist*, **119**, 151–178.
Parra, R. (1978). Comparison of foregut and hindgut fermentation in herbivores. In *Arboreal Folivores*, ed. G. G. Montgomery, pp. 205–229. Washington, DC: Smithsonian Institution Press.
Perrin, M. R. (1986). Effects of diet composition on the gastric papillae and microflora of *Mystromys albicaudatus* and *Cricetomys gambianus*. *South African Journal of Zoology*, **21**, 67–76.
Perrin, M. R., Bowland, A. E. & Faurie, A. S. (1992). Niche segregation between the blue duiker *Philantomba monticola* and the red duiker *Cephalophus natalensis*. *Proceedings Ongulés/Ungulates,* **91**, 201–204.
Perrin, M. R. & Kokkinn, M. J. (1986). Comparative gastric anatomy of *Criceteomys gambianus* and *Saccostomus campestris* (Cricetomyinae) in relation to *Mystromys albicaudatus* (Cricetinae). *South African Journal of Zoology*, **21**, 202–210.
Phillipson, A. T. & McNally, R. A. (1942). Studies on the fate of carbohydrates in the rumen of sheep. *Journal of Experimental Biology*, **19**, 119–214.
Reddy, B. S., Pleasants, J. R. & Wostmann, B. S. (1969). Effect of intestinal microflora on calcium, phosphorus and magnesium metabolism in rats. *Journal of Nutrition*, **99**, 353–362.
Reid, R. T. (1970). The future role of ruminants in animal production. In *Physiology of Digestion and Metabolism in the Ruminant*, ed. A. T. Phillipson, pp. 1–22. Newcastle-upon-Tyne: Oriel Press.
Root, R. B. (1967). The niche exploitation pattern of the blue-grey gnatcatcher. *Ecological Monographs*, **37**, 317–350.
Smith, R. H. (1975). Nitrogen metabolism in the rumen and the composition and nutritive value of nitrogen compounds entering the duodenum. In *Digestion and Metabolism in the Ruminant*, ed. I. W. McDonald & C. I. Warner, pp. 399–415. Armadale, NSW: University of New England.
Sugihara, G. (1980). Minimal community structure: an explanation of species abundance patterns. *American Naturalist*, **116**, 770–787.
van Valen, L. (1973). A new evolutionary law. *Evolutionary Theory*, **1**, 1–30.
Vorontsov, N. N. (1961). Variation in the transformation rates of organs of the digestive system in rodents and the principle of functional compensation. *Proceedings of the Academy of Sciences of the USSR* (Doklady Akademii

Nauk SSR), **136**, 1494–1497. English translation in *Evolutionary Morphology*, Biological Sciences Section, **136–138**, 49–52.

Vorontsov, N. N. (1962). The ways of food specialisation and the evolution of the alimentary system in Muroidea. In *Symposium Theriologica*, pp. 360–370. Prague: Czechoslovakian Academy of Sciences.

Vrba, E. S. (1980). Evolution, species and fossils: how does life evolve? *South African Journal of Science*, **76**, 61–84.

Vrba, E. S. (1984). Evolutionary pattern and process in the sister-group Alcelaphini-Aepycerotini (Mammalia: Bovidae). In *Living Fossils*, ed. N. Eldridge & S. M. Stanley, pp. 62–79. New York: Springer-Verlag.

Williams-Smith, H. (1967). Observations on the flora of the alimentary tract of animals and factors affecting its composition. *Journal of Pathology and Bacteriology*, **89**, 95–122.

10

Taste discrimination and diet differentiation among New World primates

BRUNO SIMMEN

An integrative approach to feeding adaptations should consider taste discrimination as one of the first steps, functioning to assess the quality of food before it is processed in the gut. My recent work with Neotropical primates has shown that, in addition to taste thresholds, the study of above-threshold responses for different chemicals allows a better understanding of subtle differences in food choices among species. Although they cannot be considered as exclusively depending on the taste channel, these global taste responses may shed light on an important component that, together with gut differentiation, aids the animal in coping with the biochemical environment.

The data presented here were obtained on primate species belonging to the family Callitrichidae, namely marmosets (*Cebuella pygmaea* and *Callithrix* spp.), tamarins (*Saguinus oedipus* and *Leontopithecus* spp.) and Goeldi's monkey (*Callimico goeldii*). Each of these small-bodied primates (120 g to 710 g) eats fruits, nectars and plant exudates (mostly gums) in addition to insects, but their diet may differ to a large extent in the proportion of these plant food categories, marmosets being the most gummivorous callitrichids (Snowdon and Soini, 1988; Soini, 1988; Stevenson and Rylands, 1988; Rylands, 1989). All these species feed opportunistically on gums that are released from branches and tree trunks, presumably as a consequence of damage inflicted to the tree by xylophagous insects, but only marmosets have a specialized dentition allowing them to gnaw barks in order to stimulate exudate flow in a relatively predictable manner.

It is usually assumed that gums, which are mainly composed of long polymers of β1,4 linked monosaccharides and glucuronic acid (Smith and Montgomery, 1959), cannot provide energy unless they go through a digestive tract with a specific form, including chambers (Chivers and Hladik, 1980) whose function involves fermentation (Langer, 1988). The analysis of gums selected by primates has shown that galactose and arabinose account for 45–80% of the polysaccharide content, although gums do contain other hexoses and pentoses as well as low amounts of nitrogen and minerals such as calcium (Anderson *et al.*, 1974; Hausfater and Bearce, 1976; Bearder and Martin, 1980; Garber, 1984).

Platyrrhine primates have different strategies to face the potentially harmful substances, such as digestibility-reducing compounds and toxins, that are present in many immature fruits, seeds, barks and that may also occur in gums (Willaman and Schubert, 1961; Glicksman, 1969; Golstein and Swain, 1963; Gartlan *et al.*, 1980; Wrangham and Waterman, 1981). Accordingly, it might be relevant to determine whether or not gum-specialist primates share similar taste characteristics, especially with regard to soluble sugars and allelochemicals, with closely-related species exhibiting a much more frugivorous tendency and showing no particular gut differentiation.

Methods

Determination of taste threshold for various compounds was conducted on *Cebuella pygmaea*, *Callithrix jacchus*, *Callithrix geoffroyi*, *Callithrix argentata*, *Callimico goeldii*, *Saguinus oedipus*, *Leontopithecus chrysomelas* and *L. rosalia*. Most of the animals tested were adult or sub-adult males and females grouped in separate cages. The number of groups per species ranged from two to seven (mode = 4) including one to nine individuals per group (mode = 2). Details on the composition of groups may be found in a previous study (Simmen, 1991).

After a training period during which fructose solutions at 0.3 M were provided, taste thresholds were determined using the method of the 'two-bottle test' derived from that of Glaser (1968) and described elsewhere (Simmen and Hladik, 1988). Following this procedure, the gustatory solution was presented simultaneously with tap water to the animals, the concentration of the stimulus being varied at random each day. In each test, the period during which the two bottles were left lasted from a total 24-h cycle – when no visible responses could be recorded at regular time intervals following the presentation of the two bottles – to a few minutes (in order to prevent overfeeding). Taste threshold for any particular compound was defined as the lowest concentration for which the mean difference of consumption between the gustatory solution and tap water was significant ($P < 0.05$ with a paired-sample t-test).

In addition, mean volumes of the gustatory solutions consumed per unit time were used to establish the profile of each species' supra-threshold responses for fructose, as well as for quinine hydrochloride. Allowing for differences in the grouping of animals, rates of consumption recorded in each cage were divided by group weight on the basis of specific body weight (Simmen, 1991).

The method of the 'two-bottle test' was further used to measure the consumption of fructose solutions at 60 and 300 mM, added with various concentrations of a gallotannin (tannic acid $C_{76}H_{52}O_{46}$, M_r: 1701.23; Fluka Chemie AG, CH-9470, Buchs, Switzerland) in *Callithrix jacchus* and *Callimico goeldii*. The binary solutions and tap water, with volumes limited to 10 ml each, were presented during 1-h tests. Alternatively, fructose solutions at 60 and 300 mM were given to reinforce the ingestive behaviour. The paired-sample *t*-test was applied to determine if the mean difference of consumption of the mixture versus tap water as well as versus pure fructose solutions was significant.

For each compound tested, the number of concentrations that were used differed between species: D-fructose, 19 to 30; D-ribose, 9 to 12; L-arabinose, 6; quinine hydrochloride, 12 to 19; tannic acid, 6, concentrations, respectively.

The consumption of the gustatory solution and tap water recorded for each species were grouped in successive concentration intervals. The taste threshold for any particular compound was considered as the lowest concentration interval, characterized by the minimum interval width, for which the mean difference of consumption of the gustatory solution versus tap water was still significant. For instance, the taste preference threshold for fructose of *Cebuella pygmaea* is between 30 and 44 mM, as illustrated in Fig. 10.1. Results of taste threshold determination are summarized in Tables 10.1 and 10.2.

Taste responses of primates

The taste responses described for various species, as presented in this chapter, result from behavioural studies conducted with the above-mentioned method. In primates, the relationship between taste threshold obtained with a behavioural method and the 'true' threshold determined by recording the chorda tympani nerve responses for gustatory stimuli applied onto the tongue has been discussed in a few papers. From studies of taste discrimination in three primate species, it appears that both methods yield similar threshold values when using sucrose or quinine hydrochloride (Glaser and Hellekant, 1977). However, contrasting results were obtained in *Microcebus murinus*, a prosimian characterized by an annual marked fattening cycle in the wild as well as in captivity. In this species, the seasonal variation of food choices and body weight is accompanied by a significant change in the taste preference thres-

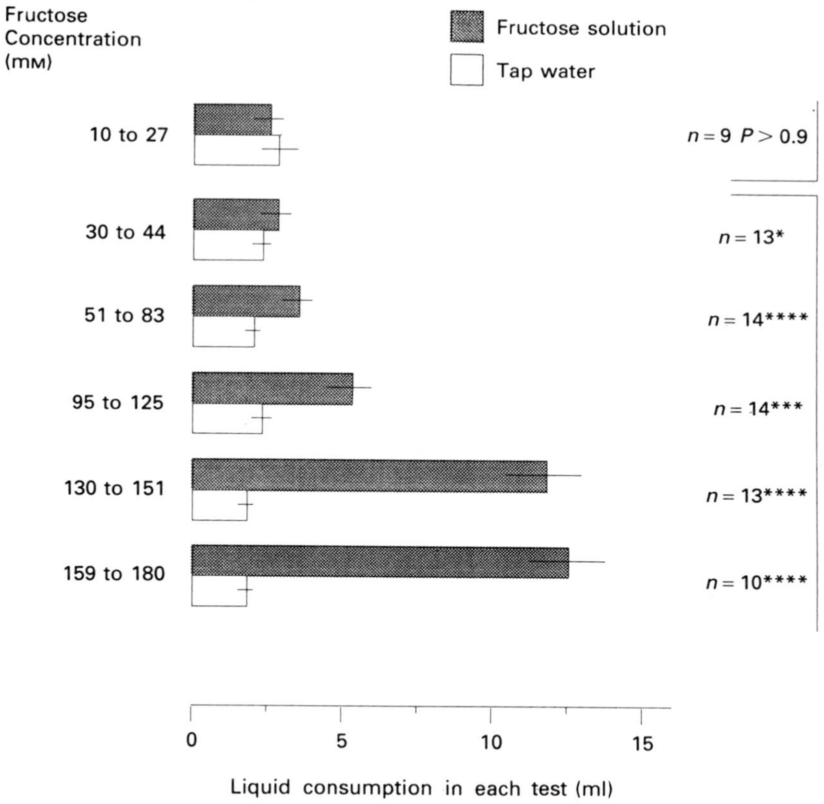

Fig. 10.1. Result of the 'two-bottle test' with D-fructose in *Cebuella pygmaea*. Mean liquid consumption (fructose versus tap water) with standard error of the mean is figured for each fructose concentration interval. The threshold is considered as the lowest interval for which the mean difference between the tasty solution and water is still significant (see Methods) $^{*}P < 0.05$, $^{***}P < 0.01$, $^{****}P < 0.001$.

hold for sucrose without any concomitant modification of the sensitivity of the chorda tympani nerve (Simmen and Hladik, 1988; Hellekant *et al.*, 1993). Obviously, more information is needed in this area, but we may presently assume that, for most species, long-term experiments based on the two-bottle test procedure yield taste threshold values comparable to the lowest concentrations perceived. In this respect, it is striking that the taste rejection threshold for quinine of *Cebuella pygmaea* as determined in our study was equivalent to the value published by Glaser for this species (Glaser, 1968). Similarly, a taste threshold at 5×10^{-5} M was found in *Saguinus midas* using both behavioural and electrophysiological methods (Glaser, 1968; Glaser and

Table 10.1. *Taste thresholds for fructose, ribose and arabinose in callitrichid species*

	D-Fructose	D-Ribose	L-Arabinose
	threshold P (mM)	threshold P (M)	threshold P (M)
Tamarins			
Saguinus o. oedipus	10–22 (18) < 0.02		0.4–0.6 (6) < 0.05
L. rosalia	12–27 (10) < 0.05		
L. chrysomelas	12–31 (17) < 0.05		
Goeldi's monkey			
Callimico goeldii	27–35 (19) < 0.05	0.2–0.25 (23) < 0.05	
Marmosets			
Cebuella pygmaea	30–44 (13) < 0.05	0.1–0.2 (16) < 0.05	≤ 0.1–0.33 (8) < 0.02
Callithrix a. argentata	12–27 (17) < 0.05		
Callithrix jacchus	27–32 (16) < 0.05	0.33–0.4 (18) < 0.001	
Callithrix geoffroyi	31–51 (16) < 0.05		

The number of tests in the threshold interval is given in parentheses and the corresponding probability (P) indicates significance of the difference between consumption of the gustatory solution and tap water by paired t-test.

Table 10.2. *Taste threshold for quinine in callitrichid species*

	Quinine hydrochloride		
	Threshold (mM)	*n*	*P*
Tamarins			
Saguinus o. oedipus	0.06–0.07	13	< 0.02
L. rosalia	0.16–0.25	18	< 0.01
L. chrysomelas	0.05–0.06	14	< 0.05
Goeldi's monkey			
Callimico goeldii	0.08	13	< 0.05
Marmosets			
Cebuella pygmaea	0.7–0.75	12	< 0.05
Callithrix a. argentata	0.0006–0.0008	12	< 0.05
Callithrix jacchus	0.5–0.55	14	< 0.05
Callithrix geoffroyi	0.33–0.4	12	< 0.05

n, the number of tests in the threshold interval.
P, the probability that the mean difference of consumption between the gustatory solution and tap water is significant.

Hellekant, 1977), which is very close to that presented here for *Saguinas oedipus*.

Responses of callitrichids to fructose

Taste thresholds for D-fructose ranged from the interval 10–22 mM for *Saguinus oedipus* to 31–51 mM for *Callithrix geoffroyi* (Table 10.1). Threshold values of some marmosets (*Callithrix geoffroyi*, *Cebuella pygmaea*) tended to be slightly higher than those of some of the tamarins (*Saguinus oedipus* and *Leontopithecus rosalia*), showing no overlap.

The above-threshold rate of consumption of the sweet solution for each species, as expressed by the mean volume of sweet solution consumed per unit time and body weight as fructose concentration increases, is shown in Fig. 10.2. In most species, the rate of ingestion increased monotonically as near-threshold concentrations were provided. However, the ingestive pattern of *Callithrix jacchus* and *Cebuella pygmaea* differed in that, although fructose solutions at 60 mM and 100 mM were discriminated by these two species respectively (2 to 2.5-fold higher than the taste threshold), they did not elicit any marked increase in the rate of consumption of the sweet liquid. Contrasting with these marmosets, the Goeldi's monkey and the golden lion tam-

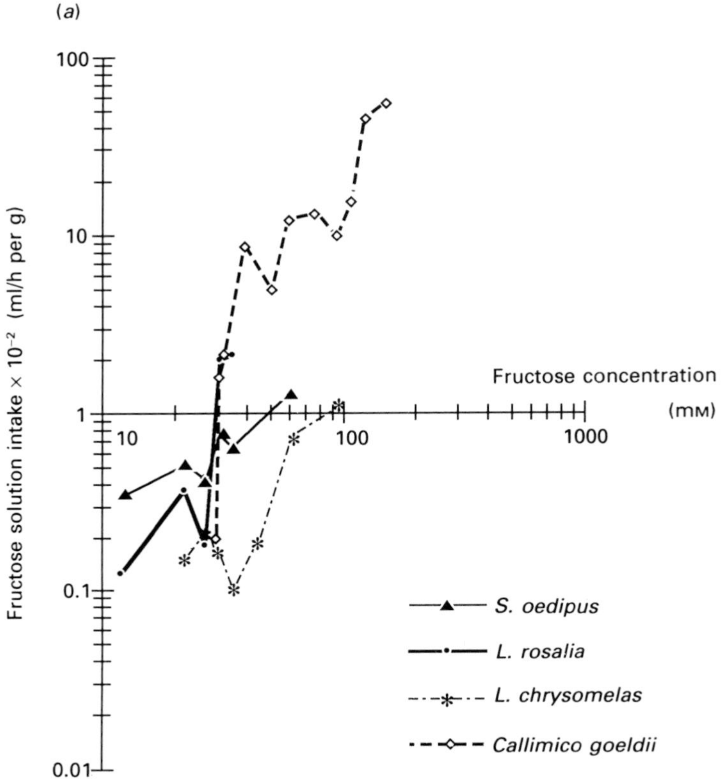

Fig. 10.2. Mean rate of consumption of sweet solutions plotted against their fructose concentration in (*a*) tamarins and Goeldi's monkey and (*b*) marmosets. Below-threshold responses are not shown.

arin, *Leontopithecus rosalia*, were the species most attracted by low fructose concentrations. Intermediate ingestive characteristics were presented by the less responsive species of the tamarin group (*Leontopithecus chrysomelas*) and the most responsive species of the marmoset group (*Callithrix argentata*).

Responses of callitrichids to ribose and arabinose

D-Ribose was avoided by *Cebuella pygmaea* and preferred both by *Callimico goeldii* and *Callithrix jacchus*, the taste thresholds ranging from 0.1–0.2 M for the former species to 0.33–0.4 M for the latter species (Table 10.1). L-Arabinose was preferred by *Cebuella pygmaea* and *Saguinus oedipus* (Table 10.1). Although mean differences of consumption between L-arabinose sol-

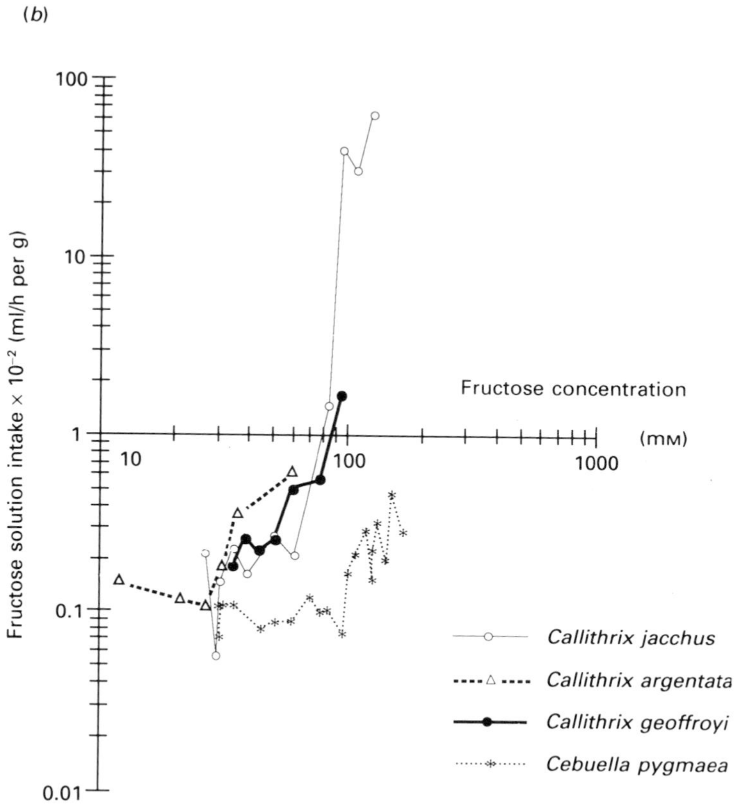

Fig. 10.2 *Cont.*

utions and tap water were all significant in *Cebuella pygmaea*, the taste threshold was presumably close to 0.1–0.33 M since the mean volume consumed of the pentose solutions at these concentrations was almost similar to that of tap water.

Responses of callitrichids to quinine

All of the eight species tested rejected quinine hydrochloride and the taste thresholds ranged from the interval 0.0006–0.0008 mM for *Callithrix argentata* to 0.7–0.75 mM for *Cebuella pygmaea* (Table 10.2). With the exception of *Callithrix argentata*, tamarins and the Goeldi's monkey discriminated quinine at the lowest concentrations and no overlap with marmoset taste thresholds was found. As shown in Figure 10.3, the slope which characterizes the rate of consumption of bitter solutions decreased more progressively in marmosets, even in *Callithrix argentata*. It may be seen, however, that in most tamarins as well as in Goeldi's monkey threshold and near-threshold concentrations, although discriminated, are tolerated to some extent (Fig. 10.3*a*).

Responses of callitrichids to tannic acid and fructose mixtures

The ingestive responses of *Callimico goeldii* and *Callithrix jacchus* for binary solutions composed of fructose (F), either at 60 mM or at 300 mM, added with tannic acid (T), were compared because both species exhibited an equivalent taste threshold for fructose (*cf.* Table 10.1).

Agreeing with the results shown in Table 10.1, pure fructose solutions at 60 mM and 300 mM (F/T: 60 mM/0% and 300 mM/0%, the tannic acid proportion being given as a percentage of the dry weight of fructose dissolved) were significantly preferred over tap water by both species (Fig. 10.4). Non-significant differences between the mixture and water intake were obtained when the mixed solution was characterized by F/T: 60 mM/≥ 2% and F/T: 300 mM/3.33% in *Callimico goeldii* on the one hand (Fig. 10.4*a*), and F/T: 60 mM/≥ 0.5% and F/T: 300 mM/3.33% in *Callithrix jacchus* on the other hand (Fig. 10.4*b*). Both species tolerated tannic acid to some extent, as indicated by the fact that respective mean differences of consumption between the gustatory solution and pure fructose solution at 60 mM were significant only when the mixture was characterized by F/T: 60 mM/4% ($P < 0.01$ with $n = 7$ in *Callimico goeldii* and $n = 6$ in *Callithrix jacchus* using the paired-sample *t*-test).

Taste and dietary adaptations

In addition to taste threshold, the ingestive pattern of sweet and bitter solutions as concentration increases provides an approximate measure of the pleasure/displeasure value of gustatory stimuli. For instance, the above-threshold consumption of fructose solutions (Fig. 10.2*a*, *b*) approximated the foot of the bell-shaped curve which characterizes affective perception of sweet solutions in man (Moskowitz *et al.*, 1974; Cabanac, 1979) as well as sucrose solution intake in rat and squirrel monkey (Pfaffmann, 1960; Michels *et al.*, 1988). However, although taste thresholds for fructose were very similar among the eight callitrichid species, large differences in the attractiveness of near-threshold concentrations were found: gummivorous marmosets such as *Cebuella pygmaea* and *Callithrix jacchus* being the least responsive. Given

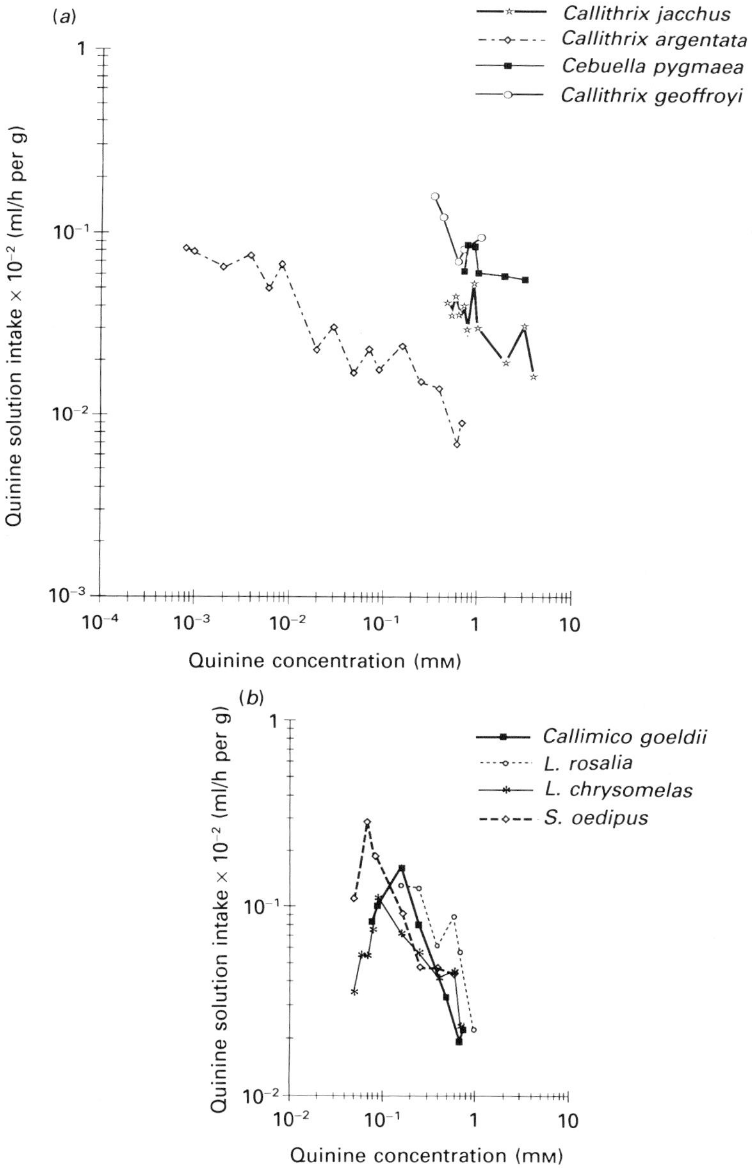

Fig. 10.3. Mean rate of consumption of bitter solutions plotted against their quinine hydrochloride concentration in (*a*) tamarins and Goeldi's monkey and (*b*) marmosets. Below-threshold responses are not shown.

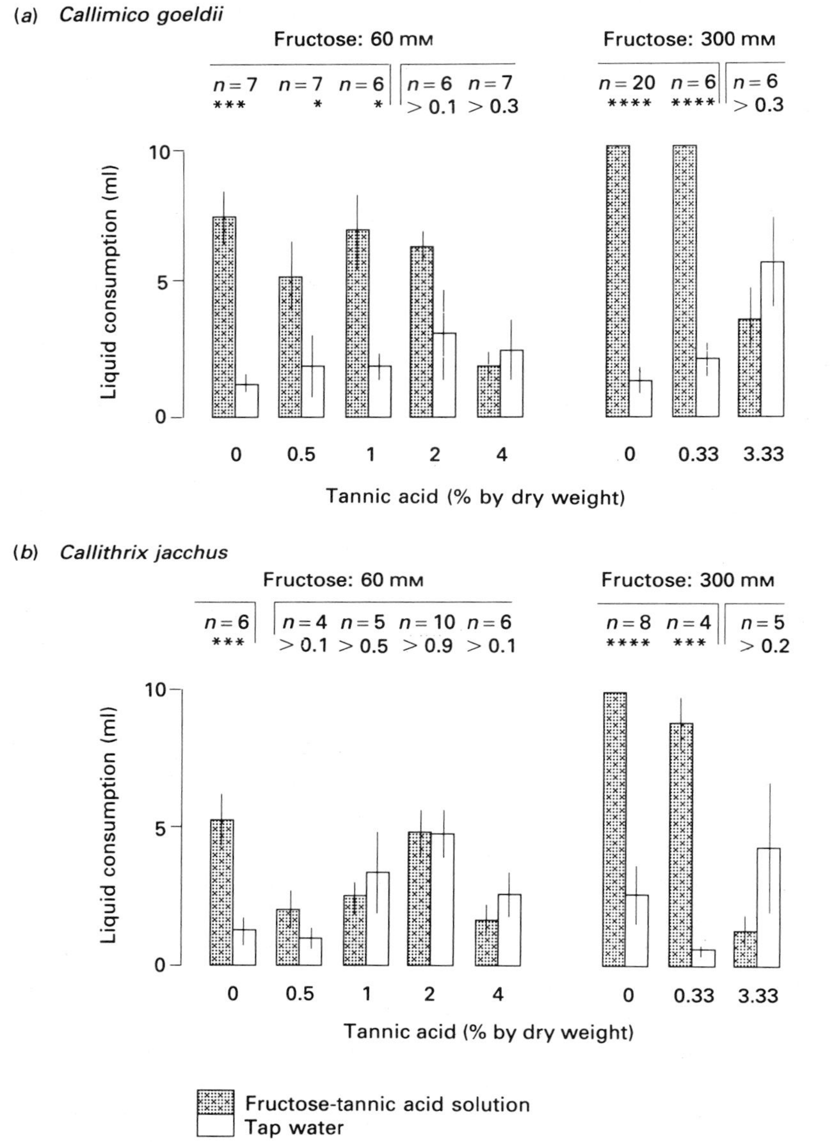

Fig. 10.4. Results of the 'two-bottle test' with mixtures of fructose + tannic acid in *Callimico goeldii* and *Callithrix jacchus*. Tannin concentration is given as a percentage of the dry weight of sugar dissolved. Mean liquid consumption (fructose + tannic acid versus tap water) with standard error of the mean is figured for each binary solution, including number of tests (n). The corresponding probability that the mean difference of consumption between the solution and water is significant is indicated: $^{*}P < 0.05$, $^{***}P < 0.01$, $^{****}P < 0.001$.

that our data for fructose were obtained using near-threshold and low caloric concentrations, it is likely that oropharyngeal stimulations predominated over post-ingestive effects such as long-term satiety in determining such differential ingestive responses.

From responses to both sweet and bitter solutions, it appears that species' discriminative abilities are specific for each compound and, where relevant to naturally occurring substances, are adapted to the diet. For instance, the high motivation required for harvesting 'unpredictable' and scattered foods such as ripe fruits and nectars in tamarins and Goeldi's monkey might be sustained by the great palatability of sweet compounds. Conversely, low taste sensitivity for soluble sugars in some marmosets is associated with a diet including lesser proportions of fleshy fruits. Feeding mainly on gums probably does not depend on any sensory reward associated with the taste of hexoses and pentoses. As was previously hypothesized for gummivorous prosimians as well as for leaf-eating primates (Hladik, 1979, 1981), feeding on polysaccharide-rich plant materials – foliage is mainly composed of cellulose and hemicelluloses, digestion of which requires fermentation in the stomach and/or in the large intestine (Bauchop and Martucci, 1968; van Soest, 1982) – might rather be maintained predominantly by a long-term conditioning based on the balance of the energetic needs. However, the possibility that exudate-eating primates may be able to taste large polymerized sugars, thus making gums palatable, should be further examined.

Given that the allelochemical composition of barks and gums used by marmosets is unknown, the interpretation of interspecific differences in taste discrimination of bitter and astringent substances is presently hypothetical. From results obtained with quinine and tannic acid (an hydrolyzable tannin), it appears that the deterrent effects on taste of toxins (e.g. alkaloids) and of digestibility-reducing compounds (e.g. tannins) differ within species as well as between species. The most gummivorous marmosets were the most tolerant of quinine hydrochloride, presenting taste rejection thresholds up to 10 times higher than those of tamarins and Goeldi's monkey and approximately 100 times higher than values obtained in the rat (see Thaw and Smith, 1992). However, adding relatively low tannin concentrations to fructose solutions resulted in a comparable inhibitory effect on the ingestive response of *Callithrix jacchus* and *Callimico goeldii*. Hence, hydrolyzable tannins are efficient in constraining food choices through repellent effects on taste. Contrary to what is believed for larger primates eating leaves (Hladik and Gueguen, 1974; Hladik, 1978), these small-bodied primates probably did not evolve efficient behavioural or physiological mechanisms, such as earth consumption and detoxification systems, to avoid the non-specific effects of tannins, i.e. sequestration of proteins and inhibition of digestive enzyme activity (Golstein and Swain, 1965; Feeny, 1969; Martin and Martin, 1982; Mole and Waterman, 1987; but see Martin *et al.*, 1987).

Allelochemicals mainly occur in immature fruits, seeds and leaves, as well as in barks of tropical trees and sometimes in gums (Willaman and Schubert, 1961; Glicksman, 1969; McKey, 1978; Gartlan *et al.*, 1980; Wrangham and Waterman, 1981; Beeson, 1984). The relatively acute discrimination of bitter and astringent tastes enables primates relying on highly digestible plant materials and having a simple digestive tract to avoid foods containing alkaloids and tannins. By contrast, low sensitivity to quinine would allow gum-specialist marmosets, which gnaw barks to obtain exudates, to accept to some extent distasteful chemical defences such as alkaloids, saponins and cyanogenic glycosides.

Surprisingly, the lowest taste threshold for quinine hydrochloride ever obtained in primates, including humans (Skramlik, 1948; Glaser, 1986; Hladik *et al.*, 1986), was found in the silvery marmoset, *Callithrix argentata*. In this species, the rate of ingestion of bitter solutions decreased very slowly as concentration increased, thus indicating a relative taste tolerance to a wide range of quinine concentrations. Taking into account the ingestive responses both for fructose and quinine solutions, this marmoset appeared to have taste characteristics relatively close to those of tamarins. According to Stevenson and Rylands (1988), this Amazonian primate, the ecology of which is poorly known, might be more frugivorous than any marmosets of the *Callithrix jacchus* group (including *C. jacchus* and *C. geoffroyi*) living in the Atlantic forest and in drier habitats of eastern Brazil; its geographical range is contiguous to areas inhabited by the most frugivorous marmoset so far studied (*Callithrix humeralifer*; Rylands, 1982, 1984) and its dentition is intermediate between the specialized gummivorous type of marmosets (V-shaped mandible as well as long lower incisors adapted to gouging) and the tamarin type (U-shaped mandible and elongated lower canines; Hershkowitz, 1977). Clearly, quantitative and qualitative data on the diet of *Callithrix argentata* are needed to define more precisely the functional significance of such an acute discrimination of the bitter taste.

Finally, the overall results presented here offer new insights into the endogenous factors involved in primate food choices and it should be emphasized that further comparison of taste threshold and supra-threshold responses between species may help us better define the role of taste perception in diet differentiation among primates.

Acknowledgements

I would like to thank C. M. Hladik and D. B. McKey for their help with the manuscript. I am also particularly grateful to R. D. Martin, D. Glaser and B. Carroll who provided me with facilities to carry out the experiments at the Anthropology Institute in Zürich and at the Jersey Wildlife Preservation Trust.

References

Anderson, D. M. W., Bell, P. C. & Millar, J. R. A. (1974). Composition of gum exudates from *Anacardium occidentale*. *Phytochemistry*, **13**, 2189–2193.

Bauchop, T. & Martucci, R. W. (1968). Ruminant-like digestion of the langur monkey. *Science*, **161**, 698–700.

Bearder, S. K. & Martin, R. D. (1980). *Acacia* gum and its use by bushbabies, *Galago senegalensis* (Primates: Lorisidae). *International Journal of Primatology*, **1**, 103–127.

Beeson, M. (1984). Blue monkeys (*Cercopithecus mitis*) in Malawi: condensed tannins and bark-stripping. *Primate Eye*, **23**, 4 (abstract).

Cabanac, M. (1979). Sensory pleasure. *Quarterly Review of Biology*, **54**, 1–29.

Chivers, D. J. & Hladik, C. M. (1980). Morphology of the gastrointestinal tract in primates: comparisons with other mammals in relation to diet. *Journal of Morphology*, **166**, 337–386.

Feeny, P. P. (1969). Inhibitory effect of oak leaf tannins on the hydrolysis of proteins by trypsin. *Phytochemistry*, **8**, 2119–2126.

Garber, P. A. (1984). Proposed nutritional importance of plant exudates in the diet of the Panamanian tamarin, *Saguinus oedipus geoffroyi*. *International Journal of Primatology*, **5**, 1–15.

Gartlan, J. S., McKey, D. B., Waterman, P. G., Mbi, C. N. & Struhsaker, T. T. (1980). A comparative study of the phytochemistry of two African rain forests. *Biochemical Systematics and Ecology*, **8**, 401–422.

Glaser, D. (1968). Geschmacksschwellenwerte bei Callitrichidae (Platyrrhina). *Folia Primatologica*, **9**, 246–257.

Glaser, D. (1986). Geschmacksforschung bei Primaten. *Vierteljahrsschrift der Naturforschenden Gesellschaft in Zuerich*, **131/2**, 92–110.

Glaser, D. & Hellekant, G. (1977). Verhaltens- und electrophysiologische Experimente über den Geschmackssinn bei *Saguinus midas tamarin* (Callitrichidae). *Folia Primatologica*, **28**, 43–51.

Glicksman, M. (1969). *Gum Technology in the Food Industry*. New York: Academic Press.

Golstein, J. L. & Swain, T. (1963). Changes in tannins in ripening fruits. *Phytochemistry*, **2**, 371–383.

Golstein, J. L. & Swain, T. (1965). The inhibition of enzymes by tannins. *Phytochemistry*, **4**, 185–192.

Hausfater, G. & Bearce, W. H. (1976). *Acacia* tree exudates: their composition and use as a food source by baboons. *East African Wildlife Journal*, **14**, 241–243.

Hellekant, G., Roberts, T. W., Hladik, C. M., Dennys, V., Simmen, B. & Glaser, D. (1993). On the relationship between sweet taste and seasonal body weight changes in a primate (*Microcebus murinus*). *Chem. Senses*, in press.

Hershkovitz, P. (1977). *Living New World Monkeys (Platyrrhini) with an Introduction to the Primates*. Chicago: Chicago University Press.

Hladik, C. M. (1978). Adaptive strategies of primates in relation to leaf-eating. In *The Ecology of Arboreal Folivores*, ed. G. G. Montgomery, pp. 373–395. Washington, DC: Smithsonian Institution Press.

Hladik, C. M. (1979). Diet and ecology of prosimians. In *The Study of Prosimian Behavior*, ed. G. A. Doyle & R. D. Martin, pp. 307–357. New York: Academic Press.

Hladik, C. M. (1981). Diet and the evolution of feeding strategies among forest primates. In *Omnivorous Primates. Gathering and Hunting in Human Evolution*, ed. R. S. O. Harding & G. Teleki, pp. 215–254. New York: Columbia University Press.

Hladik, C. M. & Gueguen, L. (1974). Géophagie et nutrition minérale chez les primates sauvages. *Comptes Rendus de l'Académie des Sciences, Paris,* **279**, 1393–1396.

Hladik, C. M., Robbe, B. & Pagézy, H. (1986). Sensibilité gustative différentielle des populations Pygmées et non Pygmées de forêt dense, de Soudaniens et d'Eskimos, en rapport avec l'environnement biochimique. *Comptes Rendus de l'Académie des Sciences, Paris,* **303**, 453–458.

Langer, P. (1988). *The Mammalian Herbivore Stomach. Comparative Anatomy, Function, and Evolution*. Stuttgart: Gustav Fischer.

Martin, J. S. & Martin, M. M. (1982) Tannin assays in ecological studies: lack of correlation between phenolics, proanthocyanidins and protein-precipitation constituents in mature foliage of six oak species. *Oecologia*, **54**, 205–211.

Martin, J. S., Martin, M. M. & Bernays, E. A. (1987). Failure of tannic acid to inhibit digestion or reduce digestibility of plant protein in gut fluids of insect herbivores: implications for theories of plant defense. *Journal of Chemical Ecology*, **13**, 605–621.

McKey, D. B. (1978). Soils, vegetation and seed-eating by black colobus monkeys. In *The Ecology of Arboreal Folivores*, ed. G.G. Montgomery, pp. 423–437. Washington, DC: Smithsonian Institution Press.

Michels, R. R., King, J. E. & Hsiao, S. (1988). Preference differences for sucrose solutions in young and aged squirrel monkeys. *Physiology and Behaviour*, **42**, 53–57.

Mole, S. & Waterman, P. G. (1987). A critical analysis of techniques for measuring tannins in ecological studies. II. Techniques for biochemically defining tannins. *Oecologia*, **72**, 148–156.

Moskowitz, H. R., Kluter, R. A., Westerling, J. & Jacobs, H. L. (1974). Sugar sweetness and pleasantness: evidence for different psychological laws. *Science*, **184**, 583–585.

Pfaffmann, C. (1960). The pleasures of sensation. *Psychology Review*, **67**, 253–268.

Rylands, A. B. (1982). The Behaviour and Ecology of Three Species of Marmosets and Tamarins (Callitrichidae, Primates) in Brazil. Ph.D. thesis, University of Cambridge.

Rylands, A. B. (1984). Exudate-eating and tree-gouging by marmosets (Callitrichidae, Primates). In *Tropical Rain Forest: The Leeds Symposium*, ed. S. L. Sutton & A. C. Chadwick, pp. 155–168. Leeds: Leeds Philosophical and Literary Society.

Rylands, A. B. (1989). Sympatric Brazilian callitrichids: the black tufted-ear marmoset, *Callithrix kuhli*, and the golden-headed lion tamarin, *Leontopithecus chrysomelas*. *Journal of Human Evolution*, **18**, 679–695.

Simmen, B. (1991). Stratégies Alimentaires des Primates Néotropicaux en Fonction de la Perception des Produits de l'Environnement. Ph.D. thesis, University of Paris XIII, Villetaneuse.

Simmen, B. & Hladik, C. M. (1988). Seasonal variation of taste threshold for sucrose in a prosimian species, *Microcebus murinus*. *Folia Primatologica*, **51**, 152–157.

Skramlik, E. von (1948). Über die zur minimalen Erregung des menschlichen Geruchs- bzw. Geschmackssinnes notwendigen Molekülmengen. *Pflügers Archiv Gesellschaft Physiologie*, **250**, 702–716.

Smith, F. & Montgomery, R. (1959). *The Chemistry of Plant Gums and Mucilages*. London: Reinhold, Chapman & Hall.

Snowdon, C. T. & Soini, P. (1988). The tamarins, genus *Saguinus*. In *Ecology and Behavior of Neotropical Primates*, Vol. 2, ed. R. A. Mittermeier, A. B. Rylands, A. Coimbra-Filho & G. A. B. Fonseca, pp. 223–298. Washington, DC: World Wildlife Fund.

Soini, P. (1988). The pygmy marmoset, genus *Cebuella*. In *Ecology and Behavior of Neotropical Primates*, Vol. 2, ed. R. A. Mittermeier, A. B. Rylands, A. Coimbra-Filho & G. A. B. Fonseca, pp. 79–129. Washington, DC: World Wildlife Fund.

Stevenson, M. F. & Rylands, A. B. (1988). The marmosets, genus *Callithrix*. In *Ecology and Behavior of Neotropical Primates*, Vol. 2, ed. R. A. Mittermeier, A. B. Rylands, A. Coimbra-Filho & G. A. B. Fonseca, pp. 131–222. Washington, DC: World Wildlife Fund.

Thaw, A. K. & Smith, J. C. (1992). Conditioned suppression as a method of detecting taste thresholds in the rat. *Chemical Senses*, **17**, 211–223.

van Soest, P. J. (1982). *Nutritional Ecology of the Ruminant*. Corvallis: O & B Books.

Willaman, J. J. & Schubert, B. C. (1961). *Alkaloid-Bearing Plants and their Contained Alkaloids*. 1234, US Department of Agriculture, Technical Bulletin.

Wrangham, R. W. & Waterman, P. G. (1981). Feeding behaviour of vervet monkeys on *Acacia tortilis* and *Acacia xanthophloea*: with special reference to reproductive strategies and tannin production. *Journal of Animal Ecology*, **50**, 715–731.

11

Potential hominid plant foods from woody species in semi-arid versus sub-humid sub-tropical Africa

CHARLES R. PETERS and
EILEEN M. O'BRIEN

One question, posed from the perspective of an ecologist studying wild-plant food species, that is appropriate to a discussion of the digestive system in mammals is 'how are potential plant food diets affected by changes in climate, in particular a drier climate versus a wetter climate with increased seasonality?'

Our interests lie in the edible flora of sub-Saharan Africa and the evolution of the hominid diet. In this chapter when we speak of edible plant foods we are referring to hominids. In its extreme form, this interest can be expressed as the African Dietary Hypothesis: it is the African sub-tropical flora in particular that has nurtured human evolution through most of the earth's recent pre-history; this is the nutritional environment to which our dietary physiology is most fundamentally adapted. An extreme position to be sure, perhaps only partially true.

To return to our initial question about climate and equitability of the nutritional environment, the climate of sub-tropical Africa is dominated by strong seasonality: summer rain systems primarily, with restricted bimodal zones (Fig. 11.1) where failure of the lesser rains often results in a seasonal pattern similar to that of the summer rain zones. Theoretically, one might expect a gradient of increasing seasonality of potential diet in the sub-tropical zones, with a notable difference, for example, between a drier climate with greater seasonal temperature extremes and a wetter climate with more equitable seasonal differences in temperature. This gradient and type of contrast should be most readily seen in Africa south of the equator, with decreasing annual precipitation and cooler winters associated with hotter summers across a broad geographic area. Fortunately this is a part of Africa where the edible wild plants are relatively well known.

The edible wild plants of sub-Saharan Africa in general are not well known. We have become increasingly aware of this since the 1970s when

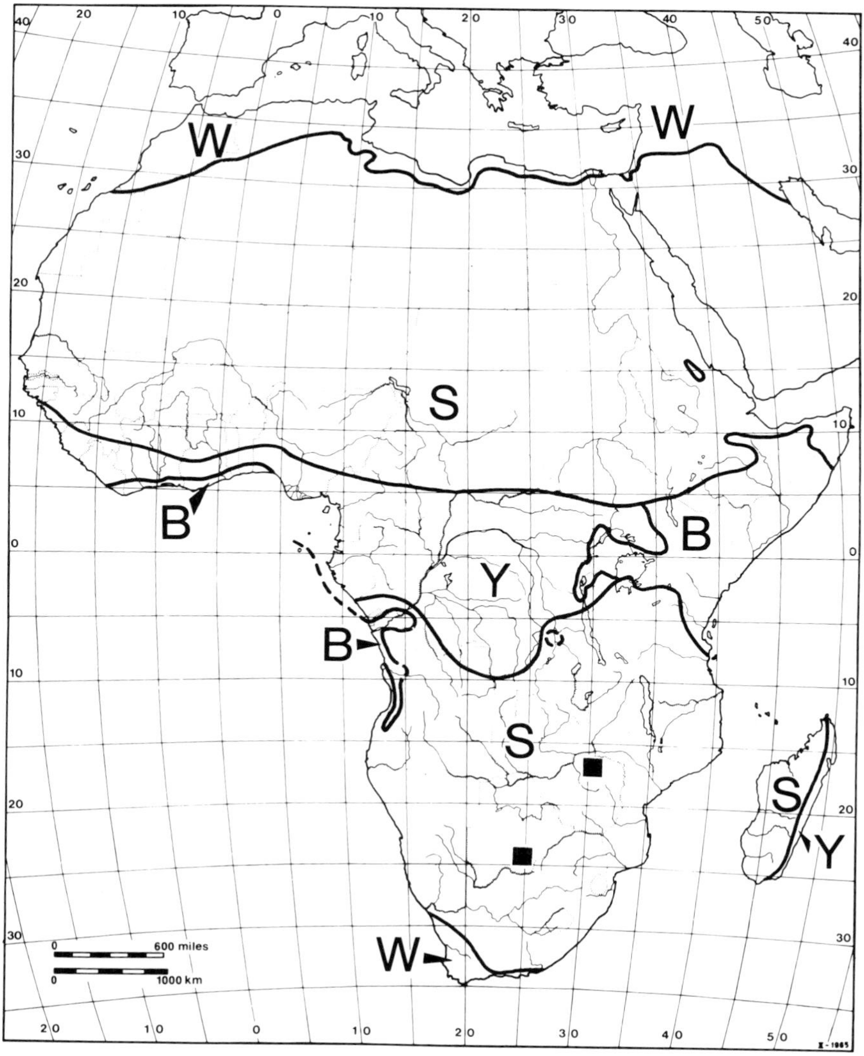

Fig. 11.1. Seasonal variation in rainfall in different climate regions of Africa: **Y**, humid-equatorial diurnal climate with rain more-or-less year-round (only a short dry season invariably occurs in most years); **B**, equatorial bimodal-rain climate with two marked dry seasons, particularly pronounced in semi-arid east Africa; **S**, tropical summer-rain climate (sub-humid to desert); **W**, winter-rain climate of the Cape and circum-Mediterranean. From Peters (1990) based upon the climate diagrams and subzonobiomes provided by Walter *et al.* (1960). The Harare grid cell (Zimbabwe) and the Gaborone grid cell (Botswana) are marked as black squares.

we began to catalogue information on their ecology. As part of that compilation, we have prepared a fairly complete checklist of the common edible wild plants of eastern and southern Africa. This species list is annotated as to the plant part(s) consumed by humans, and that information is used here for a partial model of potential hominid plant food resources. For a variety of reasons the woody species (excepting some of the suffrutices) are most reliably known. Southern Africa (the area south of Angola, Zambia and the Zambezi River) is particularly fortunate in having a compendium of botanical descriptions and small-scale distribution maps for its trees and shrubs, published by Coates Palgrave (1983). Nothing similar to this is available for the other parts of Africa.

Theoretically and empirically, a strong relationship between species richness and climate is plausible given climate's direct relationship to variations in ambient thermal range and photosynthetic potential (amount and duration of precipitation and energy) both of which are directly related to variations in the amount and duration of potential growth and reproduction. For southern Africa's wild, edible, woody plants, variations in climate appear to explain about 80% of the macro-scale spatial variation in species richness (number of species per unit equal-area) (O'Brien, 1988). Some of the results of this research can be applied in the investigation that we have initiated here.

What we have developed for this volume is the beginning of a case study, to which future contributions might be made by a variety of researchers. This study revolves around two contrasting climate areas, which can also be thought of, in a more dynamic evolutionary light, as a kind of place-for-time substitution. There are several minimum criteria we used for the selection of the two study areas. They should be examples of two common (dominant) climatic regimes within the summer rainfall zone of eastern and southern Africa, i.e. semi-arid and sub-humid, common versus rare winter frosts, respectively. Topography and elevation should be similar. Species richness necessarily varies as a function of sample area size, so to avoid confounding comparisons the area covered should be kept constant. Data on species richness and climate should be available or obtainable and the areas must have potential accessibility for future research. The two areas that we have chosen for preliminary consideration include one near Gaborone, the capital city of Botswana, and one near Harare, the capital city of Zimbabwe (see Fig. 11.1 and Table 11.1), each sample area (termed a grid cell) being about 20 000 km^2. The two areas are closely related in their flora (see Table 11.2), both being members of the Zambezian floristic region which extends northwards well into Tanzania.

Table 11.1. *Description of Gaborone and Harare grid cells*

	Gaborone	Harare
Country	Botswana	Zimbabwe
Main towns	Gaborone[a], Molepolole	Harare[a], Miami, Sipolilo
Location		
NW corner	23°48′ S, 24°12′ E	16°18′ S, 29°30′ E
SE corner	25°00′ S, 25°36′ E	17°36′ S, 31°00′ E
Approximate area	20 000 km^2	20 000 km^2
Mean elevation[b] (m)	1030	1251
Topography	Both areas are etch plains with scattered rocky hills and kopjes, underlain by crystalline and metamorphic bedrock which, in the case of the western portion of the Gaborone grid cell, is covered by a mantle of Kalahari sand	
Climate		
Type	Summer rainfall	Summer rainfall
Class (TMI)[c]	Semi-arid (–48)	Sub-humid (–10)
Climate variables: (averages)		
Precipitation[b] (mm)		
Annual	507 (546[a])	829 (840[a])
Maximum monthly	102	205
Minimum monthly	2	0
Temperature[a] (°C)		
Annual	19.7	18.5
Maximum monthly	24	22
Minimum monthly	12	14
Daily minimum, coldest month	1.5	6.7
Absolute minimum recorded	–6.7	0
Significant frost months	6	0
Potential evapotranspiration[b] (mm)		
Annual	973	921
Maximum monthly	138	109
Minimum monthly	22	36
Species richness		
All trees and shrubs	100	308
Edible (expected)[d]	60 (47)	121 (133)
Vegetation	Leguminous bushlands (*Acacia* dominated)	Leguminous woodlands (*Brachystegia* dominated)

[a] Climate stations described in Fig. 11.2, data based on Walter *et al.* (1960).
[b] Values from Thornthwaite and Mather (1962).
[c] TMI is Thornthwaite's Moisture Index (Thornthwaite and Mather, 1957).
[d] Expected values based on O'Brien (1988).

Table 11.2. *Taxonomic overlap in woody food plants between the Gaborone and Harare grid cells*

Taxonomic level	Number of plant taxa found in the grid cells			Total plant taxa
	Gaborone alone	Both grid cells	Harare alone	
Family	1 (1)	23 (9)	21 (19)	45 (29)
Genus	4 (4)	32 (15)	47 (39)	83 (58)
Species	24	36	85	145

The values in parentheses are the number of taxa that are monotypic in these areas, i.e. families with one food-providing genus, genera with one food-providing species. Therefore, most of the species that are found exclusively in the Gaborone grid cell come from genera that are found in both grid cells.

Methods and study sites

The only mammal with a diverse plant food diet for which we have anything approaching an adequate record of use is humans. They provide the plant-species data base for our climate comparison. In this paper, our universe of edible wild plants consists of native species that provide plant foods that are reported to be eaten raw by humans, or suspected of being edible raw. To our way of thinking, these plant foods are good candidates for inclusion in a model of the fundamental plant-food diet of the African hominids. Plant parts that reportedly require cooking or parts that are known to be poisonous are excluded. Many human famine foods are poisonous unless carefully processed.

There is dietary overlap between humans and other mammals, especially primates, but many of the plant parts eaten by baboons, for example, the dominant non-human primate in these environs, are poisonous to humans, e.g. the legumes of *Faidherbia albida* (syn. *Acacia albida*) and *Xeroderris stuhlmannii*. Our best guess is that baboons are capable of eating almost everything that humans eat, plus things we consider poisonous. Thus baboons can be thought of as potential primate competitors for the hominid plant foods considered here (also see Peters and O'Brien, 1981). Perhaps more importantly, the food-plant species list that we have for baboons provides an insight into how far we are from being able to develop a climate-based model for other polytrophic mammals. Of the 152 edible plant species that we consider here (see Appendix), 68 are also known to be exploited for food by baboons. The records for the other mammals with diverse plant food diets are much less systematic and much less complete. None-the-less, the preliminary conclusions that we reach here probably bear significantly on potential dietary resources for a variety of polytrophic mammals, including, in addition to the baboon, the vervet monkey, lesser bushbaby, bushpig, warthog, common duiker, impala, bushbuck, kudu, some of the smaller carnivores, and the African elephant (see Walker, 1986).

The plant universe for this model is further restricted to woody species of plants that attain, at least on occasion, a minimum height of 2.5 m, except in *Aloe* and *Protea* where the minimum height is more like 1.5 m. Most of the species are shrubs, bushes and trees. Here we use the term shrub in a rather broad sense, including arborescents (as in Peters *et al.*, 1984), i.e. shrubby forms and scandents with a less determinate growth form that can become trees when conditions permit. It is all these woody life forms, plus the bamboo *Oxytenanthera*, that Coates Plagrave (1983) covers in his broad treatment of 'trees'.

The climate model of species richness for woody plants developed by O'Brien (1988) suggests that, as the amount and duration (length of growing season) of potential plant photosynthesis decreases, species richness tends to decrease. The variables best describing climate's relationship to species richness are the minimum monthly potential evapotranspiration (a measure of incoming solar radiation, expressed as the energy demand for water) and the maximum monthly precipitation:

$$\text{species richness} = -61.28 + 2.98X_1 + 0.42X_2 \pm 20$$

where X_1 is the minimum monthly potential evapotranspiration, and X_2 is the maximum monthly precipitation.

Table 11.1 provides a general comparison of the two study areas. Both grid cells occur in the southern summer-rainfall zone. Gaborone typifies semi-arid conditions with moderately cold winter nights, Harare, sub-humid conditions with mild winters. Gaborone receives about 60% of the precipitation that Harare receives and it is colder in winter and hotter in summer than Harare. As a consequence, the potential evapotranspiration is more seasonally variable and extreme than Harare, resulting in a greater energy demand for water during the growing season. Figure 11.2 graphically portrays the pattern of annual variation in temperature and precipitation and the length of the potential growing season in each study area. Although their general patterns appear relatively similar, on closer scrutiny they differ significantly. First, Harare receives almost twice as much monthly precipitation throughout the growing season. Second, Gaborone has 6 months with frost, while Harare usually has no months with significant frost. In effect, the amount of photosynthesis is

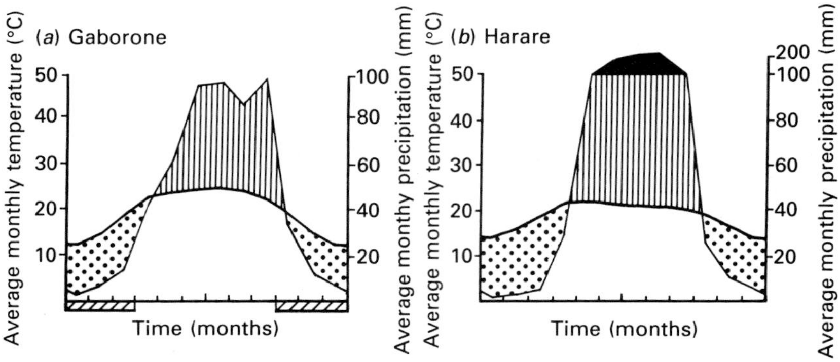

Fig. 11.2. Ecological climate diagrams for (*a*) Gaborone and (*b*) Harare (simplified from Walter *et al*., 1960). Overlapping curves of average monthly temperature and average monthly precipitation show the seasonal periods of relative drought (dotted), relative humidity (vertical shading), and the period of average monthly precipitation greater than 100 mm (the perhumid season, scale reduced to 1/10, black area). Below the abscissa in (*a*) are marked the months with absolute minimum temperatures below 0 °C (diagonally shaded).

likely to be much greater in Harare. The length of the growing season may also be longer than depicted, if there is sufficient soil moisture to sustain growth after the rains cease. Regardless of available soil moisture, the growing season in Gaborone ends with the advent of frost.

The vegetation differs structurally, and compositionally, between the study areas, although there is a certain amount of overlap. Like most of Africa outside the tropical forests, these areas are dominated by shrubs and trees of the Leguminosae. In the semi-arid area, there are bushlands where *Acacia* spp. are dominant, with broadleaf sub-dominants; in the wetter, sub-humid area there are woodlands, dry 'miombo', where *Brachystegia* spp. are dominant.

The appendix is a partially annotated list of the edible woody species that are found in the Gaborone and Harare grid cells. The lists of woody species were initially obtained using the species range maps in Coates Palgrave (1983) and an equal-area grid cell matrix to circumscribe the presence–absence of species in each area. This work was undertaken as part of O'Brien's (1988) analysis of the variations in the species richness of woody plants in relation to climate changes. The size of the grid cells, i.e. approximately 20 000 km^2 (about 140 km × 140 km), was based on two factors. First, climate is a macro-scale variable, measurable variation occurring at a minimum of 100 km (Griffiths, 1976). Second, the radius of accuracy for the species-range maps is probably within 70 kms. The lists were given a final check by soliciting the knowledgeable comments of Robert Drummond and Brian Williams during a trip to Zimbabwe in 1992. As a result, we were able to eliminate several species (especially Zambezi Valley species from the Harare list) and add a few that we had overlooked. This in turn was supplemented, in the case of the Gaborone grid cell, by additions made on the basis of species maps in Taylor and Moss (1982).

We recognize the species listed in the appendix as edible because of their appearance in our current checklist of the edible wild plants of sub-Saharan Africa, which has as its major emphasis the plants of eastern and southern Africa (Peters, O'Brien and Drummond, 1992; see footnote to Appendix). This sub-Saharan checklist is based for the most part on currently recognized scientific names, which are also adopted here in the appendix but without listing of synonyms. Further annotation is provided for the edible parts and whether they are known to be eaten raw. Generally the leaves that are reported to be eaten are cooked as 'pot herbs' or 'spinach'. Some may be edible raw but not reported as such. Those listed in the appendix as being edible raw were included after checking for poisonous effects against the information in Watt and Breyer-Brandwijk (1962) and Verdcourt and Trump (1969). Some poisonous plant parts may go unrecognized here. Certainly there is a great deal of unrecognized diversity of food quality represented here.

The appendix is further annotated with phenological and/or availability information from Coates Palgrave (1983), Palmer and Pitman (1972), Silberbauer (1981) and Scudder (1971), supplemented with personal observations and notes from van Wyk (1972/1974), Williamson (1975) and Drummond (1981, and personal communication). In these annotations the early dry season represents the period from April through June. Only notes on the season of scarcity, the dry season into earliest rainy season, are presented here. Most of the plants provide edible parts in the rainy season.

Results

The appendix presents a compilation of dietary information for the woody species occurring in the Gaborone and Harare grid cells. The data are presented in alphabetic order by family, then genus, then species; first for the monocotyledons and then for the dicotyledons. As can be seen in Table 11.2, there is significant taxonomic overlap between these two areas, even though they are about 1000 km apart. In effect, the sub-humid climate area is twice as species-rich in edible taxa as the semi-arid climate area, and this is generally reflected in the diversity of edible plant parts.

Table 11.3 is a tabulation of plant-species counts for the edible parts listed in the appendix. It is a summary of the diversity of edible plant parts for the two study areas and contains some notes on season of availability. A few plant parts not clearly assignable to the major categories have been left out of the Table. Some entries questionable on other grounds have also been

Table 11.3. *Edible plant parts from woody species in sub-tropical Africa in a semi-arid versus sub-humid climate*[a]

	Grid cell	
	Gaborone	Harare
Total number of edible plant species[b]	**60**	**121**
Nut-like oil-seeds	**1** Available early dry season	**2**(1) Available early dry season, 1(1); late dry, 1
Fleshy fruits	**33** Season available: early dry, 6(1); dry, 11; late dry, 2(1)	**81**(2) Season available: early dry, 16(8); dry, 18(7); late dry, 13(3)
Dry fruits[b]	**1**(1)	**3**(1) Season available: early dry, 1; dry, 1; late dry, 2
Other seeds	**2** Dry season, 1	**9**(1) Season available: early dry, 2(1); dry, 2(2); mid to late dry, 1
Flowers, nectar	**2** Dry season, 1; late dry, 1	**4** Late dry season, 2
Leaves, stems, shoots	**13** Includes stem, 1; shoots, 0 Late dry season, 2	**30** Includes stems, 4; shoots, 3 Late dry season, **3**
Rootstocks	**5**(1) Dry season, 2	**14** Dry season, 1
Gum[c]	**19**	**12**
Bark, cambium, wood	**3**	**7**

[a] A plant species may provide more than one edible part.
[b] Including pods and capsules.
[c] The gums appear to be available year round.
Numbers in parentheses are possibilities that need to be checked further.

omitted from the counts. The data are rough in many ways but can be thought of as an introduction to the potential plant food diets provided by the trees and shrubs of these climate areas. The presentation of results that follows is by plant part, in their order of appearance in Table 11.3.

Species of nut-like (hard endocarp) oil seeds are present in both climate areas. *Sclerocarya birrea* (Anacardiaceae) is the one species that is found in both areas although it is not common in the Harare grid cell. It is primarily a species of mid to lower elevations and is common in the Zambezi Valley and found in significant numbers in the Gaborone area. *Balanites aegyptiaca* (Balanitaceae) may occur in the Harare grid cell but is not common there or elsewhere in Zimbabwe. Its main distribution is to the north, in the hotter, drier parts of tropical Africa. *Parinari curatellifolia* (Chrysobalanaceae) is the common species providing nut-like oil seeds in the Harare grid cell. It would appear to provide a critical food resource in the late-dry/early-rainy season. *Sclerocarya*, *Parinari* and *Balanites* seeds have reported fat contents that range from 34% to 61% (Malaisse and Parent, 1985; Wehmeyer, 1986; Peters, 1988). The apparent protein values range from 28% to 31%, although, oddly, Malaisse and Parent (1985) report a protein content of only 3% for *Parinari*. Comparing the abundance of *Parinari* in the Harare grid cell (C. R. Peters, personal observation) with that of *Sclerocarya* in the Gaborone grid cell (Taylor and Moss, 1982; C. R. Peters, personal observation) it appears likely that the former is not only more common but reaches higher densities locally.

Fleshy fruit is the most species-rich category in Table 11.3. Although there is a significant drop in their availability during the dry season, this is still the major dry season food provided by trees and shrubs. The climate-related pattern for the dry season taxa has at least three aspects: (a) the genera that characterize the drier climate are also present in the wetter climate with only rare exceptions; (b) few of the dry-climate, dry-season genera provide more than one food species in that season, with *Rhus* (Anacardiaceae) and especially *Grewia* (Tiliaceae) notable exceptions; and (c) the dry-season genera of the wetter climate are also primarily monotypic in that season, with high family diversity and an overall species richness nearly as great as the wet season taxa of the drier climate. The late dry season/early rainy season is the time when fewest species are in fruit: only one identified in the drier climate (*Ochna pulchra*), but most of the late dry-season species in the sub-humid climate appear to provide fruit into the early rainy season. Notable plants that can act in this way in the sub-humid climate include *Ficus* (Moraceae), *Myrica* (Myricaceae) the wax berry, with a fat content reported to be 20% (*syn. M. cordifolia* in Watt and Breyer-Brandwijk, 1962), *Parinari* (see notes on nut-like oil seeds above), and *Strychnos* (Loganiaceae). We still need to identify the dominant edible species and the major producers in each climate area.

Dry fruits consisting of pods and capsules have a very limited role to play with the possible exception of *Piliostigma* (Leguminosae–Caesalpinioideae) whose thick, woody, tough pods might be an important late dry-season famine food in the sub-humid climate area. There is one additional very important type of 'dry fruit' that is not counted as such in Table 11.3. This is the thinly fleshy and/or fibrous indehiscent fruit that dries out and remains edible well into the dry season. *Grewia* spp. are notable in this regard, and are eaten by humans, pyrenes and all (Peters, 1993). *Ziziphus mucronata* (Rhamnaceae) is similar and these species might play a critical dietary role in the drier climate.

The group 'other' seeds is another species-poor category with a few possibly important members. For the drier climate, *Bauhinia petersiana* (Leguminosae–Caesalpinoideae) provides a proteinaceous seed (Wehmeyer, 1986), apparently in the dry season. Three food species seem to provide proteinaceous seeds during the dry season in the sub-humid climate area: *Adansonia* (Bombacaceae) an oil-seed, *Bauhinia petersiana* and *Piliostigma* (Wehmeyer, 1986).

Flower nectar is a category that appears to be of particular interest to children, and the two species noted for the drier climate provide nectar that is reported to be eaten by children. One of these, an *Aloe* (Aloeaceae), is said to have its main blooming period in the mid-dry season. One of the four species catalogued for the sub-humid climate area, *Combretum paniculatum* (Combretaceae), may also be of special interest because its flowers are reported to be sucked by children, presumably for the nectar (as in related species), and it is recorded to flower at the end of the dry season through early rainy season.

In the category leaves/stems/shoots it is leaves that are the most common plant part eaten. Second to fleshy fruits this is the most species-rich category, particularly in the sub-humid climate. Because leaves are usually cooked as 'pot herbs' the records provide few additional details. A few are noted to be *young* leaves. Rarely it is noted that they may be eaten raw. Further study of those reported to be eaten as young leaves might examine their potential importance in the early wet season, when leaf-flush generally occurs. Only a few species are known to have leaf-flush just before the rains.

Edible rootstocks is a relatively species-poor category for trees and shrubs. A few are known to be eaten raw and they appear to be significant sources of moisture: *Commiphora* spp. (Burseraceae) and *Bauhinia petersiana*. Information is lacking to judge the potential significance of others.

Gums is a relatively species-rich category in the drier climate. Lee (1979) notes that, in the northern Kalahari, several of the most conspicuous tree species (*Acacia*, *Combretum* and *Terminalia* spp.) exude non-toxic edible resins that are collected through most of the year by men, women and children. They are eaten in small amounts, not highly rated as food and claimed to be 'hard on the digestion'. Silberbauer (1981) also notes some species of gum as available all year round. It is likely, however, that during the coldest months of the winter/dry season production is reduced significantly.

Bark/cambium/wood is a species-poor category. Two *Acacia* spp. (Leguminosae–Mimosoideae) found in both study areas and *Ficus sur* (Moraceae) in the sub-humid area, have an inner bark that can be chewed for thirst. As sometimes depicted in photoessays on the African elephant, the soft wood of the baobab tree *Adansonia* (Bombacaceae) can also be chewed for its moisture content. The potential significance of the other species is unclear.

Discussion

Under the climate model introduced in O'Brien (1988), the expected values for edible species in the Gaborone and Harare grid cells are close to the observed values, 47 versus 60 species and 133 versus 121 species, respectively. Therefore, it would appear that, primarily because of differences in climate, the Harare grid cell has twice the number of woody food-species ('edible' trees and shrubs) of that found in the Gaborone grid cell. However, the difference in total number of species of trees and shrubs is almost three times that of the Gaborone area. This means that the relative proportion (but not necessarily abundance) of edible species is greater (60% compared with 39%) under the more seasonably harsh climate of the Gaborone area. In other words, although a great number of species drop-out as we move to the harsher climate, a larger portion of those remaining are edible. It also appears that, in spite of this climate difference, the basic food categories do not change. What changes is the absolute number of species providing those resources, and perhaps potential competition, as well as reliable seasonal availability, other qualities and crop abundance. Food plant abundance does not necessarily change, but decreased productivity is likely to be associated with the drier climate. This is in addition to the loss of important food species *per se*.

The food species that remain in the drier climate probably take on increased importance for the variety of polytrophic mammals that include leaves and fruit, in particular, in their diet. Potential competition for fruit is also probably higher in the harsher climate. Hamilton *et al.* (1976) note, for example, that in the northern Kalahari, even on the floodplains of the Okavango Swamp, there is no excess of ripe fruit on the important fruit-producing trees. As the crops of fruit ripen they are 'consumed completely by baboons, hornbills, fruit pigeons, impala, warthogs, mongooses, vervet monkeys, elephants and other animals'. To put it another way, more of the potential competition is probably realized in the scramble for food in the harsher climate. Only lower quality plant foods or those that require extensive processing are likely to be subject to less competition. Perhaps there is less competition for gums, in which case the lesser bushbaby may benefit.

Broadly speaking, trees and shrubs (*sensu lato*) are the dominant plant life form providing wild-plant foods in Africa north of the Cape floristic region. Forbs, particularly deciduous forbs, are the second most important ecophysiognomic type of life form (Peters *et al.*, 1984). (Although the grasses are diverse and abundant across eastern and southern Africa, few appear to provide food items for humans.) Formal treatment of the relationship of climate to species richness in the forbs has not been developed due to the lack of species distribution maps. Perennial forbs are particularly important, however, as sources of edible rootstocks.

Our phenology notes indicate a pattern of seasonal availability for wild plant foods characterized by a preponderance of fruits/seeds and leaves in the rainy season, fruits/seeds in the dry season and a few rootstock species available perhaps more or less all year round. This is similar to the pattern reported by Peters *et al.* (1984), but, in addition, they note the potential importance of rootstocks provided by the perennial forbs, particularly in the dry season. What this current analysis forces us to consider, however, is the apparently marginal nature of the drier, more seasonal climate area. The diversity of edible species and plant parts drops off precipitously. The question remains as to whether the perennial forbs and their potentially important rootstocks follow a similar pattern.

For woody plants there is a gradient of loss in species richness going from north-east Zimbabwe through to south-west Botswana (O'Brien, 1988; O'Brien & Peters, 1991). A number of the species in the Harare grid cell are found to the north and/or east but drop out to the south-west, well before the appearance of Kalahari Sand in west-central Zimbabwe. The pattern recurs further to the south-west in the Gaborone grid cell. A number of species found in the east-north-east portion of the Gaborone grid cell drop out further west, this time approximately where the sandveld or deep Kalahari Sand begins. These differences in soil, plus topography and perhaps elevation probably account for most of the difference between the predicted number of species based on climate alone and the observed number of species in each grid cell.

Before we turn our attention to some speculations about early hominid adaptations we should emphasize a few points about what it is that we would like to know next, some of which is research that must be carried out by others. We need to learn which species are dominant and which could provide potential staples. We need to clear up the phenology notes. The physical and chemical characteristics of the plant foods need to be analysed. Judgements about the potential digestibility and nutritional significance for children would perhaps be the most interesting application to pursue. At a finer scale of spatial resolution, we would like to know something of the landscape correlates of distribution for edible species. Possibly, for example, there are larger areas without significant food plants in the drier climate area: larger areas of non-habitat from the plant food point-of-view.

With regard to early hominids, including *Paranthropus* and *Homo erectus*, the preliminary findings presented here suggest greater within- and between-species dietary diversity is possible in the sub-humid climate. The semi-arid climate narrows dietary choices severely as the background diversity of numerous monotypic food taxa is eliminated and taxa that remain take on enhanced importance. Specialized capabilities for food processing presumably are increasingly relied upon as the resource base shrinks. The bottom line, from a regional perspective, is that the semi-arid climate areas may have been marginal habitats, occupied by hominids but functioning as demographic sinks. It has previously been suggested (O'Brien and Peters, 1991) that the ultimate source area, in southern Africa, for the Transvaal early hominids was probably to the north-east in Zimbabwe.

References

Coates Palgrave, K. (1983). *Trees of Southern Africa*. Cape Town: Struik.

Drummond, R. B. (1981). *Common Trees of the Central Watershed Woodlands of Zimbabwe*. Harare: Department of Natural Resources.

Griffiths, J. F. (1976). *Climate and the Environment*. Boulder: Westview Press.

Hamilton, W. J., Buskirk, R. E. & Buskirk, W. H. (1976). Defense of space and resources by chacma (*Papio ursinus*) baboon troops in an African desert and swamp. *Ecology*, **57**, 1264–1272.

Lee, R. B. (1979). *The !Kung San*. Cambridge: Cambridge University Press.

Malaisse, F. & Parent, G. (1985). Edible wild vegetable products in the Zambezian woodland area: a nutritional and ecological approach. *Ecology of Food and Nutrition*, **18**, 43–82.

O'Brien, E. M. (1988). Climate correlates of species richness for woody 'edible' plants across southern Africa. *Monographs in Systematic Botany from the Missouri Botanical Garden*, **25**, 385–402.

O'Brien, E. M. & Peters, C. R. (1991). Ecobotanical contexts for African hominids. In *Cultural Beginnings: Approaches to Understanding Early Hominid Life-ways in the African Savanna*, ed. J. D. Clark, pp. 1–15. Bonn: Dr Rudolf Habelt Gmbh.

Palmer, E. & Pitman, N. (1972). *Trees of Southern Africa*, Vol. 1–3. Cape Town: Balkema.

Peters, C. R. (1988). Notes on the distribution and relative abundance of *Sclerocarya birrea* (A. Rich.) Hochst. (Anacardiaceae). *Monographs in Systematic Botany from the Missouri Botanical Garden*, **25**, 403–410.

Peters, C. R. (1990). African wild plants with rootstocks reported to be eaten raw: the monocotyledons, part 1. *Mitteilungen aus dem Institut für Allgemeine Botanik Hamburg*, **23**, 935–952.

Peters, C. R. (1993). Shell strength and primate seed predation of non-toxic species in eastern and southern Africa. *International Journal of Primatology*, **14**, 315–344.

Peters, C. R. & O'Brien, E. M. (1981). The early hominid plant-food niche: insights from an analysis of plant exploitation by *Homo*, *Pan*, and *Papio* in eastern and southern Africa. *Current Anthropology*, **22**, 127–140.

Peters, C. R., O'Brien, E. M. & Box, E. O. (1984). Plant types and seasonality of wild-plant foods, Tanzania to southwestern Africa: resources for models of the natural environment. *Journal of Human Evolution*, **13**, 397–414.

Scudder, T. (1971). *Gathering Among African Woodland Savannah Cultivators. A Case Study: The Gwembe Tonga*, Zambian Papers 5. Lusaka: University of Zambia, Institute for African Studies.

Silberbauer, G. B. (1981). *Hunter and Habitat in the Central Kalahari Desert.* Cambridge: Cambridge University Press.

Taylor, F. W. & Moss, H. (1982). Maps: density and distribution of 111 plants investigated during survey of 83 village areas around Botswana. *Final Report on the Potential for Commercial Utilization of Veld Products*, Vol. II, Appendix I. Gaborone: The Government Printer.

Thornthwaite, C. W. & Mather, J. R. (1957). Instructions and tables for computing potential evapotranspiration and the water balance. *Publications in Climatology*, Vol. 10 (1). Centerton, NJ: Laboratory of Climatology.

Thornthwaite, C. W. & Mather, J. R. (1962). Average climate water balance data of the continents, Part 1: Africa. *Publications in Climatology*, Vol. 15 (2). Centerton, NJ: Laboratory of Climatology.

van Wyk, P. (1972/1974). *Trees of the Kruger National Park*, Vols. 1, 2. Cape Town: Purnell.

Verdcourt, B. & Trump, E. C. (1969). *Common Poisonous Plants of East Africa.* London: Collins.

Walker, C. (1986). *Signs of the Wild: Field Guide to the Spoor and Signs of the Mammals of Southern Africa.* Cape Town: Struik.

Walter, H., Lieth, H. & Rehder, H. (1960). *Klimadiagramm-Weltatlas.* Jena: Gustav Fischer.

Watt, J. M. & Breyer-Brandwijk, M. G. (1962). *Medicinal and Poisonous Plants of Southern and Eastern Africa*, 2nd edn. Edinburgh: E. & S. Livingstone.

Wehmeyer, A. S. (1986). *Edible Wild Plants of Southern Africa: Data on the Nutrient Contents of Over 300 Species.* Pretoria: CSIR, National Food Research Institute.

Williamson, J. (1975). *Useful Plants of Malawi*, Revised and extended edn. Zomba: University of Malawi.

Appendix to Ch. 11 *Potential wild plant foods for humans: woody species found in the Gaborone and Harare grid cells*[a]

Plant name		Edible parts	Grid cell
Monocotyledones			
Aloeaceae	*Aloe marlothii* Berger	Flower nectar (eaten by children), main bloom, mid dry season	G
Arecaceae	*Phoenix reclinata* Jacq.	Fruit, raw; shoot, young; heart of crown ('cabbage'); sap	H
Poaceae	*Oxytenanthera abyssinica* (A. Rich.) Munro	Fruit, grain; shoot, young	H
Dicotyledones			
Anacardiaceae	*Lannea discolor* (Sond.) Engl.	Fruit, raw (ripen in the late dry season, earliest rainy season)	H
	L. schweinfurthii (Engl.) Engl.	Fruit, raw	H
	Rhus lancea L.f.	Fruit	Both
	R. leptodictya Diels	Fruit (ripen in early dry season)	H
	R. longipes Engl.	Fruit (eaten by children) ripen in late dry season, earliest rainy season	H
	R. pyroides Burch.	Fruit, raw (may last into earliest dry season)	Both
	R. tenuinervis Engl.	Fruit, raw (available into the early dry season)	Both
	Sclerocarya birrea (A. Rich.) Hochst.	Fruit pulp and seed kernel, raw (nuts unspoiled lasting into the dry season)	Both

Appendix to Ch. 11 *Continued*

Plant name		Edible parts	Grid cell
Annonaceae	*Friesodielsia obovata* (Benth.) Verdc.	Fruit, raw (ripen in early dry season and dry fruit can be found on the branch almost any time of the year)	H
Apocynaceae	*Carissa edulis* (Forssk.) Vahl	Fruit, raw; root (boiled)	H
	Diplorhynchus condylocarpon (Muell. Arg.) Pich.	Gum	H
Aquifoliaceae	*Ilex mitis* (L.) Radlk.	Fruit (ripen in the early dry season) Root (boiled)	H
Araliaceae	*Cussonia arborea* A. Rich.	Fruit (ripen in the early rainy season)	H
	C. spicata Thunb.	Fruit (ripen through the dry season) root	H
Asteraceae	*Vernonia amygdalina* Del.	Leaf stem (chewstick); root (chewstick, bark removed)	H
Balanitaceae	*Balanites aegyptiaca* (L.) Del.	Fruit pulp and seed kernel (nuts unspoiled lasting into the dry season)	H[b]
Bignoniaceae	*Kigelia africana* (Lam.) Benth.	Seed (roasted) available into the early dry season	H
Bombacaceae	*Adansonia digitata* L.	Fruit pulp raw and seed kernel raw (fruit mature in earliest dry season and remain on branch through the dry season); leaf; shoot from germinating seed; flower; root (very young baobab); wood (source of water)	H
Boraginaceae	*Cordia africana* Lam.	Fruit (ripen mid to latter part of the dry season)	H

	Cordia sinensis Lam.	Fruit, raw (some ripen in the early dry season); root, raw	H
	Ehretia amoena Klotzsch	Fruit	H
	E. rigida (Thunb.) Druce	Fruit, raw	G
Burseraceae	*Commiphora africana* (A. Rich.) Engl.	Fruit; leaf; stem (chewstick); root pith (chewed for sweet juicy sap); gum; bark	H
	C. mossambicensis (Oliv.) Engl.	Root, raw (eaten by children)	H
	C. pyracanthoides Engl.	Root pith, raw (juice) available year-round; gum	G
Capparidaceae	*Boscia albitrunca* (Burch.) Gilg & Bened.	Fruit, raw; leaf, raw; root (famine)	G
	B. angustifolia A. Rich.	Fruit (available in the dry season); leaf	H
	B. foetide Schinz	Fruit, raw (sweetish oily pulp)	G
	Capparis tomentosa Lam.	Fruit	H
	Maerua triphylla A. Rich.	Fruit (available into the early dry season?); root (both fruit and root may be poisonous?)	H
Celastraceae	*Cassine aethiopica* (Thunb.	Fruit (ripen in dry season)	H
	C. transvaalensis (Burtt Davy) Codd	Fruit (ripen in the later dry season)	H
	Maytenus heterophylla (Eckl. & Zeyh.) N. Robson	Fruit aril (available in the dry season plus early rainy season and sporadically most times of the year?)	H
	M. senegalensis (Lam.) Exell	Root (boiled)	Both
	M. undata (Thunb.) Blakelock	Bark (relish)	Both

Appendix to Ch. 11 *Continued*

Plant name		Edible parts	Grid cell
Chrysobalanaceae	*Parinari curatellifolia* Benth.	Fruit pulp and seed kernel, raw (fruits ripen from the mid-dry season into the early rainy season; nuts, unspoiled lasting into the late rainy season)	H
Clusiaceae	*Garcinia buchananii* Bak.	Fruit (ripen into the early dry season?)	H
Combretaceae	*Combretum apiculatum* Sond.	Gum	Both
	C. collinum Fresen.	Gum	H
	C. erythrophyllum (Burch.) Sond.	Gum	Both
	C. hereroense Schinz	Leaf; gum	Both
	C. imberbe Wawra	Leaf; gum raw (year round)	G
	C. molle G. Don	Root (tea and soup)	Both
	C. paniculatum Vent.	Leaf; flower nectar eaten by children (end of the dry season and early rainy season)	H
	C. zeyheri Sond.	Gum	Both
	Terminalia sericea DC.	Leaf, raw (young) flush just before rains; gum	Both
Ebenaceae	*Diospyros kirkii* Hiern	Fruit, raw (ripen from early to late dry season?)	H
	D. lycioides Desf.	Fruit, raw (ripen in the latter part of the rainy season and early dry season, can remain on tree through the dry season)	Both
	D. mespiliformis A. DC.	Fruit pulp and seed, raw (fruit ripen in the dry season?)	H

	Euclea crispa (Thunb.) Gürke	Fruit, raw (ripen in the dry season?), may be strongly purgative; leaf (appetizer)	H
	E. divinorum Hiern	Fruit, raw (often purgative), ripen in rainy season and early dry season; leaf, young	H
	E. natalensis A. DC.	Fruit (ripen into the dry season)	Both
	E. racemosa Murr.	Fruit, raw (ripen into and through the dry season?)	H
	E. undulata Thunb.	Fruit (often purgative), ripen in the dry season	G
Euphorbiaceae	*Antidesma venosum* Tul.	Fruit, raw (sporadically available in the early dry season)	H
	Bridelia cathartica Bertol. f.	Fruit, raw (ripen in the dry season?)	H
	B. micrantha (Hochst.) Baill.	Fruit, raw (eaten by children)	H
	B. mollis Hutch.	Fruit (eaten by children), may last into earliest dry season	G
	Flueggea virosa (Willd.) Voigt	Fruit, raw	Both
	Margaritaria discoidea (Baill.) Webster	Fruit (dry capsule), cooked	H
	Pseudolachnostylis maprouneifolia Pax	Fruit, raw (ripen in the dry season and can remain on the branch into the early rainy season)	H
	Uapaca kirkiana Muell. Arg.	Fruit, raw (ripen in the early rainy season)	H
	U. nitida Muell. Arg.	Fruit, raw (ripen from the dry season into the early rainy season?)	H
Flacourtiaceae	*Flacourtia indica* (Burm.f.) Merr.	Fruit, raw (ripen into the mid dry season)	H
	Oncoba spinosa Forssk.	Fruit, raw (ripen in the early to mid dry season?)	H[b]
	Scolopia zeyheri (Nees) Harv.	Fruit (ripen in the dry season)	G

Appendix to Ch. 11 *Continued*

Plant name		Edible parts	Grid cell
Leguminosae–Caesalpinioideae	*Afzelia quanzensis* Welw.	Leaf, young	H
	Bauhinia petersiana Bolle	Fruit seed raw (ripe and green), mature in the dry season; flower nectar, raw (eaten by children), flowering from the late dry season through most of the rainy season; root, raw (water source, also baked and pounded), used in the dry season	Both
	B. tomentosa L.	Leaf, young	H[b]
	Burkea africana Hook.	Gum	Both
	Cassia abbreviata Oliv.	Leaf (purgative)	H
	Piliostigma thonningii (Schumach.) Milne-Redh.	Fruit pod ('biscuit-like inside') and seed (ripen in the mid to late dry season); leaf, young (chewed to relieve thirst)	H
	Schotia brachypetala Sond.	Fruit seed, roasted (available into the earliest dry season)	H
	Senna singueana (Del.) Lock	Fruit pod, raw (young pods available in the late dry season, mature pods in the rainy season into the early dry season); leaf (new leaves available just before the rains)	H
Leguminosae–Mimosoideae	*Acacia amythethophylla* A. Rich.	Leaf	H
	A. erioloba E. May	Gum	G
	A. erubescens Oliv.	Gum	G

	A. fleckii Schinz	Gum	G
	A. hebeclada DC.	Gum	G
	A. karroo Hayne	Gum; bark (fibres beneath the coarse outer bark chewed to alleviate thirst)	Both
	A. luederitzii Engl.	Gum	G
	A. mellifera (Vahl) Benth.	Fruit; pod (famine); leaf; gum, raw (year round)	G
	A. nilotica (L.) Del.	Gum, raw (year round)	G
	A. polyacantha Willd.	Gum	H
	A. tortilis (Forssk.) Hayne	Leaf; gum; bark (fibres beneath the coarse outer bark chewed to alleviate thirst)	Both
	Albizia anthelmintica (A. Rich.) Brongn.	Leaf, young	G
	Dichrostachys cinerea (L.) Wight & Arn.	Leaf; gum	Both
Leguminosae-Papilionoideae	*Dalbergia melanoxylon* Guill. & Perr.	Leaf (flush with the first rains and fall very early in the dry season)	H
	Ormocarpum kirkii S. Moore	Leaf	H
Loganiaceae	*Strychnos cocculoides* Bak.	Fruit pulp, raw (ripen in the early dry season?)	Both
	S. innocua Del.	Fruit pulp, raw (seed toxic), ripen into the early dry season	H
	S. madagascariensis Poir.	Fruit pulp, raw and seed (questionably toxic?), ripen in the dry season on into the early rainy season	H
	S. pungens Solered.	Fruit pulp and seed, raw	G
	S. spinosa Lam.	Fruit pulp, raw (ripen in the dry season); leaf (questionably toxic)	H

Appendix to Ch. 11 *Continued*

Plant name		Edible parts	Grid cell
Malvaceae	*Azanza garckeana* (F. Hoffm.) Exell & Hillcoat	Fruit pulp, raw, green, mature the outer covering is furry, hard and woody but when chewed like gum the flesh is glutinous and sweet (available through the dry season)	H
Meliaceae	*Ekebergia benguelensis* C. DC.	Fruit, raw	H
	E. capensis Sparrm.	Fruit, raw (may ripen into the early dry season?)	H
	Turraea nilotica Kotschy & Peyr.	Fruit (thinly woody capsule whose black shiny seeds are partially covered by a red aril), available in latest dry season and early rainy season	H
Moraceae	*Ficus abutilifolia* (Miq.) Miq.	Fruit, raw	G
	F. cordata Thunb.	Fruit (ripen from the late dry season through the rainy season into the earliest dry season)	Both
	F. ingens (Miq.) Miq.	Fruit (ripen from the mid dry season into the mid rainy season)	H
	F. stuhlmannii Warb.	Fruit (ripen from the mid rainy season through the dry season); leaf	H
	F. sur Forssk.	Fruit (ripen in the latest dry season on into the rainy season); leaf/shoot; root (young aerial roots); bark (inner bark chewed for thirst)	H

	Ficus sycomorus L.	Fruit, raw (ripen in the late dry season on into the early rainy season; some at almost any time of year); leaf, young	H
	F. thonningii Blume	Fruit, raw (available almost anytime of year?); leaf	H
Myricaceae	*Myrica serrata* Lam.	Fruit (ripen in latest dry season/earliest rainy season)	H
Myrtaceae	*Syzygium cordatum* Krauss	Fruit, raw	H
	S. guineense (Willd.) DC.	Fruit, raw (ripen in the latter part of the rainy season just into the earliest dry season)	H
Ochnaceae	*Ochna pulchra* Hook.	Fruit, raw (pulp contains some fat), mature latest dry season, early rainy season	Both
Olacaceae	*Ximenia americana* L.	Fruit pulp, raw	G
	X. caffra Sond.	Fruit, raw (available mid rainy season)	Both
Oleaceae	*Olea europaea* L. subsp. *africana* (Mill.) P. S. Green	Fruit (ripen in the latest rainy season through the mid dry season); leaf (astringent); flower (rainy season)	H
Polygalaceae	*Securidaca longipedunculata* Fresen.	Leaf, young (flush in the latter part of the dry season)	Both
Proteaceae	*Protea gaguedi* Gmel.	Bark (infusion mixed with soup as an appetizer)	H
Rhamnaceae	*Berchemia discolor* (Klotzsch) Hemsl.	Fruit, raw (ripen into the earliest dry season)	H
	Ziziphus mucronata Willd.	Fruit, raw (ripen in the early dry season; can remain on the tree through the dry season)	Both

Plant name		Edible parts	Grid cell
Rubiaceae	*Canthium lactescens* Hiern	Fruit (ripen in the early dry season or later?)	H
	C. mundianum Cham. & Schlecht.	Fruit (can remain on the tree into the dry season)	G
	Gardenia ternifolia Schumach. & Thonn.	Fruit (ripen into the early dry season?)	H
	G. volkensii K. Schum.	Fruit pulp, needs further checking (can ripen and remain on the branch into the dry season)	G
	Tapiphyllum parvifolium (Sond.) Robyns	Fruit, raw	G
	Vangueria apiculata K. Schum.	Fruit (ripen into earliest dry season?)	H
	V. infausta Burch.	Fruit, raw; leaf	Both
	Vangueriopsis lanciflora (Hiern) Robyns	Fruit, raw (may ripen as early as the latest dry season?)	H
Rutaceae	*Zanthoxylum chalybeum* Engl.	Leaf ('fresh') flush with the rains, apparently not before	H
Sapindaceae	*Allophylus africanus* Beauv.	Fruit; leaf	H
	Dodonaea angustifolia L.f.	Seed (ripen in the dry season?)	H
	Haplocoelum foliolosum (Hiern) Bullock	Fruit, raw	H
	Pappea capensis Eckl. & Zeyh.	Fruit, raw (may last into early dry season); leaf	Both
	Zanha africana (Radlk.) Exell	Fruit, raw	H
	Z. golungensis Hiern	Fruit, raw	H

Sapotaceae	*Bequaertiodendron magalismontanum* (Sond.) Heine & J. H. Hemsl.	Fruit, raw (ripen in the rainy season)	Both
	Manilkara mochisia (Bak.) Dubard	Fruit, raw	H[b]
	Mimusops zeyheri Sond.	Fruit, raw (ripen in the dry season)	Both
Simaroubaceae	*Kirkia acuminata* Oliv.	Root	H
Sterculiaceae	*Dombeya rotundifolia* (Hochst.) Planch.	Fruit (needs further checking, the fruit is a capsule), mature latest dry season/early rainy season; stem	Both
	Sterculia africana (Lour.) Fiori	Seed (ripen into the early dry season?)	H
Tiliaceae	*Grewia bicolor* Juss.	Fruit, raw (ripen into the early dry season, can remain on the branch through the dry season)	Both
	G. flava DC.	Fruit, raw (remain on the branch into the dry season); leaf; gum	G
	G. flavescens Juss.	Fruit, raw (stay on the branch through the dry season)	Both
	G. micrantha Boj.	Fruit (available into the early dry season)	G[c]
	G. monticola Sond.	Fruit (ripen into the early dry season and can remain on the branch well into the dry season)	Both
	G. pachycalyx K. Schum.	Fruit, raw (ripen into the mid dry season?)	H[b]
	G. praecox K. Schum.	Fruit, raw	H[b]
	G. retinervis Burret	Fruit, raw (may ripen into the early dry season, especially with late rains, and remain on the branch well into the dry season)	G

Appendix to Ch. 11 *Continued*

Plant name		Edible parts	Grid cell
Ulmaceae	*Trema orientalis* (L.) Blume	Leaf	H
Urticaceae	*Pouzolzia mixta* Solms-Laub.	Leaf	Both
Verbenaceae	*Vitex payos* (Lour.) Merr.	Fruit, raw (ripen in late rainy season through the early dry season)	H
Vitaceae	*Rhoicissus revoilii* Planch.	Stem juice (used in times of water shortage)	H
	R. tridentata (L.f.) Wild & Drummond	Fruit, raw (dry season)	Both

[a] Notes on the edibility of these plant species are drawn from various sources throughout sub-Saharan Africa, with annotations on the edible parts and their phenology added for purposes of this analysis (see text). The identification and nomenclature of the species follow that in Peters, C. R., O'Brian, E. M. and Drummond, R. B. (1992). *Edible Wild Plants of Subsaharan Africa: an Annotated Checklist, Emphasizing the Woodland and Savanna Floras of Eastern and Southern Africa, Including the Plants Utilized for Food by Chimpanzees and Baboons.* Kew, UK: Royal Botanical Gardens.

[b] Not common (questionable occurrence; possible azonal occurrence).

[c] Mapped as a very distant outlier in Coates Palgrave (1983); unlikely to really be there (R. B. Drummond, 1992 personal communication).

G, Gaborone grid cell; H, Harare grid cell.

Part III

Form

12

The form of selected regions of the gastro-intestinal tract

GÖRAN BJÖRNHAG and PETER LANGER

In this section of five chapters, the authors present information on the morphology and functional anatomy of the mammalian gastro-intestinal tract.

It is the purpose of the digestive tract to:

1. Reduce food particle size mechanically
2. Degrade macromolecules of the nutrient chemically
3. Absorb food constituents as well as endogenous matter through specialised structures in certain regions of the tract's wall into the blood and lymph circulatory system
4. Excrete indigestible components of the food as well as some of the endogenous matter
5. Protect the internal milieu of the body against disturbances from the external environment.

Although there is considerable morphological as well as functional diversity in the different parts of the digestive tract, a general sub-division according to functional needs can be found in practically all mammals.

In the **oral cavity** food is transferred from the outside world into the digestive tract. In most animal classes, dentition (if present) serves for catching and holding the food but does not necessarily reduce particle size to any considerable degree. However, in many Mammalia, especially herbivores, particulate food is reduced in size by dentition as described in Ch. 13 by Lucas. In some mammals, such as in humans, enzymatic digestion of starch starts in the oral cavity.

The **pharynx** and the **oesophagus** rapidly transport the swallowed bolus into the **stomach**, which in most mammals is the beginning of the region of the digestive tract where enzymatic digestion takes place. Digesta are retained in this organ for very different periods of time. The digestive enzymes are generally produced by the mucosal lining of the stomach and are secreted into the lumen to digest food **autoenzymatically**; however, there is also the possibility of making use of enzymes produced by microbial symbionts (**alloenzymatic** digestion). This occurs, in such cases, in the proximal part of the stomach and separated from the part producing stomach juices. This type of digestion has remarkable ontogenetic variation as described in Ch. 16 by Langer.

The **small intestine** follows the stomach and has three parts, the **duodenum**, the **jejunum** and the **ileum**, which are sites of transport as well as of retention, secretion and absorption. Most of the digestive enzymes are brought into the small intestine from the **pancreas**. The **bile** helps in the digestion of lipids by means of emulsification. Some enzymes are situated in the mucosal epithelium of the jejunum and act by splitting molecules just before absorption.

The **large intestine** follows and consists of three parts, the **caecum**, the **colon**, and the **rectum**. The first two can be highly differentiated and demonstrate a remarkable functional diversity including alloenzymatic digestion, absorption and separation mechanisms. Several animal species require a different retention time and passage pattern for constituents of different physical characteristics. The anatomical structures and their functions needed for this separation are described by Björnhag in Ch. 17.

The importance of the cross-mural transport makes it necessary to obtain a precise quantitative idea about internal surfaces in different regions of the gastro-intestinal tract. The anatomical structures responsible for internal surface enlargement vary considerably, not only in form but also in functional effect. It is therefore of practical importance to determine the internal surface of different regions of the digestive tract carefully. It is a surprising fact that both in Champaign, Illinois and Giessen, Germany, two investigators independently applied similar techniques to tackle this problem. Young Owl (Ch. 14) measured with semi-macroscopic methods the intestinal surface of the stomach and small and large intestines. With a wider range of methods and scales of magnification Snipes (Ch. 15) investigated the small and large intestines.

13

Categorisation of food items relevant to oral processing

PETER W. LUCAS

Most terrestrial vertebrates use their teeth to guide food particles into the mouth and then swallow this food quickly. However, this ingestion procedure has been elaborated in mammals such that the teeth act together with soft tissues for longer periods to fracture food particles inside the mouth prior to swallowing. This process is called chewing or mastication and involves rhythmic cycles of muscle contraction and jaw movement (Hiiemae and Crompton, 1985). Overviews of the process that are based on the understanding of neuromuscular activity and its control (Otten, 1991) cannot as yet be integrated with the fracture mechanics of foods (Jeronimidis, 1991). So only one aspect is examined in this chapter: the contact of teeth with food. Firstly, a general theoretical outline of mastication is given, much of the supporting physiological evidence being derived from studies on humans. An attempt is then made to define the direction in which an efficient dentition would adapt to changes in the food input. Efficiency can be defined in two ways, either in terms of the rate at which the process runs (there are three possible definitions of masticatory rate) or energy expenditure. The predictions made here require much more study of the physical properties of the foods of mammals. This is difficult to do in the field. Most of the evidence discussed here relates to the diets of anthropoid primates because they are close relatives of the human and have been well-studied. Since these primates are largely herbivorous, only plant foods are treated in any detail.

Why chew?

Three reasons are offered for chewing food. Firstly, the fracture of food particles exposes new surfaces which increases the rate at which digestive enzymes and cellulytic symbionts can act lower down the gut. Mammals need a faster rate of digestion than other terrestrial vertebrates because of their higher metabolic rates. The alternative to mechanical breakdown would be a great increase in gut length. However, the cost of mastication is probably lower than the cost of carrying around very large quantities of food. This is the only generally acceptable reason to evolve a chewing mechanism, being independent of the chemical structure and initial size of any food object. The relevant masticatory rate is the rate of production of new food surface area.

Fracture of food particles by chewing usually involves fragmentation into separate pieces, the characteristic particles produced at the end of mastication being in the low millimetre range (Rensberger, 1973; Lillford, 1991). It is possible that a second reason for chewing in some mammals is in order to make particles small enough to pass easily down the pharynx and larynx. This is not a very credible evolutionary argument because mammals evolved from reptiles, which can often swallow very large particles. However, it may apply to carnivorous mammals that eat rapidly digestible tissue but also can prey on animals larger than themselves. The rate to consider here is the rate of particle size reduction.

Thirdly, mammals may also chew to remove a chemically-impervious outer coat from a food object. This is relevant to the evolution of herbivores and applies at various levels, e.g. to plant cell walls (Laca and Demment, 1991) or to seed coats. If the objective is to open these casings prior to swallowing the contents, then the efficiency is simply the number of chews per unit volume of food necessary to do this.

Only the last of these three reasons gives a well-defined, if idealised, end-point for the process. Otherwise, there must be a trade-off between the time spent on chewing or on feeding (Alexander, 1991 and Ch. 4; Laca and Demment, 1991). There have been many studies of the sensory cues that might be employed by humans to trigger swallowing and, in general, these do not favour food particle size. Particle size is important, particularly for harder foods, but it is probably the largest particles that influence swallowing and not the average or most common particles which are, however, more usually measured (Lucas and Luke, 1986). Hutchings and Lillford (1988) describe what they call breakdown pathways for different foods during mastication in humans. They envisage two thresholds, one which relates to food particle size and the other which refers to the lubrication of particle surfaces. Some foods need to be reduced in particle size to be swallowed; others simply need to be lubricated by saliva (fracture being incidental). Yet other foods are fractured merely to liberate juice which serves to wet them. Towards the end of mastication, the particles are often coagulated into a bolus just before swallowing. However, it should be pointed out that the process of swallowing in man is different from that in many other mammals where the airway can be separated from the foodway (Hiiemae and Crompton, 1985).

Description of the process of mastication

Mastication is described most easily as a particle size reduction (or comminution) process. A mammal chewing on a homogeneous food, presented as a group of particles of similar size, would produce particle size distributions such as those shown in Fig. 13.1*a*. The skew of these distributions changes markedly in the first few chews but, thereafter, develops a long tail pointing towards larger particle sizes. Rensberger (1973) shows such a particle size distribution produced by a horse chewing grass. Distributions are usually obtained from sieving and examined in a cumulative form. Two methods of characterising these distributions have been employed: the Rosin–

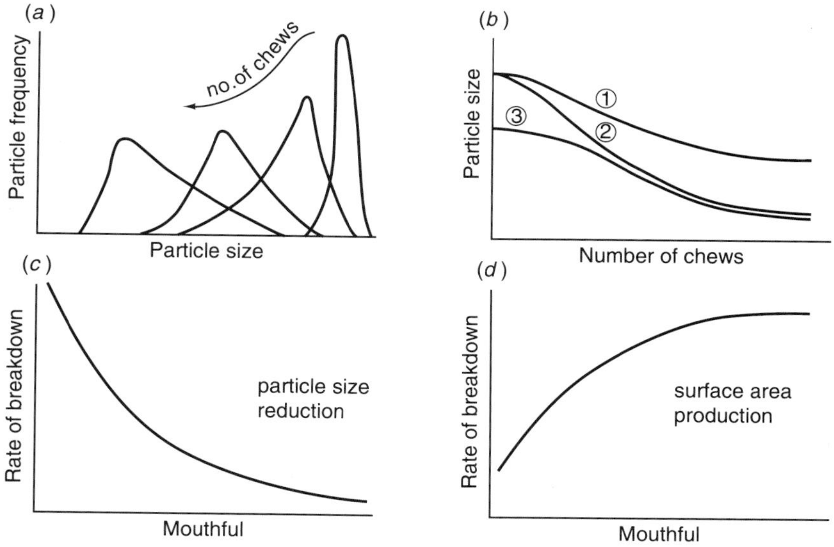

Fig. 13.1. Comminuted output for a hypothetical homogeneous food produced by mammalian chewing (based on data on humans). (*a*) Trends in particle size distributions with increased chewing; (*b*) reduction in median particle size as a function of initial particle size (curves 2 and 3) or the mouthful of food (curves 1 and 2). When the initial particle size is greater for the same total volume (curve 2 versus 3), the initial rate of breakdown of particles differs but converges with increased chewing. The rate of particle size reduction is slowed with a larger total volume in the mouth (curve 1 versus 2). (*c*) and (*d*) The rate of total surface area production as a function of the mouthful size.

Rammler equation (Olthoff *et al.*, 1984) and a normal curve fitted to square-root transformed particle size data (Voon *et al.*, 1986). However, it is agreed that the most appropriate measure of central tendency when the skew of a distribution is so variable is the median particle size. Several studies have shown that the breakdown curves of median particle sizes are exponential after a few chews (Sheine and Kay, 1982; Lucas and Luke, 1983; Olthoff *et al.*, 1984).

The production of these distributions can be understood, at least partially, in terms of two analytically distinct variables. One is the probability of a food particle being fractured and the other is the distribution of fragment sizes produced by that fracture (Epstein, 1947). These comminution functions are very difficult to measure but provide a simple analytical framework by which the activity of the mouth can be understood. Each may depend on various factors: particle size, the number and distribution of particles of other sizes and on how far chewing has progressed. Of these, it is important for this paper that the chance of a particle being fractured has been found to depend strongly on particle size raised to a power of between 1.3–2.6 (Lucas and Luke, 1983; van der Bilt *et al.*, 1987; van der Glas *et al.*, 1987, 1992).

An increase in the total volume of food particles present in the mouth appears to reduce the probability of fracture of all particles in the same proportion (Lucas and Luke, 1984a). The distribution of particle sizes present at any particular moment in chewing does not influence the chance of fracture of any particle size under consideration. Van der Glas *et al.* (1992) explain this by a model in which particles 'compete' to occupy sites on the dentition where they are broken. It is also clear that, for the test foods used, the distribution of fragment size of individual particles follows a distribution curve of a class originally derived by Gaudin and Meloy (1962) for the single fracture of a particle into a random number of fragments. Using these results, the process has been modelled using computers, the unit step of action of the process being taken to be the chewing cycle. Two published studies (Voon *et al.*, 1986, van der Glas *et al.*, 1992) have obtained results similar to those shown in Fig. 13.1*b* where the three curves represent the breakdown of median particle sizes under different conditions. The total volume of particles in the mouth is larger in curve 1 than in curve 2 and this slows the rate of particle size reduction. Curves 2 and 3 show the breakdown of particles of different initial size but with the same total volume. These curves converge.

The total volume of particles in the mouth is better thought of as the proportion of the volume of intra-oral space that is occupied by food and can be termed the mouthful (Lucas and Luke, 1984a). The rate of particle size reduction declines in an exponential manner with increase in the mouthful

(Fig. 13.1*c*). However, assuming a constant particle shape, the rate of surface area production increases with small mouthfuls and then either peaks or levels off (Fig. 13.1*d*). Since a 100% filling of the oral cavity cannot be chewed, there must be an optimum mouthful. Though this optimum is difficult to establish experimentally, it is likely to be only a small fraction (say 10–20%) of the capacity of the mouth. To understand this, it is important that food particles are treated by the mouth in a statistical fashion. The tongue and cheeks attempt to move particles towards the teeth so that they can be fractured but they cannot impose this inflexibly. The teeth themselves occupy only a small proportion of the oral volume and cannot accommodate many particles in any given chew. It is remarkable, then, that bimodal distributions are not reported from masticatory studies. The explanation must be that the teeth are not designed to retain particles on their working surfaces for more than one or two chews. Instead, fragments slip from these surfaces to be recirculated by the tongue so as to promote an even breakdown of particles (Lucas and Luke, 1984b).

Characterisation of the food input

Chewing is largely a mechanical process and so both the input and output must be described in physical terms. The physical properties of foods which affect the rate at which oral activities can proceed can be separated into two sets. The first set defines the properties of the external surface of the food at any point (**external physical characteristics**) and consists largely of geometrical properties such as food particle shape and size and the mouthful. The stickiness of particles (to each other and to the teeth) is also important. The abrasiveness of a food acts over time to reduce the probability of fracture and can also be considered a surface property in this context. The second set describes properties that act variably to resist the formation of new surface under load. These are **internal mechanical properties** and consist of the elastic modulus, strength and fracture toughness. The last is understood by some materials scientists to refer to the critical stress intensity factors (Ashby, 1989) and by others to the critical strain energy release rate and/or the work of fracture (Atkins and Mai, 1985). I use it here to mean the latter, which is measured as the work needed to create unit area of new surface by cracking (units of kJ m^{-2}). This description of toughness is convenient because its definition in energetic terms is applicable to major ecological theories such as optimal foraging (Choong *et al.*, 1992) and also allows wide comparisons between solid materials of any construction.

A wide range of methods for obtaining the toughness of plant tissues is given by Vincent (1990). One aspect of testing for toughness requires comment, which is that the properties are derived from force–displacement curves. In mechanical tests, one of these variables (force or displacement) is controlled and the other quantity allowed to vary. Toughness is awkward to measure unless displacement is controlled (which is difficult to achieve in field tests). Under load-control, there can be large energy losses which are unrelated to fracture and lead to an overestimate of toughness (Vincent, 1990). However, mammals probably use 'load-control' and so toughness values multiplied by the fractured area of food do not represent the cost of chewing but rather the minimum cost required.

Strength, which is the stress (force per unit area) at which a solid either yields or fractures, provides resistance to the initiation of cracks. Yield strength is thought to be a fundamental property. However, fracture strength, particularly compressive strength, is a function of particle size (Atkins and Mai, 1985). Toughness provides resistance to cracks propagating through the structure. In a homogeneous solid, even very small cracks, notches or flaws markedly reduce strength. In many composite materials, strength is not impaired by such cracks. The area necessary to support a given load is reduced but the strength of the remaining material is the same. Such materials are called notch-insensitive.

Adaptation of the dentition

External physical characteristics

The external physical characteristics of foods, being surface characteristics, affect the probability of food particle fracture. Since tooth surfaces act on food surfaces, the most logical adaptation of the dentition to a low probability of fracture would be change in the proportion of the surface area of the oral cavity, that is the working surface of the post-canine dentition, i.e. change in tooth size. An increase in post-canine tooth size increases the rate of processing.

Particle size

There is an upper size limit to the entry of particles into the oral cavity that is governed by the size of the mouth slit. This is, in turn, restricted by the size of the post-canine dentition because that has to be covered by the cheeks in order for particles not to be lost out of the mouth during chewing. Large mouth slits are found in mammals with large canine teeth (Herring, 1975) and which tend to have mesio-distally long tooth rows (Lucas *et al.*, 1986a). The incisors can regulate initial food particle size, as for example when a felid eats a prey item which is larger than itself. However, inside the mouth, it might seem that the smaller the initial particles, the less likely that they would be fractured and the slower the rate at which the process would run. However, smaller particles are closer to or below a size threshold for swallowing and particle size reduction curves from particles of different sizes also converge provided that the mouthful is constant (Fig. 13.1*b*).

Particle size only seems to be an important obstacle to the rate of mastication if the purpose of chewing is for chemical access and several particles are ingested. Then, the rate of reduction of average particle sizes is irrelevant; swallowing must be delayed until all particles are broken. The smaller the particles are, the longer the delay. This delay could be cut by increasing post-canine tooth size.

The above prediction would not apply if the central nervous system had sufficient sensory feedback to be able to get all particles directed towards the teeth rapidly. However, this is difficult with small particles. There is an absolute limit for stereological perception involving the soft tissues, of the order of 1–2 mm in humans (Ringel and Ewanowski, 1965). Though there are many oral and circum-oral receptors in humans (Heath and Lucas, 1988), there is evidence that these do not operate in an ideal manner. Owall and Vorwerk (1974) found that the discrimination of interocclusal distances (i.e. food particle size) by human subjects was over an order of magnitude poorer during normal chewing than during a static bite.

Mouthful size

The volume of food in the mouth has a marked effect on the rate of processing. Though rates of particle size reduction decline with an increase in the mouthful (Fig. 13.1*c*), the rate of surface area production is higher (Fig. 13.1*d*). Therefore, if small volumes of food are ingested regularly for mastication, an increase in tooth size would increase the rate of food processing.

Particle shape

Forces act over surfaces and so the presentation of food surface to the teeth is important. This depends on particle shape. Three extremes could be envisaged: a sheet, a rod and a sphere. The smallest dimension of each of these objects will have the greatest effect on the amount of new surface produced. Therefore, the initial advantage that a sheet material like a leaf, for example, might present by virtue of its large surface area is lost in comparison with a spherical seed, the fracture of which would expose a much greater new area. There are no easy generalisations. If particle size is judged in terms of particle volume, then it might be best to consider some leaves as small objects.

Stickiness

The stickiness of a particle affects whether fragments separate after fracture and also affects the formation of a bolus. The advantage of particle fragments sticking together is that a greater proportion of small particles gets broken per chew. However, this ball of fragments would probably be distributed over a very limited area of the dentition. The tongue throws this ball laterally in the opening phase of a chewing cycle (Hiiemae and Crompton, 1985). There is, therefore, an advantage in having a mesio-distally short tooth row with the bucco-lingual dimension expanded in the centre. In contrast, the fracture of non-sticky foods would benefit from a longer tooth row of more even width.

Abrasiveness

Abrasion is a mechanical process by which very small particles get fractured from much larger ones. It is responsible for wear of the dentition. It obviously depends strongly on the internal mechanical properties of a solid. We do not attempt to explain this complex process here but, instead, ask what problems abrasive foods (or abrasive particles attached to foods) pose for the dentition.

It can conveniently be assumed that there are certain sites or positions on the dentition at which food particles can be broken (Lucas *et al.*, 1985, 1986b). These sites can be identified with the breakage sites of the model of van der Glas *et al.* (1992). When there is abrasion, every fracture of a particle at a breakage site incurs wear and eventually that site is lost. By increasing the number of such breakage sites, each is used less often. An increase in the area of the working surface of the teeth (i.e. tooth size) could, therefore, be an adaptation to counteract abrasion.

The teeth also wear against each other during chewing because protective mechanisms that switch off muscle activity only operate after tooth–tooth contact (Otten, 1991). This type of wear cannot be reduced by increasing tooth surface area because the same parts of the teeth always bite against each other. The only adaptation that makes sense is to thicken the enamel, because this tissue is wear-resistant.

Table 13.1. *Predicted associations between different external physical characteristics and post-canine tooth size*

Food characteristic	Small teeth	Large teeth
Initial particle size	Large	Small (if chemically sealed)
Mouthful	Large	Small
Abrasiveness	Low	High
Stickiness	Yes	No
Bolus formation	Yes	No

The combination of external physical characteristics that favours large or small post-canine teeth is summarised in Table 13.1.

Internal mechanical properties

Internal mechanical properties are, by definition, not exposed to the surface. Rather, these properties define the resistance of any food particle to the formation of new surfaces. However, unlike surface properties, mechanical properties act together in resisting fracture and cannot be considered individually. This aspect of the food input is largely adapted for by changes in tooth shape, which are designed to reduce the energetic cost of chewing. Energy is expended in three ways: friction, flow and fracture.

Friction can work to the advantage of a mammal, such as when the anterior teeth act to grip a food item so as to resist an external tensile or torsional force, but inside the mouth, no such grip is available. When the upper and lower post-canine teeth are pressed onto a food object, friction between the working surface of the teeth and the food may be disadvantageous in that it may increase the mean compressive stress necessary to fracture the food item; it will certainly do so if there is any indentation (Tabor, 1985). The smooth surfaces of the teeth and the lubricating effect of saliva are important in reducing friction.

Flow refers to permanent changes in the dimensions of a food object, either by plastic flow, densification (e.g. collapse of cells or loss of air spaces between cells) or localised slippage at weak structural junctions. Such yielding is inevitable in most tougher foods when they are chewed and is usually the cause of toughness. Flow can be much more expensive than the cost of the surface exposed by fracture.

Fracture should cost very little (the chemical surface energies of solids are very low) but several mechanisms act in biological solids to inflate this cost. These include the pull-out of fibres (e.g. the bundles of sclerenchyma in some leaves pulling out of surrounding tissues) and the elastic blunting of cracks (Vincent, 1983), such as seen in vertebrate skin (Atkins and Mai, 1985). One mechanism, that of crack-stopping (Cook and Gordon, 1964), does not necessarily increase cost but frustrates fracture direction and therefore fragmentation. The following is a suggested logical derivation of the shapes of the working parts of teeth (written in evolutionary shorthand).

Consider a flat compressive tooth surface, of higher modulus than any food, pressed onto a solid food particle supported from below by another flat tooth surface. This arrangement imposes far more strain energy on the particle than is necessary simply to obtain fracture. The opposite extreme, that of localised loading by a pointed cusp, will fracture the food much more cheaply but will fragment it only if a crack spreads out across the solid. A crack can only do this if the food particle is rigid (high elastic modulus) and brittle (low toughness). These solids are shatterable and include bone, seed shells and beetle cuticles. They are generally strong and, therefore, the pointed cusp is best designed blunt to avoid its fracture (Preuschoft, 1989). There is little energetic sacrifice in blunting the cusp because of the small area of loading between rigid solids. In addition, the stresses that produce cracks spreading from the indented area may actually be lower with blunt than with sharp cusps. The lower supporting surface can be made concave to support the fragments formed, giving the blunt cusp a higher probability of hitting fragments and reducing them further in size. This arrangement is the essence of a pestle and mortar.

With high modulus foods, the attainment of the fracture stress is important. In contrast, the fracture of low modulus foods depends on producing a displacement adequate to provide a required strain. Pliant or floppy food particles (of low modulus and/or with one particle dimension very small) may be fractured by a pestle and mortar but are not likely to be fragmented because critical strains cannot be reached. This is exacerbated by any great toughness in the food and also by any significant heterogeneity in the food (at the structural level of about 1 mm or more), which will tend to stop the crack. To produce fragmentation, the pointed cusp must be extended in one direction to form a ridge (or wedge), the result being sub-division of the particle into two. The sharpness of this ridge is critical to reduce the strain energy imparted to the food and to keep the loading area small; if these are increased, then toughening and crack-stopping mechanisms will tend to inhibit fragmentation. Because the tip of the ridge stays close to the crack, it must completely traverse the particle and end up contacting the lower supporting surface. The resulting wear could be avoided by designing the lower surface as another (inverted) ridge of identical direction and orienting it so as just to miss its opponent at the end of fragmentation. This is the essence of all double-bladed devices such as guillotines and pairs of scissors, a joint being essential to preserve the alignment of the blades.

Features such as pestles and mortars and pairs of blades can be found on post-canine dentitions arranged in arrays. In most bladed dentitions, such as tribosphenic molars (Crompton and Sita-Lumsden, 1970) and those of large herbivores (Fortelius, 1985), the blades are curved so that there is point contact. Usually, the ends of the blades meet before the centres (Savage, 1977; Fortelius, 1985). This traps pliant or thin (easily bent) food particles that need large deformations before failure and that also may have high Poisson's ratios.

The combination of internal mechanical properties that is predicted to favour particular tooth shapes is summarised in Table 13.2.

Teeth and diet

Most biological foods have a cellular composition that often, in addition, contains fibres. The understanding of such composite materials (which is how tough biological tissues may be viewed) is advancing (Gibson and Ashby, 1988; Jeronimidis, 1991) but beyond the scope of this chapter. However, if this heterogeneity is on a large enough scale, then different teeth could be used to process different parts of the structure. For example, mammals themselves consist of soft tissues of very low modulus supported by bones of high modulus. Carnivores that fragment both types of tissues, like some canids and hyaenids, use different teeth for different tissues. The outermost layer of the prey, the skin, represents the first major hurdle for these mammals. Fractures

Table 13.2. *Predicted association of particular combinations of internal mechanical properties with tooth shapes*

Food property	Dental adaptation
High elastic modulus Homogeneous structure Low toughness	Blunt cusps
Low elastic modulus Heterogeneous structure High toughness	Sharp blades

made by pointed incisors and canines do not extend through the tissue and they act primarily to slow the prey down. Mammalian skin has a very low elastic modulus but a very high toughness. Rat skin has been tested and possesses a toughness greater than 10 kJ m^{-2} that is dependent, apparently, on collagen fibre direction and position of the body (Purslow, 1983). Though some of the toughness is explained by the pulling out of fibres from the ground substance of the dermis, a substantial portion is due to the crack blunting that results when highly extensible materials are deformed. Carnivores have, fairly uniformly, circumvented this problem by developing upper and lower pairs of relatively sharp-bladed post-canine teeth, such as the carnassials of felids. Carnassials are also appropriate for fracturing skeletal muscle tissue which is low in modulus and has a toughness that is highly influenced by the distribution of connective tissue within it (Purslow, 1991). Hyaenids fracture bones with the blunt premolars in front of the carnassials whereas canids use the molars behind the carnassials.

A brief discussion of some plant food types follows with indications of how certain dentitions may or may not accord with the theoretical approach advocated here.

Fruits

A simple classification of fruits distinguishes between those that have a flesh and those that are dry.

Fleshy fruits

There is not enough known of the properties of the flesh itself to characterise it. However, the outer covering of these fruits is important. They may possess a thin skin, which separates with difficulty from the flesh, or a thick layer that peels easily from the flesh and is, therefore, best called a peel. Only fleshy fruits are intended for consumption by vertebrates. In the South American rain-forest, Janson (1983) found that skinned fruits were consumed mainly by birds whereas anthropoid primates appeared to concentrate on peely fruits.

It is quite possible that the spatulate incisor, distinctive of anthropoid primates, evolved in the common ancestor for opening peely fruit (Lucas, 1989a). Evidence from South America (Janson, 1983) and South-east Asia (Leighton, 1993) supports this as do evolutionary arguments (Collinson and Hooker, 1991), but evidence from West Africa does not (Gautier-Hion *et al.*, 1985). Where the posterior teeth act on fruit flesh, the low blunt cusps of frugivores act to break open cells to liberate their contained juice over a wide area (Lucas and Luke, 1984b).

Dry fruits

Squirrels (e.g. in Leighton and Leighton, 1983) and some New World anthropoid primates in the sub-family Pitheciinae (Kinzey and Norconk, 1990) appear to specialise in dry fruits as do some of the Old World colobine monkeys (Davies, 1991). The object of feeding on dry fruits is to consume the seeds. In order to get at these, it may be necessary to remove a pod and also a thick shell (even if endocarp, this is included under seed shells below). Mature pod tissue can be very tough, of the order of 5–6 kJ m^{-2} (P. Lucas, unpublished data), though mammals apparently often eat podded fruits when they are unripe.

Seeds

Seeds are not generally intended for consumption by mammals. Mechanical protection is provided by coverings that surround the embryo and its food supply. These coverings can be derived from the seed coats or the fruit endocarp (regardless, the whole object is called a seed here). Seed shape is very variable, partly linked to the mode of dispersal from the parent plant, but mostly unexplained.

Seeds can be broadly classified into small or large. This dichotomy is supported by the bimodal distribution of seed sizes in a south-east Asian tropical rain-forest (Corlett and Lucas, 1990) with peaks at about 1 mm and 6–8 mm in terms of maximum seed width (a measurement presumed relevant to swallowing). Large seeds are of two types: (a) those with thick shells which protect them mechanically and (b) those with thin outer coverings which are thought to be laden with protective secondary compounds.

Small seeds

Small seeds rarely have shells. Vincent (1990, 1991) has proposed that small seeds are protected from seed predators because very small objects are stronger than much larger ones and, below a certain size, will yield rather than crack under compressive loads (Atkins and Mai, 1985). This is probably relevant mostly to seeds in dry fruits. Those inside fleshy fruits probably usually avoid predation by mammals by not being detected in the mouth. For example, the maximum width of seeds in the diet of *Macaca fascicularis*, an Old World monkey, ranges from about 0.3 mm to more than 30 mm. *Macaca fascicularis* seems to try to spit out most of the seeds that it takes into the mouth inside fruit flesh (Corlett and Lucas, 1990) but, below a distinct threshold of about 2–3 mm in maximum seed width, it swallows seeds intact probably because it cannot detect them (see above). These seeds appear to pass through the gut undamaged. The successful gut passage of intact small seeds has also been reported in fruit bats (Phua and Corlett, 1989) and treeshrews (Emmons, 1991; Emmons *et al.*, 1991), which usually spit seeds, and in a rat *Bolomys lasiurus* (Magnusson and Sanaiotti, 1987), which normally destroys them.

Thick-shelled large seeds

Some seeds have thick fibrous shells but this is the exception rather than the rule and seems to be restricted to larger seeds with a lipid energy source (Peters, 1987). Virtually all seed shells appear to have a weakness, which can be thought of as a notch, that is probably necessary for successful germination of the seed (Lucas *et al.*, 1991a). The shells are mostly composed of fibrous sclerenchyma, the lumina of which are very small. Their density can be 90% or more of the cell wall material (Jennings and MacMillan, 1986). The density of seed shells should act in their favour to promote high toughness (Jeronimidis, 1980; Ashby *et al.*, 1985) but, in fact, shells are brittle (Vincent, 1990; Table 13.3). The orientation of fibres in the shell is very variable but is generally such that they are stiff and strong but easily split. Fractures tend to run through the middle lamella, which is (cellulose) fibre-free, and are associated with a much lower energy cost than fracture through the cell wall (Ashby *et al.*, 1985).

Table 13.3. *Properties of a thick seed shell compared with a thin covering*

Species	Thickness of covering (mm)	Elastic modulus (GPa)	Fracture strength (MPa)	Toughness ($J\ m^{-2}$)
Mezzettia parviflora (Annonaceae)[a]	3–4	7	67	2000
Milletia atropurpurea (Leguminosae)[b]	0.5	0.34	5	330

Data from: [a] Lucas *et al.* (1991a); [b] P. W. Lucas and M. F. Teaford (unpublished data).

There is good evidence that some mammals, such as peccaries (Kiltie, 1982; Bodmer, 1989) and some anthropoid primates (Peters, 1987), prey on this class of seed. Orang-utans have also been observed to do this (Galdikas, 1982) though this may not be a general dietary pattern (Leighton, 1993). All these mammals have low blunt-cusped molars.

Many squirrels (Payne, 1980; Leighton and Leighton, 1983; Galetti *et al.*, 1992) and other rodents (Forget, 1990) also open these seeds but appear almost entirely to use their incisors.

Thinly covered large seeds

Thin seed coverings are rarely fibrous though an outer palisade layer is often seen (Corner, 1976). Their properties are very different from shells and are qualitatively similar to fruit peels (Table 13.3).

Seeds that are swallowed whole generally pass through the mammalian digestive tract unharmed. However, Bodmer (1989) has shown that intact, thinly covered seeds can be digested by ruminants and has proposed that the origin of the ruminant stomach may have been an adaptation for digesting whole seeds. The fermenting stomach evolved in small animals before the advent of wide-spread grasslands (Langer, Ch. 2). Small forest deer eat fruit, some of which contain thinly covered seeds. Bodmer (1989) showed that these would be at least partially digested. Microbes in the fermenting stomach would help to detoxify secondary compounds. However, it is strange that these mammals do not break up the seeds in their mouths. Several species of colobine Old World monkeys, e.g. *Presbytis rubicunda* (Davies, 1991), eat seeds, some of which are so large that they must be broken down by the teeth before swallowing. Preliminary analysis of one species, *Milletia atropurpurea* (Leguminosae), has shown that the food storage tissue forming the bulk of this seed is very difficult to fragment by compression with flat plates. Cracks formed by pointed indenters also do not tend to spread. The fracture toughness is high (about 1 kJ m^{-2}, similar to that reported for *Zea mays*; Vincent, 1991) and the material yields extensively before failing. Sharp wedges are required to fragment such tissue and the ridges or blades on colobine teeth may act like this when they break such foods. They seem in experiments to be capable of imparting sufficient strain energy to start cracks that can split these seeds into two (Lucas and Teaford, 1993).

Most mammalian frugivores, such as fruit bats and gibbons, have a small dentition set in a wide arch, disposed so that much of the dentition acts at the front of the mouth (Freeman, 1988). The posterior teeth tend to be small. However, some frugivorous primates have large posterior teeth, such as

Macaca fascicularis. This primate seems to remove the flesh from individual seeds retrieved from its cheek pouches. This is a slow process, the posterior teeth are involved and the time necessary to do this is probably reduced by an increase in posterior tooth size.

Leaves

Many leaves are eaten by mammals before the leaves have matured. Very young leaves are small objects (p. 204) in the form of flat, relatively floppy, sheets. Cracks that are made by pointed indenters on such sheets do not spread to fragment the sheet. Therefore, even though many such leaves might consist only of friable parenchyma with very thin cell walls (of toughness < 0.1 kJ m^{-2}), bladed dentitions are inevitable for breaking them down.

Kursar and Coley (1991) have found that the mechanical properties of leaves change very rapidly early on in development. However, they employed a punch-and-die measuring device which unfortunately can confuse tissue toughness with lamina thickness (Choong *et al.*, 1992). Toughness in the mature leaves of dicotyledons is associated with low-order leaf veins. If, during the development of a leaf, the xylem of the veins thickens and sclerenchyma develops alongside the veins, then the leaf becomes very tough and almost completely notch-insensitive (Vincent, 1982, 1983). The toughness of the veins can average 6 kJ m^{-2}, which contrasts with an average of 0·2–0·3 kJ m^2 for other leaf tissues (Lucas *et al.*, 1991b). Toughness is strongly correlated with the crude fibre content (Vincent, 1990) and with the sclerophylly index (crude fibre content ÷ 6.25 (nitrogen content); Choong *et al.*, 1992). These are both measures which have been used to examine the selection of leaves by many mammalian herbivores, which seem, in general, to favour a high nitrogen but low fibre content (Cork and Foley, 1991). Since the fibre which produces toughness is tasteless, it is possible to argue that the ability to perceive toughness in the mouth would allow herbivores to forage more rapidly and efficiently rather than waiting for some digestive clue long after ingestion (Choong *et al.*, 1992).

A mammal requires ridged molars, which act like blades, to break leaves down. The extent of these ridges in anthropoid primate dentitions has been measured by Kay (1975, 1978) and Kay and Covert (1984) where it was shown that the more folivorous a primate is, the longer are the ridges on its second molar teeth. The post-canine dentition of leaf-eating primates are large (Kay, 1975, 1978), particularly mesio-distally (Lucas *et al.*, 1986b), possibly because browsing on the young leaves of dicotyledonous plants leads to the ingestion of small amounts of these small objects. The incisors of these primates tend to be small (Hylander, 1975), which may reflect their use in notching of the petiole or young (not mature) laminae prior to pulling them off the plant by movements of the head and hands.

Rod and sheet particles, such as leaves and petioles, tend, because of the low modulus of most cellular tissues, to deflect very easily (their stiffness in bending is proportional to the cube of their thickness). This makes it difficult for the tongue and cheeks to move such particles around the mouth. Herbivores appear to get round this by chewing a 'compressed' clump of such particles (Osborn and Lumsden, 1978).

Discussion

The comparative dental literature generally uses a complex terminology – grinding, crushing, cutting and shearing – to describe the effects of teeth on food (the properties of which are rarely considered). The distribution of stresses set up in a food particle by the post-canine teeth must be complex. However, the most important stresses are those at the point of crack initiation in the food particle or, once a crack has been initiated, at a short distance ahead of the crack tip. With reference to the long axis of such a crack, propagation may involve crack opening due to tensile stresses (I). Alternatively, it may involve the sliding (shearing) of the two surfaces on either side of the crack in the direction of the long axis of the crack (II) or at right-angles to it, tending to twist the material out of this plane (III). Each of these directions of crack propagation is referred to as a mode of fracture and given the Roman numeral shown in the brackets. Cracks produced in rigid brittle foods by 'pestle and mortar' type dentitions are probably largely in mode I, which is tensile. Bladed dentitions, because of their reciprocal curvature (Crompton and Sita-Lumsden, 1970) and scissors-like arrangement, may produce mode III shear failure in thin floppy structures that do not store much strain energy (e.g. young leaves).

The difficulties of measurement in biological tissues often mean a limited choice of tests. It is difficult to make any fundamental tests under field conditions but, since most physical properties change with storage, it is only through the development of field testing that sufficient data will be collected to test hypotheses such as the predictions given above. Those field tests that are made currently, e.g. Coley (1983) for leaves, Peters (1987) and Kinzey and Norconk (1990) for fruits and seeds, usually describe the results in terms of hardness and toughness. These can be confused but are very different qualities. Hardness is defined as the ability of a solid to resist being indented (usually a permanent change in shape following removal of the indenter). It is not a fundamental property but its measurement can yield much information and, importantly, can be measured under load-control.

References

Alexander, R. M. (1991). Optimization of gut structure and diet for higher vertebrate herbivores. *Philosophical Transactions of the Royal Society of London*, **B333**, 249–255.

Ashby, M. F. (1989). Overview no. 80. On the engineering properties of materials. *Acta Metallurgica*, **37**, 1273–1293.

Ashby, M. F., Easterling, K. E., Harrysson, R. & Maiti, S. K. (1985). The fracture and toughness of woods. *Proceedings of the Royal Society of London*, **A398**, 261–280.

Atkins, A. G. and Mai, Y. W. (1985). *Elastic and Plastic Fracture*. Chichester: Ellis Horwood.

Bodmer, R. E. (1989). Frugivory in Amazonian artiodactyla: evidence for the evolution of the ruminant stomach. *Journal of Zoology*, **219**, 457–467.

Choong, M. F., Lucas, P. W., Ong, J. S. Y., Pereira, B., Tan, H. T. W. & Turner, I. M. (1992). Leaf fracture toughness and sclerophyllly: their correlations and ecological implications. *New Phytologist*, **121**, 597–610.

Coley, P. D. (1983). Herbivory and defensive characteristics of tree species in a lowland tropical forest. *Ecological Monographs*, **53**, 209–233.

Collinson, M. E. & Hooker, J. J. (1991). Fossil evidence of interaction between plants and plant-eating mammals. *Philosophical Transactions of the Royal Society of London*, **B333**, 197–208.

Cook, J. & Gordon, J. E. (1964). A mechanism for the control of crack propagation in all-brittle systems. *Proceedings of the Royal Society of London*, **A282**, 508–520.

Cork, S. J. & Foley, W. J. (1991). Digestive and metabolic strategies of arboreal mammalian folivores in relation to chemical defenses in temperate and tropical forests. In *Plant Defenses against Mammalian Herbivory*, ed. R. T. Palo & C. T. Robbins, pp. 133–166. Boca Raton, FA: CRC Press.

Corlett, R. T. & Lucas, P. W. (1990). Alternative seed-handling strategies in primates: seed-spitting by long-tailed macaques (*Macaca fascicularis*). *Oecologia*, **82**, 166–171.

Corner, E. J. H. (1976). *The Seeds of Dicotyledons*, Vol. 1. Cambridge: Cambridge University Press.

Crompton, A. W. & Sita-Lumsden, A. G. S. (1970). Functional significance of the therian molar pattern. *Nature*, **227**, 197–199.

Davies, A. G. S. (1991). Seed-eating by red leaf monkeys (*Presbytis rubicunda*) in dipterocarp forests of northern Borneo. *International Journal of Primatology*, **12**, 119–145.

Emmons, L. H. (1991). Frugivory in treeshrews (*Tupaia*). *American Naturalist*, **138**, 642–649.

Emmons, L. H., Nais, J. & Biun, A. (1991). The fruit and dispersers of *Rafflesia keithii* (Rafflesiaceae). *Biotropica*, **23**, 197–199.

Epstein, B. (1947). The mathematical description of certain breakage mechanisms leading to the logarithmico-normal distribution. *Journal of the Franklin Institute*, **244**, 471–477.

Forget, P. M. (1990). Seed dispersal of *Vouacapaia americans* (Caesalpinaceae) by

caviomorph rodents in French Guiana. *Journal of Tropical Ecology*, **6**, 459–468.
Fortelius, M. (1985). Ungulate cheek teeth: developmental, functional and evolutionary interrelations. *Acta Zoologica Fennica*, **180**, 1–76.
Freeman, P. W. (1988). Frugivorous and animalivorous bats (Microchiroptera): dental and cranial adaptations. *Biological Journal of the Linnaean Society*, **33**, 249–272.
Galdikas, B. (1982). Orang-utans as seed dispersers at Tanjung Puting, Central Kalimantan: implications for conservation. In *The Orang-Utan: its Biology and Conservation*, ed. L. E. M. de Boer, pp. 285–298. The Hague: W. Junk.
Galetti, M., Paschoal, M. & Pedroni, F. (1992). Predation on palm nuts (*Syagrus romanzoffiana*) by squirrels (*Sciurus ingrami*) in south-east Brazil. *Journal of Tropical Ecology*, **8**, 121–123.
Gaudin, A. M. & Meloy, T. P. (1962). Model and a comminution distribution equation for repeated fracture. *AIME Transactions*, **223**, 43–50.
Gautier-Hion, A., Duplantier, J. M., Quris, R. *et al.* (1985). Fruit characters as a basis of fruit choice and seed dispersal in a tropical forest vertebrate community. *Oecologia*, **65**, 324–347.
Gibson, L. J. & Ashby, M. F. (1988). *Cellular Solids*. Oxford: Pergamon Press.
Heath, M. R. & Lucas, P. W. (1988). Oral perception of texture. In *Food Structure – its Creation and Evaluation*, ed. J. M. V. Blanchard & J. R. Mitchell, pp. 465–481. London: Butterworths.
Herring, S. W. (1975). Adaptations for gape in the hippopotamus and its relatives. *Forma e Functio*, **8**, 85–100.
Hiiemae, K. M. & Crompton, A. W. (1985). Mastication, food transport, and swallowing. In *Functional Vertebrate Morphology*, ed. M. Hildebrand, D. M. Bramble, K. F. Liem & D. B. Wake, pp. 262–290. Harvard: Belknap Press.
Hutchings, J. B. & Lillford, P. J. (1988). The perception of food texture – the philosophy of the breakdown path. *Journal of Texture Studies*, **19**, 103–115.
Hylander, W. L. (1975). Incisor size and diet in anthropoids with special reference to the Cercopithecoidea. *Science*, **189**, 1095–1097.
Janson, C. H. (1983). Adaptation of fruit morphology to dispersal agents in a neotropical rain forest. *Science*, **219**, 187–189.
Jennings, J. S. & MacMillan, N. H. (1986). A tough nut to crack. *Journal of Materials Science* **21**, 1517–1524.
Jeronimidis, G. (1980). The fracture behaviour of wood and the relations between toughness and morphology. *Proceedings of the Royal Society of London*, **B208**, 447–460.
Jeronimidis, G. (1991) Mechanical and fracture properties of cellular and fibrous materials. In *Feeding and the Texture of Food*, ed. J. F. V. Vincent & P. J. Lillford, pp. 1–17. Cambridge: Cambridge University Press.
Kay, R. F. (1975). Functional adaptations of primate teeth. *American Journal of Physical Anthropology*, **43**, 195–216.
Kay, R. F. (1978). Molar structure and diet in extant cercopithecoids. In *Development, Function and Evolution of Teeth*, ed. P. M. Butler & K. A. Joysey, pp. 309–339. London: Academic Press.
Kay, R. F. & Covert, H. H. (1984). Anatomy and behaviour of extinct primates. In *Food Acquisition and Processing in Primates*, ed. D. J. Chivers, B. A. Wood & A. Bilsborough, pp. 467–508. New York: Plenum Press.
Kiltie, R. F. (1982). Bite force as a basis for niche differentiation between rain forest peccaries. *Biotropica*, **14**, 188–195.
Kinzey, W. G. & Norconk, M. A. (1990). Hardness as a basis of food choice in

two sympatric primates. *American Journal of Physical Anthropology*, **81**, 5–15.

Kursar, T. A. & Coley, P. D. (1991) Nitrogen content and expansion rate of young leaves of rain forest species: implications for herbivory. *Biotropica*, **23**, 141–150.

Laca, E. A. & Demment, M. W. (1991). Herbivory: the dilemma of foraging in a spatially heterogeneous food environment. In *Plant Defenses against Mammalian Herbivory*, ed. R. T. Palo & C. T. Robbins, pp. 29–44. Boca Raton, FA: CRC Press.

Leighton, M. (1993). Modelling dietary selectivity by Bornean orangutans: evidence for integration of multiple criteria in fruit selection. *International Journal of Primatology*, in press.

Leighton, M. & Leighton, D. R. (1983). Vertebrate responses to fruiting seasonality within a Bornean rain forest. In *Tropical Rain Forest: Ecology and Management*, ed. S. L. Sutton, T. C. Whitmore & A. C. Chadwick, pp. 181–196. Oxford: Blackwell.

Lillford, P. J. (1991). Texture and acceptability of human foods. In *Feeding and the Texture of Food*, ed. J. F. V. Vincent & P. J. Lillford, pp. 231–243. Cambridge: Cambridge University Press.

Lucas, P. W. (1989a). A new theory relating seed processing by primates to their relative tooth sizes. In *The Growing Scope of Human Biology, Proceedings of the Australasian Society for Human Biology 2*, ed. L. H. Schmitt, L. Freedman & N. W. Bruce, pp. 37–49. Perth, WA: Centre for Human Biology, University of Western Australia.

Lucas, P. W. (1989b). Significance of *Mezzettia leptopoda* fruits eaten by orang-utans for dental microwear analysis. *Folia Primatologica*, **52**, 185–190.

Lucas, P. W. & Luke, D. A. (1983). Methods of analysing the breakdown of food in human mastication. *Archives of Oral Biology*, **28**, 821–826.

Lucas, P. W. & Luke, D. A. (1984a). Optimum mouthful for food comminution in human mastication. *Archives of Oral Biology*, **29**, 205–210.

Lucas, P. W. & Luke, D. A. (1984b). Chewing it over: basic principles of food breakdown. In *Food Acquisition and Processing in Primates*, ed. D. J. Chivers, B. A. Wood & A. Bilsborough, pp. 283–301. New York: Plenum Press.

Lucas, P. W. & Luke, D. A. (1986). Is food particle size a criterion for the initiation of swallowing? *Journal of Oral Rehabilitation*, **13**, 127–136.

Lucas, P. W. & Teaford, M. F. (1993). Functional morphology of colobine teeth. In *Colobine Monkeys: Their Evolutionary Ecology*, ed. A. G. Davies & J. F. Oates. Cambridge: Cambridge University Press, in press.

Lucas, P. W., Corlett, R. T. & Luke, D. A. (1985). Plio-Pleistocene hominid diets: an approach combining masticatory and ecological analysis. *Journal of Human Evolution*, **14**, 187–202.

Lucas, P. W., Corlett, R. T. & Luke, D. A. (1986a). Sexual dimorphism of tooth size in anthropoids. *Human Evolution*, **1**, 1–23.

Lucas, P. W., Corlett, R. T. & Luke, D. A. (1986b). Postcanine tooth size in anthropoid primates. *Zeitschrift für Morphologie und Anthropologie*, **76**, 253–276.

Lucas, P. W., Lowrey, T. K., Pereira, B., Sarafis, V. & Kuhn, W. (1991a). The ecology of *Mezzettia leptopoda* (Hk. f. et Thoms.) Oliv. (Annonaceae) seeds as viewed from a mechanical perspective. *Functional Ecology*, **5**, 545–553.

Lucas, P. W., Choong, M. F., Tan, H. T. W., Turner, I. M. & Berrick, A. J. (1991b). The fracture toughness of the leaf of the dicotyledon *Calophyllum inophyllum* L. (Guttiferae). *Philosophical Transactions of the Royal Society of London*, **B334**, 95–106.

Magnusson, W. E. & Sanaiotti, T. M. (1987). Dispersal of *Miconia* seeds by the rat *Bolomys lasiurus*. *Journal of Tropical Ecology*, **3**, 277–278.

Olthoff, L. W., van der Bilt, A., Bosman, F. & Kleizen, H. H. (1984). Distribution of particle sizes in food comminuted by human mastication. *Archives of Oral Biology*, **29**, 899–903.

Osborn, J. W. & Lumsden, A. G. S. (1978). An alternative to 'theogosis' and a re-examination of the ways in which mammalian molars work. *Neues Jahrbuch für Paläontologie und Geologie*, **156**, 371–392.

Otten, E. (1991). The control of movements and forces during chewing. In *Feeding and the Texture of Food*, ed. J. F. V. Vincent & P. J. Lillford, pp. 123–141. Cambridge: Cambridge University Press.

Owall, B. & Vorwerk, P. (1974). Analysis of a method for testing oral tactility during chewing. *Odontologisk Revy*, **25**, 1–10.

Payne, J. B. (1980). Competitors. In *Malayan Forest Primates*, ed. D. J. Chivers, pp. 261–277. New York: Plenum Press.

Peters, C. R. (1987). Nut-like oil seeds: food for monkeys, chimpanzees, human and probably ape-men. *American Journal of Physical Anthropology*, **73**, 333–363.

Phua, P. B. & Corlett, R. T. (1989). Seed dispersal by the lesser short-nosed fruit bat (*Cynopterus brachyotis*, Pteropodidae, Megachiroptera). *Malayan Nature Journal*, **42**, 251–256.

Preuschoft, H. (1989). Biomechanical approach to the evolution of the facial skeleton of hominoid primates. *Fortschritte der Zoologie*, **35**, 421–431.

Purslow, P. P. (1983) Measurement of the fracture toughness of extensible connective tissues. *Journal of Materials Science*, **18**, 3591–3598.

Purslow, P. P. (1991). Measuring meat texture and understanding its structural basis. In *Feeding and the Texture of Food*, ed. J. F. V. Vincent & P. J. Lillford, pp. 35–56. Cambridge: Cambridge University Press.

Rensberger, J. M. (1973). An occlusion model for mastication and dental wear in herbivorous mammals. *Journal of Palaeontology*, **47**, 515–528.

Ringel, R. L. & Ewanowski, S. J. (1965). Oral perception: I. Two point discrimination. *Journal of Speech and Hearing Research*, **8**, 389–397.

Savage, R. J. G. (1977). Evolution in carnivorous mammals. *Palaeontology*, **20**, 237–271.

Sheine, W. S. & Kay, R. F. (1982). A model for comparison of masticatory effectiveness in primates. *Journal of Morphology*, **172**, 139–149.

Tabor, D. (1985). Indentation hardness and its measurement: some cautionary comments. In *Microindentation Techniques in Materials and Engineering*, ed. P. J. Blau & B. R. Lawn, pp. 129–159. ASTM Special Technical Publication 889, International Metallographic Society.

van der Bilt, A., Olthoff, L. W., van der Glas, H. W., van der Weelen, K. & Bosman, F. (1987). A mathematical description of the comminution of food during mastication in man. *Archives of Oral Biology*, **32**, 579–586.

van der Glas, H. W., van der Bilt, A., Olthoff, L. W. & Bosman, F. (1987). Measurement of selection chances and breakage functions during chewing in man. *Journal of Dental Research*, **66**, 1547–1550.

van der Glas, H. W., van der Bilt, A. & Bosman, F. (1992). A selection model to estimate the interaction between food particles and the post-canine teeth in human mastication. *Journal of Theoretical Biology*, **155**, 103–120.

Vincent, J. F. V. (1982). The mechanical design of grass. *Journal of Materials Science*, **17**, 856–860.

Vincent, J. F. V. (1983). The influence of water content on the stiffness and fracture properties of grass. *Grass and Forage Science*, **38**, 107–111.

Vincent, J. F. V. (1990). Fracture properties of plants. *Advances in Botanical Research*, **17**, 235–287.

Vincent, J. F. V. (1991). Texture of plants and fruits. In *Feeding and the Texture of Food*, ed. J. F. V. Vincent & P. J. Lillford, pp. 19–33. Cambridge: Cambridge University Press.

Voon, F. C. T., Lucas, P. W., Chew, K. L. & Luke, D. A. (1986). A simulation approach to understanding the masticatory process. *Journal of Theoretical Biology*, **119**, 251–262.

14

A direct method for measurement of gross surface area of mammalian gastro-intestinal tracts

MARCUS YOUNG OWL

In early literature, gastro-intestinal tracts (GIT) are described with subjective terms such as 'large', 'small', 'long', 'short', or 'capacious' (e.g. Böker, 1932; Hill, 1958). Comparison of the size of the GIT of a particular species requires a reliable quantitative method. Length of the small intestine, caecum and colon has been used by ornithologists interested in GIT morphology (e.g. Davis, 1961; Levin, 1963; Moss, 1972, 1974; Ankey, 1977; Pulliainen, 1981; Pulliainen *et al.*, 1981; Pulliainen and Tunkkari, 1983) and by a few researchers in mammalogy (Barry, 1977; Schieck and Millar, 1985; Woodall, 1987), but length does not reliably indicate volume and, therefore, the capacity of a GIT.

Measurements of volume have been difficult to obtain. Chivers and Hladik (1980) attempted to measure volume of GIT regions using three different techniques. Filling a gut region with water was deemed inaccurate because of the tendency for GIT tissue to expand and, thus, give an inaccurate measurement. A second method, used for the stomach only, relied on a mathematical estimation based upon the length of the greater curvature in which this length was assumed to be equal to the circumference of a sphere. The third method relied on measurements of surface area for the stomach, which was assumed to be a sphere, and measurements of lengths and widths for the small intestine and colon, which were assumed to be cylinders. In their paper, the volume data were reported by use of the latter method. Chivers and Hladik (1980) recognized the potential problems concerned with the inaccuracy of these measurements.

Surface area gives a truer picture of the size of a GIT than does length and can be measured more accurately than volume. Chivers and Hladik (1980) note that Magnan (1912) measured surface area of mammalian GIT but his data were not published and are now lost. Therefore, Hladik (1967) was the first to report surface areas of the GIT of mammals (primates).

The indirect method for determining GIT surface area employed by Hladik (1967), Charles-Dominique and Hladik (1971), Chivers and Hladik (1984), MacLarnon *et al.* (1984) and Martin *et al.* (1985) is described in Chivers and Hladik (1980). Four regions of the GIT were identified: stomach, small intestine, caecum and colon. These regions of the GIT were opened longitudinally and flattened, usually on a dissecting tray under water, and measurements of length and width were made. Surface area for the intestine and colon were estimated from lengths and a series of widths. The methods for estimating stomach and caecum surface area relied on geometric formulae and, in the case of the stomach, covering surface areas of irregular pieces with aluminium foil of a known area to mass ratio, which was then weighed to obtain surface area. Because the directions for estimating the surface area of the stomach and caecum are complicated and the accuracy of the indirect method is unknown, I developed an alternative technique for direct measurements of surface area.

The aims of this investigation are to present a more standardized method for direct measurement of GIT surface area and to duplicate as closely as possible the indirect method of Chivers and Hladik on the same material for comparison. This will allow comparison of direct measurements to the indirect method employed by Chivers and Hladik (1980).

Methods

Samples of mammalian GIT (stomach $n = 37$, small intestine $n = 46$, caecum $n = 39$, and colon $n = 46$), preserved in 15% formaldehyde, were measured using both the Chivers and Hladik indirect method (modified slightly) and my newly-developed direct technique. The taxon, specimen number and the measurements from both techniques are presented in Table 14.1.

I divided the GIT into four sections (stomach, small intestine, caecum and colon) by severing it at the following anatomical markers: oesophagus–cardiac stricture, pyloric sphincter–duodenum, ileo-caecal junction, and caecum–colon.

Whenever possible stomachs were dissected longitudinally into two halves. This was not always possible, and complicated stomachs (such as found in colobines, sloths and macropod marsupials) were cut into a number of pieces in order to flatten them. Small intestines were dissected along the line of mesenteric attachment. Caeca and colons were dissected along the taeniae (bands consisting of either longitudinal muscle or connective tissue) if they were present, resulting in two halves. As with the stomach, caeca and colons which possessed well developed haustra (recesses formed by contraction of circular muscle) were cut into a number of pieces in order to flatten them and account for the surface area contained in the haustra. The pieces from stomachs and caeca were pinned onto paper, while intestines and simple colons were splayed back and pinned onto paper (Fig. 14.1). Long small intestines were cut into shorter segments in order to fit them onto the paper for tracing their outlines. Care was taken to avoid stretching the tissue. GIT parts were kept moist throughout the procedure. Outlines of the specimens were then drawn onto paper (Fig. 14.2). An Apple graphics tablet (Figs 14.3*a*, *b*) on an Apple II computer was used to determine surface area. The outlines of the GIT were traced with the graphics tablet light pen and the computer calculated the surface area automatically. This method is similar in principle to the method used by Dawson and Hulbert (1970) who used a geographer's planimeter for estimating the surface area of skin on marsupials.

The indirect method followed Chivers and Hladik (1980) with the following modifications: (a) the lengths of small intestines and colons were divided into tenths and the widths were measured at these points, summed and averaged to obtain a mean width; (b) the instructions for obtaining initial measurements for the stomach and caecum were difficult to follow on fresh tissue under water, so their method of calculating surface area was applied to the outlines of the stomach and caecum that had been traced on paper.

To calibrate the Apple graphics tablet a series of ten measurements was taken on a 100 cm^2 (10 cm $\times$ 10 cm). The mean surface area for the ten measurements was 99.59 cm^2 with a standard error of 0.246. This variability is probably as much a reflection of the steadiness of hand when using the graphics tablet light pen as it is a reflection of the graphics tablet's accuracy.

After measuring the four GIT regions by both the direct method and the indirect method, the two techniques were subjected to statistical tests by functional regression. The great difference in the body size of the specimens used in this study required that the data be compared by linear regression. The statistical model chosen was functional regression (also referred to as geometric mean regression). Functional regression gives the preferred description of the relationship between variables when no prediction is required. Functional regression is described in Krebs (1989) and is obtained using the parameters obtained from simple linear regression,

$$\hat{v} = \frac{\hat{b}}{r}$$

where $\hat{v}$ is the estimated slope of the functional regression; $\hat{b}$ is the estimated slope of the regression of Y onto X and r is the correlation coefficient between

Table 14.1. *Comparison of direct (DM) and indirect methods (ID) for estimating mammal gastro-intestinal tract surface area*

			Surface area (cm^2)							
			Stomach		Intestine		Caecum		Colon	
Taxon	Specimen no.	Size (g)	DM	ID	DM	ID	DM	ID	DM	ID
Metatheria: Marsupialia										
Didelphidae										
Caluromys philander (woolly opossum)	UCLA 11	325	21	23	106	80	25	24	71	65
Potoridae										
Potorous tridactylis (long-nosed potoroo)	LAZ 34	85	9	8	39	37	2	2	8	7
Phascolarctidae										
Phascolarctos cinereus (koala)	LAZ 31		13	9	146	143	56	51	132	136
Eutheria: Primates										
Lorisidae										
Galago senegalensis	CIT 06	162	10	10	52	47	19	17	16	18
(Senegal bush-baby)	CIT 08	218	16	13	86	81	42	35	83	81

Callitrichidae										
Callithrix jacchus (common marmoset)	UCLA 02	210			82	76			65	60
Saguinus imperator (emperor tamarin)	LAZ 02	900	33	29	191	171	25	16	65	60
	LAZ 03	39	4	4	11	11	1	1	5	5
	LAZ 09	170	10	11	60	56	9	9	29	30
	LAZ 10	328	38	36	120	106	16	15	61	44
	LAZ 11	384	15	15	68	68	7	7	61	65
	LAZ 14	490	11	9	95	89	6	6	45	40
	LAZ 18	35			10	9	1	1	5	5
	LAZ 20	615	14	19	115	109	10	9	55	35
Saguinus oedipus (cotton-top tamarin)	LAZ 07	365	23	21	96	91	8	8	60	58
	LAZ 08	349	12	12	83	81	10	7	49	42
	LAZ 13	455	12	10	74	66	6	5	22	20
Leontideus rosalia (golden lion tamarin)	LAZ 01	430	8	7	118	110	10	9	32	40
	LAZ 06	680	22	25	196	184	8	11	59	40
	LAZ 16	57	4	3	19	17	1	1	4	4
	LAZ 17	35	3	2	19	14	3	2	4	4
Callimico goeldii (Goeldi's marmoset)	LAZ 15	45	4	3	19	18	3	2	8	4
Cebidae										
Aotus trivirgatus (owl monkey)	CIT 01	785			171	155			62	61
	CIT 02	1008			344	301			169	159
	CIT 03	565			227	210			74	65
	CIT 04	854			403	402			208	194
	CIT 05	797	90	88	386	401			101	87

Table 14.1. Continued

Taxon	Specimen no.	Size (g)	Surface area (cm^2)							
			Stomach		Intestine		Caecum		Colon	
			DM	ID	DM	ID	DM	ID	DM	ID
Ateles paniscus (black spider monkey)	LAZ 04	2200	163	160	638	699	49	39	235	131
Lagothrix lagothricha (common woolly monkey)	LAZ 12	482	27	22	110	94	12	8	48	43
Cercopithecidae										
Macaca mulatta (Rhesus macaque)	UCLA 03	13700	190	189	396	388	144	152	583	613
	UCLA 04	6000	96	93	582	528	110	106	496	513
	UCLA 05	12500	178	170	952	922	77	86	715	714
	UCLA 06	6500	152	159	827	861	78	57	581	574
	UCLA 07	15455	118	110	958	913	55	52	316	324
	UCLA 08	8280	116	135			177	192	811	852
	UCLA 09	3500			602	573				
Cercopithecus aethiops (West African green monkey)	UCLA 10	5910	251	235	581	570	93	96	507	542
	UCLA 13	5400	125	118	945	938	136	135	498	502

Mandrillus sphinx (mandrill)	LAZ 05				166	160			57	54
Colobus guereza (eastern black-and-white colobus monkey)	LAZ 36				614	625			361	347
Eutheria: Carnivora										
Canidae										
Vulpes vulpes (red fox)	LAZ 25		29	25	108	119	5	4	18	16
Nyctereutes procyonoid (raccoon-dog)	LAZ 24	3750	126	118	489	497	22	21	32	32
Otocyon megalotis (bat-eared fox)	LAZ 33	3200	87	91	245	247	22	17	63	61
Procyonidae										
Nasua nasua (coati)	UCLA 01	4500			386	429				
Felidae										
Felis manul (Pallas's cat)	LAZ 32	1589	60	52	182	175	8	6	47	45
Eutheria: Rodentia										
Sciuridae										
Sciurus griseus (western grey squirrel)	LAZ 27	520	44	39	209	210	17	14	79	51
Caviidae										
Dolichotis patagonum (mara)	LAZ 19	432	22	23	185	183	7	5	43	35

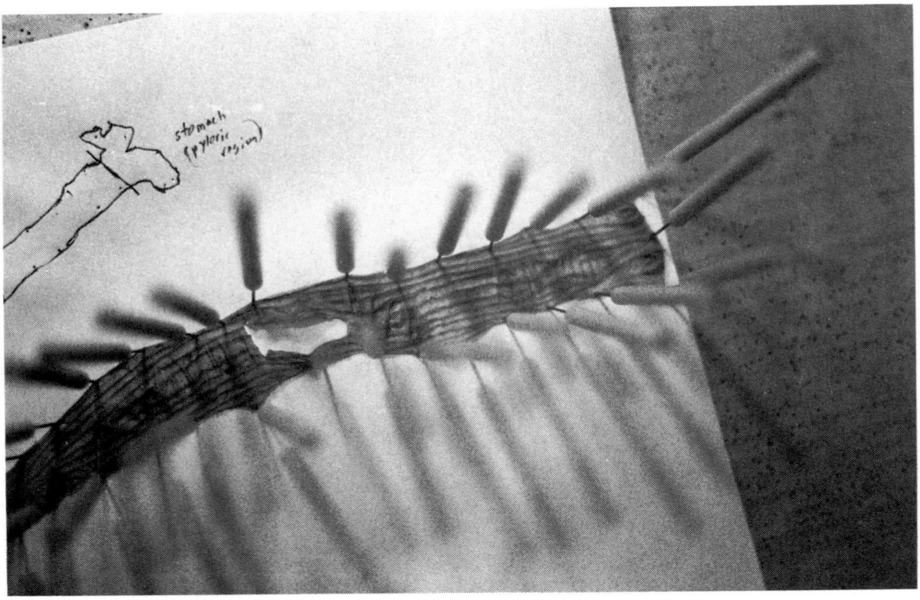

Fig. 14.1. Intestine pinned to paper prior to being outlined. The hole in the intestine is where tissue was removed for necropsy.

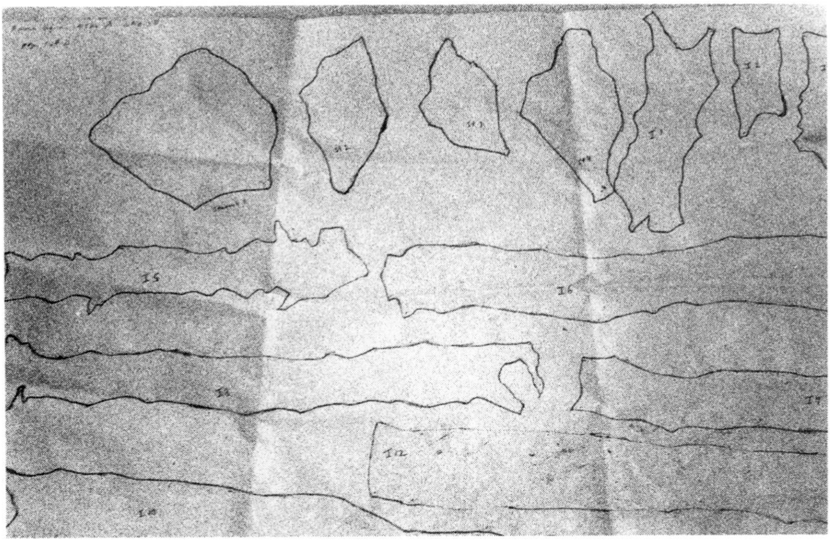

Fig. 14.2. Outlines of *Macaca nigra* gastro-intestinal tract. From the upper left and moving towards the right are the four pieces of the stomach. The remainder of the material shown is small intestine.

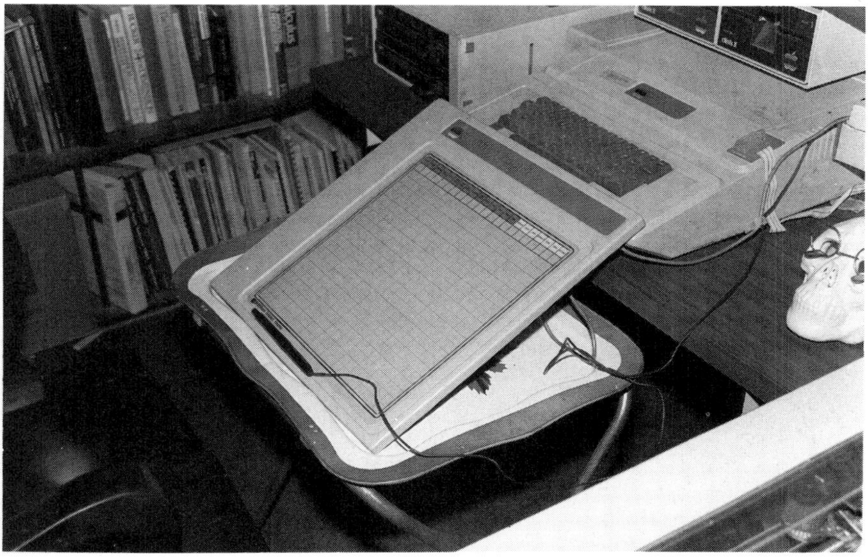

Fig. 14.3(*a*) The Apple graphics tablet (photo by Dr Robert J. Russell).

X and *Y*. The *Y*-intercept of the estimated slope was calculated by the formula, $\hat{a} = \bar{Y} - v\bar{X}$ where $\hat{a}$ is the estimated intercept of the functional regression.

In this study the measurements obtained for the indirect method were plotted onto the measurements obtained for the direct method and the estimated slope and estimated intercept of the functional regression calculated. Confidence limits (CL) for the slope and intercept were calculated following Krebs (1989). A functional regression slope of 1.0 and an intercept of 0.0 would indicate perfect agreement between the two methods.

Results

Comparison of the indirect method with the direct method by functional regression yielded slopes very close to 1.0, with only the regression coefficient of the surface area for colon differing significantly (1.034). *Y*-intercepts were very close to zero for the stomach and small intestine and marginally significant for the caecum and colon (Table 14.2). Although not significant, there was a tendency for the indirect method to underestimate GIT surface area. This underestimation is more pronounced for smaller GITs at any region.

Fig. 14.3(*b*) The gastro-intestinal tract is measured using the Apple Graphics tablet (photo by Dr. Robert J. Russell).

Table 14.2. *Surface area measurements by the direct and indirect methods. Results of functional regression: data derived from the indirect method plotted onto data derived from the direct method*

	Stomach	Intestine	Caecum	Colon
Regression coefficient (slope) ± 95% CL	0.978 ± 0.026	0.999 ± 0.022	1.041 ± 0.042	1.034 ± 0.028
Intercept ± 95% CL	−0.509 ± 2.457	−5.365 ± 8.110	−2.532 ± 2.223	−7.759 ± 7.149
Correlation coefficient (r)	0.997	0.997	0.994	0.996
P	0.0001	0.0001	0.0001	0.0001
Number of observations	37	46	39	46

Discussion

The results of functional regressions suggest that the indirect method and the direct method achieve similar results with the exception of colon surface area. However, an analysis of the regression slope indicates that there is a tendency for the indirect method to underestimate the GIT surface area of smaller GITs for any GIT region. In the case of both the caecum and colon, this underestimation becomes a slight overestimation in larger caeca (over 50 cm^2) and colons (over 230 cm^2).

These results suggest data obtained by both methods are compatible and also confirm the accuracy of the indirect method, previously unknown, for the stomach and small intestine. Compatibility between the two methods for measurement of caecum surface area is marginal, while the difference between the two methods are significant for the colon. The reason for the incompatibility of the two methods when applied to the colon may be due to the presence of semi-lunar folds between the haustra which contain surface area that is not accounted for by the formula used to obtain surface area with the indirect method.

The direct method for measurement of GIT surface area has several advantages over the indirect method. First, the direct method eliminates the uncertainty associated with calculations of geometric formulae as employed with the indirect method. Second, the procedure for measuring GIT surface area by the direct method is highly standardized and should be easy for other researchers to replicate. The major source of error is most likely to occur in differences between researchers in the tracing of GIT outlines on paper. Third, the direct method can be utilized by a single researcher, whereas the indirect method usually requires two people to hold and obtain measurements from intestines and colons. Fourth, the GIT outlines form permanent records and measurements can be taken any number of times by any number of researchers, whereas the indirect method, which obtains measurements from fresh tissue, can only be made once.

There are, however, some disadvantages to the direct method for measurement of GIT surface area. First, it is more time consuming. Depending on the size of the GIT, the direct method took 1 to 4 h longer than the indirect method. A second disadvantage is that the procedure for dissecting and pinning GIT tissue, especially the intestine, onto paper becomes difficult with very small mammals. In particular I have experienced difficulty with the small intestines and colons of voles (*Microtus* spp., body mass 30–40 g) and rice rats (*Oryzomys* spp., body mass 48–64 g). Finally, the direct method requires a computer and a digitizing graphics tablet with the capability of measuring surface area, or a geographer's planimeter. It must also be noted that this method does not consider microscopic structures such as microvilli or papillae.

Regardless of which method is used to measure GIT surface area there is a tendency for the GIT to stretch, which introduces inaccuracy into any measurement. Another complicating factor with the GIT is that it contains muscle which, when living, contracts and relaxes. There is also mounting evidence that the GIT changes size in the living mammal depending on quality of diet, season of the year, or reproductive status (Smith *et al.*, 1980; Gross *et al.*, 1985; Woodall, 1987; Sibley *et al.*, 1990; Hammond and Wunder, 1991). Thus, any measurement of the GIT, even a direct measurement, is only an estimate of the value for living tissue.

Despite these limitations, the direct method of measurement described in this paper has many applications. In addition to measuring GIT surface area, future applications may include measurement of external surface area of animals, surface area of feathers in birds, and other biological variables for which the amount of gross surface area is desired.

Acknowledgements

The author would like to thank Dr William Britt, Los Angeles County Veterinarian, and Ms Fran Woods, Los Angeles Zoo, for their assistance in procuring mammalian gastro-intestinal tracts. Helpful criticisms and comments on various versions of this manuscript were provided by Robert J. Russell, George O. Batzli, Kenneth A. Nagy, Raymond Berger, Peter Langer, Robert Snipes and Jared M. Diamond. Mammal gastro-intestinal tracts were obtained from the Los Angeles Zoo, University of California, Los Angeles, and California Institute of Technology. This research was supported by University of California Graduate Fellowships and a University of California Dissertation Fellowship.

References

Ankey, C. D. (1977). Feeding and digestive organ size in breeding Lesser Snow Geese. *Auk*, **94**, 275–282.

Barry, R. E., Jr (1977). Length and absorptive surface area apportionment of segments of the hindgut for eight species of small mammals. *Journal of Mammalogy*, **58**, 419–420.

Böker, H. (1932) Beobachtungen und Untersuchungen an Säugetieren während einer biologisch-anatomischen Forschungsreise nach Brasilien im Jahre 1928. b. Über den Magen und den Darmkanal Früchte- und Blätter-fressender Südamerikaaffen. *Gegenbaurs morphologisches Jahrbuch*, 70, 53–55.

Charles-Dominique, P. & Hladik, C. M. (1971). Le Lepilemur du sud de Madagascar: ecologie, alimentation et vie sociale. *Revue d'Ecologie (Terre et Vie)*, **1**, 3–66.
Chivers, D. J. & Hladik, C. M. (1980). Morphology of the gastrointestinal tract in primates: comparisons with other mammals in relation to diet. *Journal of Morphology*, **166**, 337–386.
Chivers, D. J. & Hladik, C. M. (1984). Diet and gut morphology in primates. In *Food Acquisition and Processing in Primates*, ed. B. A. Wood & A. Bilsborough, pp. 213–230. New York: Plenum Press.
Davis, J. (1961). Some seasonal changes in morphology of the rufous-sided towhee. *Condor*, **63**, 313–321.
Dawson, T. J. & Hulbert, A. J. (1970). Standard metabolism, body temperature, and surface areas of Australian marsupials. *American Journal of Physiology*, **218**, 1233–1238.
Gross, J. E., Wang, Z. & Wunder, B. A. (1985). Effects of food quality and energy needs: changes in gut morphology and capacity of *Microtus ochrogaster*. *Journal of Mammology*, **66**, 661–667.
Hammond, K. A. & Wunder, B. A. (1991). The role of diet quality and energy need in the nutritional ecology of a small herbivore, *Microtus ochrogaster*. *Physiological Zoology*, **64**, 541–567.
Hill, W. C. O. (1958). Some points in the enteric anatomy of the great apes. *Proceedings of the Zoological Society of London*, **119**, 19–32.
Hladik, C. M. (1967). Surface relative du tractus digestif de quelques primates, morphologie des villosites intestinales et correlations avec le alimentaire. *Mammalia*, **31**, 120–147.
Krebs, C. J. (1989). *Ecological Methodology*. New York: Harper and Row.
Levin, V. (1963). Reproduction and development of young in a population of California quail. *Condor*, **65**, 249–278.
MacLarnon, A. M., Chivers, D. J. & Martin, R. D. (1984). Gastrointestinal allometry in primates and other mammals including new species. In *Primate Ecology and Conservation*, ed. J. G. Else & P. C. Lee, pp. 75–85. Cambridge: Cambridge University Press.
Magnan, A. (1912). Le regime alimentaire et la longeur de l'intestin chez les mammiferes. *Comptes-rendus de l'Académie des Sciences*, **154**, 129–131 (cited in Chivers & Hladik, 1980).
Martin, R. D., Chivers, D. J., MacLarnon, A. M. & Hladik, C. M. (1985). Gastrointestinal allometry in primates and other mammals. In *Size and Scaling in Primate Biology*, ed. W. L. Jungers, pp. 61–89. New York: Plenum Press.
Moss, R. (1972). Effects of captivity on gut lengths in red grouse. *Journal of Wildlife Management*, **36**, 99–104.
Moss, R. (1974). Winter diets, gut lengths, and interspecific competition in Alaskan ptarmigan. *Auk*, **91**, 736–746.
Pulliainen, E. (1981). Weights of the crop contents of *Tetra urogallus, Lyrurus tetrix, Tetrates bonasia* and *Lagopus lagopus* in Finnish Lapland in autumn and winter. *Ornis Fennici*, **58**, 21–28.
Pulliainen, E., Helle, P. & Tunkkari, P. (1981). Adaptive radiation of the digestive system, heart and wings of *Turdus pilaris*, *Bombycilla garrulus*, *Sturnus vulgaris*, *Pyrrhula pyrrhula*, *Pinicola enucleater* and *Loxia pytyopsittacus*. *Ornis Fennici*, **59**, 21–28.
Pulliainen, E. & Tunkkari, P. (1983). Seasonal changes in the gut length of the willow grouse (*Lagopus lagopus*) in Finnish Lapland. *Annales Zoologici Fennici*, **20**, 53–56.

Schieck, J. O. & Millar, J. S. (1985). Alimentary tract measurements as indicators of diets of small mammals. *Mammalia*, **49**, 93–104.

Sibley, R. M., Monk, K. A., Johnson, I. K. & Trout, R. C. (1990). Seasonal variation in gut morphology in wild rabbits (*Oryctolagus cuniculus*). *Journal of Zoology* (London), **221**, 605–619.

Smith, R. L., Hubart, J. & Shoemaker, R. L. (1980). Seasonal changes in weight, cecal length, and pancreatic function of snowshoe hares. *Journal of Wildlife Management*, **44**, 719–724.

Woodall, P.F. (1987). Digestive tract dimensions and body mass of elephant shrews (Macroscelididae) and the efforts of season and habitat. *Mammalia*, **51**, 537–545.

15

Morphometric methods for determining surface enlargement at the microscopic level in the large intestine and their application

ROBERT L. SNIPES

The fact that animals react differently to experimental treatment (**biological variability**) necessitates numerical security when evaluating the **effect** of the treatment. If all animals reacted identically only one animal would be needed to measure the effect of a treatment. To state this differently, **variability** is the *noise* in a computer system, whilst the **effect** is the *signal*. Due to the inherent *noise* of the biological system, the mean signal can only be identified when the signal is made strong enough to be detected. Alternatively, by reducing the noise the effect also becomes clearer. Strengthening the signal can be attained by repetition of the experiment on different animals. Reducing the noise is achieved by use of statistical methodology. Obviously, the idea is to arrive at a favourable relationship between signal and noise.

In morphology, there is not necessarily an effect in the causal experimental sense. However, as an example, when considering the effect of different foodstuffs on an organism in an ambient environment or attempting to determine the adaptability of the intestinal tract to different dietary situations one is most certainly interested in a repetition of the signal on the test object. Obviously, more than one animal has to be observed. This statement would have been more than obvious in the early 1970s. However, in today's scientific world of government control and strict regimentation for the conservation of animals, this aspect can no longer be taken lightly. It goes without saying that the conservation of the species and rigid control of animal experimentation in those places where it is sorely needed merits not only acknowledgement but also recognition. As a consequence, sampling from an unlimited number of animals is a thing of the past.

The problem of the signal/noise relationship has to be solved by decreasing the noise component, i.e. methodologically. This problem has been approached in the past by various investigators (Weibel, 1979; Agnati and Fuxe, 1985; Snipes and Kriete, 1991) and is the subject of one of the presentations in this symposium (Young Owl, Ch. 14).

Recently the complex interplay of various external factors on the form and function of the digestive tract and their possible and plausible adaptive influence have been considered in detail (Perrin and Curtis, 1980; Chivers, 1989; Langer and Snipes, 1991). Such factors as ecology and the limitation or availability of organic matter play an enormous role in forming dietary habits and may have their morphological correlates (see Hofmann, 1989). Paleontological, socioecological, phylogenetic, ontogenetic, as well as nutritional physiological and biochemical determinants, have been discussed in the above-cited works and cannot be ignored in any coverage of this subject. All these factors are in different ways connected with the nutritional strategy that an animal employs to fulfil its vital functions, that is to say, at least to ensure an existence level and, more optimally, to ensure a reserve of energy for all necessary functions, including spurts of activity over and above the basal metabolic level. Energy is derived externally in the form of foodstuff, which must be broken down mechanically and enzymatically – either with its own or foreign enzymes (auto- and allo-enzymatically; Langer, 1987) – and then taken up by the animal in the process of absorption. The latter (together with secretion) is the primary function of the intestines. The major site of absorption in almost all animals is the small intestine. The fact that additional sites exist for absorption has excited the curiosity of many researchers in the past and today. Both stomach and large intestine are sites of volatile fatty acid uptake in many animals (Imoto and Namioka, 1978; Leng, 1978; McNeil *et al.*, 1978; Lochmiller *et al.*, 1989). In addition, especially in the large intestine, the primary and indispensable functions of water and ion homeostasis occur (Debongnie and Phillips, 1978; Vernay, 1986).

The question in the past has arisen as to how body size is scaled to metabolic rate (Kleiber, 1961; Schmidt-Nielsen, 1975). It is well known that, for thermodynamic reasons, small animals with relatively greater surface areas must have higher metabolic rates (Karasov and Diamond, 1985, 1988). An increased demand for energy to fulfil this requirement requires an increase in surface area of the absorptive surface or an increase in the efficiency of the absorptive function. The mechanisms to increase an absorptive surface include (a) elongation of a tubular gastro-intestinal (GI) tract (b) mucosal surface increase by folds and plicae (second-order enlargement) and/or villi (40 villi/mm^2 in humans) and, at the ultrastructural level, microvilli (20-fold increase; 2000–3000 microvilli/cell in humans) and (c) formation, at the molecular level, of a glycocalyx and branched filaments of glycoproteins and glycolipids (Fawcett, 1986).

A certain principle of design at all levels of organization appears to exist, namely that certain architectural devices increase the surface area without increasing the size of the organism. This principle is dramatically illustrated by the example of the GI tract. Nature strives to increase the efficiency of the organismal systems and thereby the metabolic machinery with a minimum increase in body mass (Karasov and Diamond, 1985).

What contribution can be expected by such an approach to elucidate the structure and function of the GI tract? Chivers and co-workers (Chivers and Hladik, 1980, 1984; Martin *et al.*, 1985; Stark *et al.*, 1987; Chivers, 1989) have made pioneering strides in this area by correlating morphometric parameters to body size and by grouping animals on this basis by means of multiple analysis methodology into certain dietary categories. These studies have taken into consideration the very dynamic and functionally significant measurable parameters of area and volume in relationship to body weight. To go a step further, the present analysis has coupled this morphometric methodology with an anatomical approach at the light, transmission and scanning electron-microscopic levels. Heretofore, the microscopic level has not been brought into comparative morphometric studies. The present contribution is an attempt to offer a simple methodology for obtaining morphometric data, especially for use in allometric considerations at the microscopic level. Although primarily designed for the alimentary tract it could conceivably also be used for other organs especially those with a lumen. This presentation will limit its scope to the description of a simple technique for obtaining a **surface enlargement factor** at the microscopical level. This factor should be used in combination with the area determinations obtained by methods for determining macroscopic basal or ground areas (Snipes, 1991; Snipes and Kriete, 1991; Young Owl, Ch. 14). To date, most morphometric analyses have concerned solely the macroscopically determined areas and volumes derived either by direct or indirect methods (Chivers and Hladik, 1980). Few investigations have been concerned with determining how much of the area is increased over and above this basal area, i.e. by microscopically visible structures such as villi in the small intestine (Osawa, 1990; Osawa and Woodall, 1990; Woodall and Skinner, 1993) or folds and special arrangements in the large intestine. Granted the enlargement in the small intestine due to villi is stated in almost every textbook of histology (for example, 8-fold according to Bucher and Wartenberg, 1989; also Fawcett, 1986). However, the fact that the large intestine also has a surface enlargement is rarely mentioned or noted. This has been investigated by the present author in a number of different animals, many of widely varying dietary habit (Snipes, 1978, 1979a, b, 1981, 1982a, b, 1984a, b, 1991; Snipes *et al.*, 1982, 1988, 1990; Snipes and Thiele, 1989, Snipes and Kriete, 1991) (see also Fig. 15.2).

Methods

The following account can be used for small animals without modification. Animals with more voluminous intestines will be handled separately. For a complete description of the methods for determining the ground or basal areas of intestines, see Snipes (1991), Snipes and Kriete (1991) and Young Owl (Ch. 14).

Animals

In our institute we have come to the consensus that three animals will have to suffice for all morphological and morphometric studies. For studies of different age groups or developmental stages, the number three has also been accepted. This consideration was adopted due to the situation in science today. When no experimentation in the broad sense is carried out, often obtaining permission from authorities for animal experiments can be waived. This, of course, differs in the various countries throughout the world.

Fixation

Various fixation protocols have been attempted from intracardiac perfusion to immersion. In corroboration with the results of Fenwick and Kruckenberg (1987), the intraluminal injection of fixative has been adopted as fixative procedure of choice. Unfixed material for macroscopic determination of basal areas and volume (as previously described in this symposium) is not recommended due to the tendency for tissue to spread or expand when pressure is applied. For the present consideration, the fixation of tissue as early *post-mortem* as possible is a requirement for microscopy. Fixative of choice in the field is buffered formol, whilst in the laboratory Bouin's fluid is preferred. Tissue fixed by the former method is usually postfixed in Bouin's fluid when possible. Injection of fixative should be done with ease and care should be taken not to distend the intestines.

If the intestine is rather long then the entire caecum or colon does not have to be injected, rather the predetermined regions (see below) for sampling can be fixed by ligating an appropriate segment a few centimetres in length before injecting the fixative.

Sampling

This is the most controversial point and requires some pre-knowledge of the intestines to be quantified. The ideal situation is to carry out an extensive morphological study of the intestine in a previous, separate study. Thorough knowledge of the anatomical structures, both macroscopically as well as microscopically, beforehand is the most advantageous situation. The danger in sampling an unknown large intestine is overlooking or missing some specialization of the gut wall. Granted these should or could be included in a random sampling method as purposed by most stereological investigations. However, the intestine is not as homogeneous as often thought. Good examples are the occurrence of the plicae circulares of the human small intestine and the spiral fold of the rabbit or mole rat caecum. Sampling of these structures alone would not necessarily give a true indication of their extent. Moreover, they could be completely missed in the macroscopic sampling. The present author excised and measured these structures directly for basal area as well as taking a small sample for microscopy (see p. 239). Both structures account for an enormous amount of the surface enlargement (30-fold for plicae and 6-fold for the spiral fold; Snipes and Kriete, 1991). Thus, these structures must be sampled as separate units. But many other similar structures (folds, mounds, finger-like extensions), not readily visible macroscopically, must also be accounted for. Therefore, the following pilot experiment can be helpful. To sample, it is necessary to open the intestine along its longitudinal axis after a short fixation time. This should always be done along the mesenterial side for sake of consistency. Strips of intestines across the width of the intestine are taken for light microscopic examination. In this way, any differences or specializations in the circumferential dimension can be visually detected and noted macroscopically. Strips of the length of the intestine can likewise be taken to determine any specializations in the oral-aboral direction.

Longitudinal sampling

Originally, in small animals, 20 strips of the width were taken at equidistant intervals along the colon. Pilot measurement of the areas and comparison of neighbouring samples showed that samples 1–5 could be considered as one population. As well, no major differences were generally observed between samples 5 to 10. Thus, sample 1 was always taken as the first colon specimen (C1), whereas sample 5 was considered as the second area in the length to be sampled (C2). Correspondingly, sample 10 was C3, sample 15 was C4 and sample 20 was C5. Thus, in the longitudinal dimension five equidistant sampling areas were determined as the standard sampling method.

Width sampling

To sample the width dimension, a cork borer was chosen to excise the samples for reasons discussed below. The diameter of the borer was chosen to incorporate at least 3/4 of the circumference of the opened and flattened intestines. Larger diameters necessitated taking two or more samples with the borer to accommodate this amount of the width. A method is described (p. 240) using a formula to determine how many samples are needed for measurement. This formula can be used to compare the homogeneity of the samples across the width of the intestine. Eventually when they show homogeneity fewer samples can be taken in the width dimension. Thus, for the colon, 3/4 of the circumference was sampled at five equidistant segments.

Sample systems

A point of caution is perhaps needed in line with the statement made above about sampling what one sees. Obviously, if the colon is constructed similarly to, for example, the rabbit colon where morphologically defined zones have been described (Yamasaki and Komatsu, 1971; Snipes, 1978; Snipes *et al.*, 1982), these areas should be sampled as well. For example, the proximal colon of the rabbit is highly differentiated (as it is in other lagomorphs and many rodents). It would be folly to ignore these structures in the false ardour to obey statistical and stereological rules. One cannot ignore the fact that this organ is heterogeneous. Here a new scheme should be worked out to sample each of the obviously different morphologically definable areas.

Caecum Numerous studies in various animals have shown that apex, corpus and ampulla (if not so structured then caeco-colical area) represent definable areas for sampling (Gorgas, 1967; Behmann, 1973). Routinely these areas were sampled. Initial observation and pilot measurements have shown that one sample in the middle of the circumference is sufficient since there is very little structural difference in this dimension across the circumference of the caecum. Care should be taken to excise any macroscopically visible folds for determination of their basal area. Some animals, such as nutria, possess a mound running longitudinally in the colon (Snipes *et al.*, 1988). This area was always sampled as well as an adjacent area in each sampling area; it was not ignored!

Large animals An alternative sampling method especially for huge and voluminous guts (e.g. elephant) is to use stereological principles of sampling

(Weibel, 1979; Baddeley *et al.*, 1986). The opened gut can be divided into appropriate segments (five) in the longitudinal dimension. Each of these segments can be further divided in the length and width dimension into ten equal segments. By use of random number tables or numbers from a telephone book (two digits) a coordinate can be chosen randomly for the area to be sampled. For example, 74 would indicate that the sample should be taken from the coordinate 7 in the width dimension and 4 in the length dimension. The inherent danger of missing important structures by this method is obvious with large guts when only five longitudinal segments are taken. In cases of huge colons, such as in the cow and horse, the following modification has been used with some feeling of reassurance. The colon was divided into ten equidistant segments. In segment 1 the coordinate 1–1 was sampled. In longitudinal segment 2, coordinate 2–2 was taken, etc. In this way over ten longitudinal segments, a relatively acceptable blanketing sample in the width of these very voluminous guts can be attained. Obviously, macroscopically visible folds must be excised and their basal areas determined because they would not be fully included in the probes taken in the above-described manner.

Tissue embedding

Samples were purposely taken as discs to assure the proper orientation while embedding and to assure that different planes of section through the sphere are made from one sample to the next. First, the disc should be embedded standing on edge. In this way, the plane of section is always perpendicular to the mucosa so that the entire gut wall will be included in the section. Secondly, to fulfil stereological requirements, the different discs should be cut at randomly determined angles from one sample to the next. This can be predetermined with random numbers or left to chance that each disc is oriented differently when embedded anyway.

Measurement

In order to calibrate the measuring system the following procedure was performed. A pilot measurement of up to 50 sections from one disc was performed to determine the *accuracy* of measurement. For this purpose the formula suggested by Baur (1969) was used: $n = 3$ [(standard deviation $\times$ 100)/mean].

The number of sections required for measurement (n) is a function of the standard deviation of the measurement and the mean. This formula can also be used in determining how many samples should be taken from an organ.

However, because we sample almost the entire width of the intestine within one probe, this was not necessary. According to our experience a range of three to 12 sections should be measured from each disc sample depending on the measuring method (see below). The disc should be sectioned at 10 μm and starting approximately one third from the middle of the disc. Routine haematoxylin–eosin staining gives good contrast.

Measurement device

We have three modes of measurement available.

Method 1 Automatic image analyser (IBAS) from Kontron (Oberkochen, Germany). A special programme for this purpose was written by Dr Kriete of our institute (Snipes and Kriete, 1991). The image analysing programme was developed to facilitate an automatic length measurement of both the flat tissue basal boundary and the corresponding folded surface boundary (i.e. the surface mucosa relief) from light-microscopical images. Consequently two values per individual field of view were measured. Assessing these values with a programme written in Fortran allowed us to calculate a **surface enlargement factor** (SEF) at the microscopical level by dividing the length of the folded surface boundary (i.e. the microanatomical enlargements of the mucosa) by the flat basal boundary (**reference line**).

Method 2 Semi-automatic videoplan analysing system (Kontron). This is a routine computer-assisted device for determining areas, lengths, etc. In combination, a slide projector was fitted with a stage and mirror to project the images of the microscopical sections on slides at appropriate magnifications without an ocular onto the digitizing tablet. The image could then be traced with the cursor. The software programme allowed direct tabulation and combined statistical analysis. Again the ratio of the folded mucosa to the reference line was determined as the surface enlargement factor.

Method 3 Manual optical planimeter (MOP, Kontron). This procedure is similar to the aforegoing. The MOP is a manual planimeter with cursor and digitized tablet. Before the projector was available for direct measurement as described above, the sections were traced onto transparent paper via a visopan multiviewer screen microscope and then the outlines measured with the MOP. This is a less elegant procedure than the computer-assisted programme but gives excellent results as well. Spherical aberration of the mirror system used in methods 2 and 3 was 0.3%, which is negligible in the biological sense.

Baur's formula was used to determine how many sections should be measured per disc. Routinely 10–12 sections are measured despite the fact that with method (2), according to the results of measurement and calculation, only three are necessary.

Protocol

The following protocol can thus be compiled from these techniques. Figure 15.1 shows a flow sheet for the protocol.

1. Inject fixative intraluminally
2. Open gut longitudinally
3. Sample with cork borer 3/4 of circumference or its equivalent
   a. Colon: five equidistant segments in the longitudinal dimension
   b. Caecum: apex, body, caeco-colical region
4. Fix disc on flat surface
5. Process for routine histology
6. Embed disc 'standing on edge'
7. Section disc perpendicular to mucosa: 10–12 sections, each 10 μm thick from middle of disc
8. Measurement: measure each section three times, measure at least ten sections, and measure the length of the folded mucosal surface on the microscope slide using an available morphometric system, e.g.
   a. Automatic image analyser
   b. Semi-automatic videoplan analysing system
   c. Manual optic planimeter
9. Calculate the surface area by drawing a flat reference line and determining the ratio of surface enlargement of the mucosa to the reference line (Fig. 15.1).

Results

Second-order enlargements

Morphological differentiations at the second-order microscopical level show the most striking differences among the various mammals. What is actually meant by this level? When viewing the cellular entities of the mucosae of the large intestines of different animals, it is impossible to differentiate the various species based on their particular type of enterocyte. At least to date, obvious differences at the cellular level are not evident or have not yet been

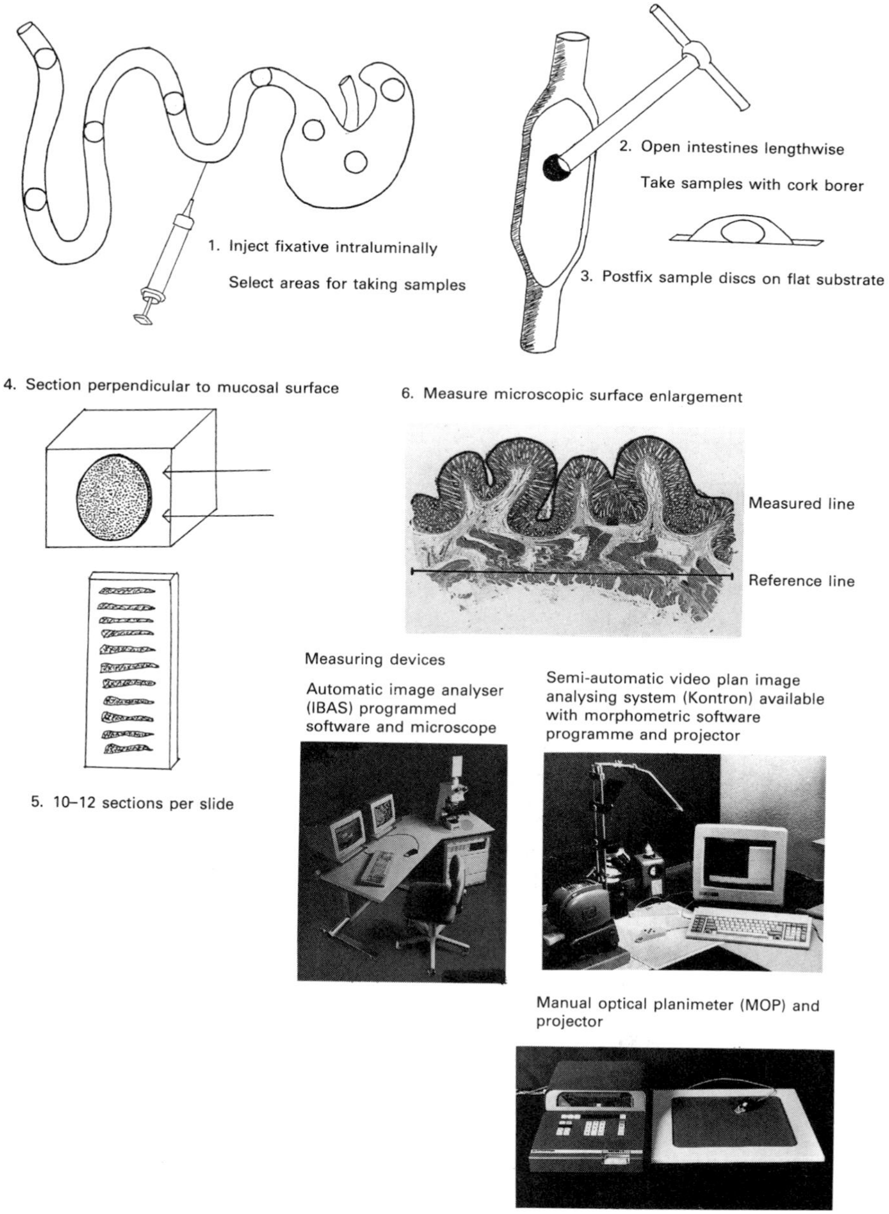

Fig. 15.1. Flow sheet showing the protocol used to measure surface area.

recorded. As described in classical textbooks (e.g. Fawcett, 1986) and by Specht (1977) in a specific monograph on intestinal cells, the enterocyte is a characteristic columnar cell with numerous microvilli on the apical surface. The content of organelles is typical of actively metabolizing cells. The difference in enterocytes of the small intestine to those in the large intestine is mostly in enzymatic machinery, which is not readily evident without special techniques. Therefore, the cellular entity appears to be uniform and not necessarily specific either to species or to small or large intestine. The latter statement can, of course, be challenged with respect to subtle differences such as height of columnar cells, numbers of microvilli (C. S. Kuehn, personal communication), content of organelles (perhaps in relationship to body weight) or changes due to external factors such as season (Ludwig, 1986); however, the basic plan is strikingly similar throughout the taxa (Gustafsson and Maunsbach, 1971; Hendrikson, 1973; Ono, 1976; Kraus *et al.*, 1977; Mora-Galindo and Martínez-Paloma, 1977; McKenzie, 1978; Bell and Williams, 1982; Sohma, 1983; Bruorton and Perrin, 1988).

With the combined use of light microscopy and scanning electron microscopy, an obvious difference in the surface architecture of the caecum/colon became apparent. The classical description of the topography of the large intestine is again taken from the human situation and has been borrowed for numerous species. This description imparts a more or less flat appearance to the large intestinal surface studded with 'pits' which represent openings to crypts of Lieberkühn. As is well known, villi almost always are missing from the large intestine. This situation can best be appreciated by a micrograph from scanning electron microscopy (SEM). The light micrographs of mucosal surfaces can best be evaluated in combination with SEM, avoiding laborious and tedious three-dimensional reconstructional processes with serial sections (Snipes and Thiele, 1989). 'Side' views in SEM micrographs often reveal the form of crypts as they extend down into the lamina propria. In the human they are simple tubes, usually more or less uniform from base to neck or slightly widened basally with a narrow neck. This situation, typical for a human, does not necessarily afford the intestinal surface with much enlargement of its surface area. In the small intestine, the surface area is increased largely by villi and, at another level in humans, the plicae circulares. These latter structures are not widely distributed among animals. Therefore, there is little possibility of a surface enlargement over and beyond what can be expected from the geometrical situation of a tube. How then do certain mammals increase the surface of their large intestines at the microscopic level?

In the simplest of cases (and as seen in many small rodents), the openings to crypts are widened. This exposes a greater amount of the crypt to luminal content or *vice versa*, the luminal content can more readily be attained and influenced by, for example, secretions produced by cells in the crypts. This can functionally be of great importance as the upper half to upper third of the crypt is populated by mature enterocytes. Therefore, not only can surface cells actively participate in the functional process but also the upper-crypt enterocytes. The SEM appearance of such a situation resembles a 'rolling hills' landscape. The intercrypt distances are not large and the circumference of the openings are somewhat raised. The 'pits', as described above, are not visible in the SEM view, but rather a configuration of elevated ridges of soft contours often in looped configurations appears. This situation is interpreted as showing a form of crypts not as long tubes, but rather as narrow channels or ravines, often in tortuous labyrinthine formations. Here is a situation where the surface mucosa is strikingly changed compared with the flat situation, creating a mucosa with a completely different topography and producing, by comparison with the flat situation, a microclimate not directly exposed to the 'currents' of intestinal flow and, thus, protected from influences in the main stream of the lumen. This could be very important functionally for situations of alloenzymatic digestion, for the vitality of bacteria, or for attaining a favourable milieu preparatory to absorptive events at the apical surface of enterocytes. Fig. 15.2 shows those animals with a 'flat' mucosa, together with a schematic drawing of their caecum and the correlated SEM and light microscopic appearance of the mucosa. In Fig. 15.3 a similar illustration is made for animals possessing surface enlargements of their mucosa (see also Snipes and Thiele, 1989).

It should be emphasized here that by conventional methods for processing tissue for microscopy, an integral part of the interphase cell/lumen is removed, namely, the mucous layer. Its functional importance in various intestinal processes and along the length of the large intestine has been emphasized by Sakata and von Engelhardt (1981) and should always be called to mind in a final analysis of the function of the intestines – even though it is often ignored or neglected due to its disappearance upon fixation.

The elevated ridges in looped configurations develop in the extreme form to deep clefts within which the openings to crypts are located (and not on the surface). This latter situation can be found in lagomorphs (Yamasaki, 1971; Yamasaki and Komatsu, 1971; Snipes, 1978; Snipes *et al.*, 1982; Snipes and Thiele, 1989). These architectural constructions, visible only at the microscopical level, are what have been termed **second-order surface enlargements**. Cellular specializations such as microvilli constitute first-order enlargements. The mucosa can be thrown into longitudinal, circular or

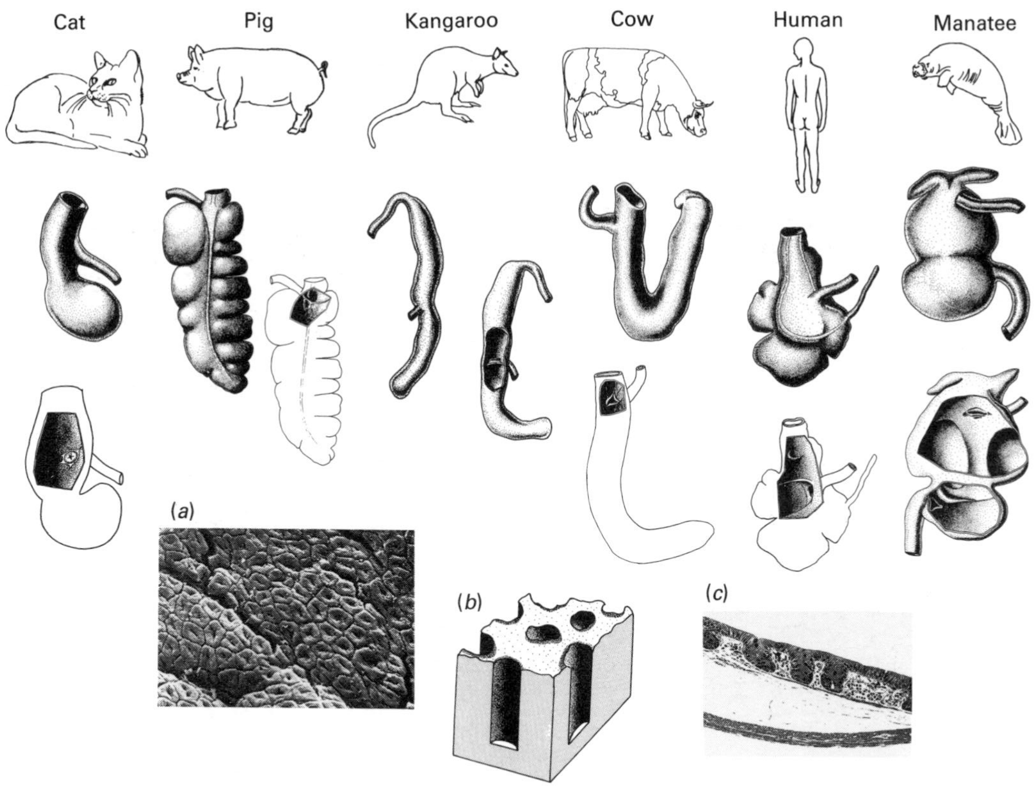

Fig. 15.2. Pictorial presentation of the macroscopic appearance of caeca with a flat mucosal surface in a variety of different mammals with widely varying dietary habits. The animals are shown at the top with below them the external morphology of the caeca with windowing in the figure below to expose the ileo-caecal area. At the bottom of the figure is (*a*) a scanning electron micrograph of the caecal mucosal surface; (*c*) a corresponding light micrograph and (*b*) a schematic representation of the mucosal type.

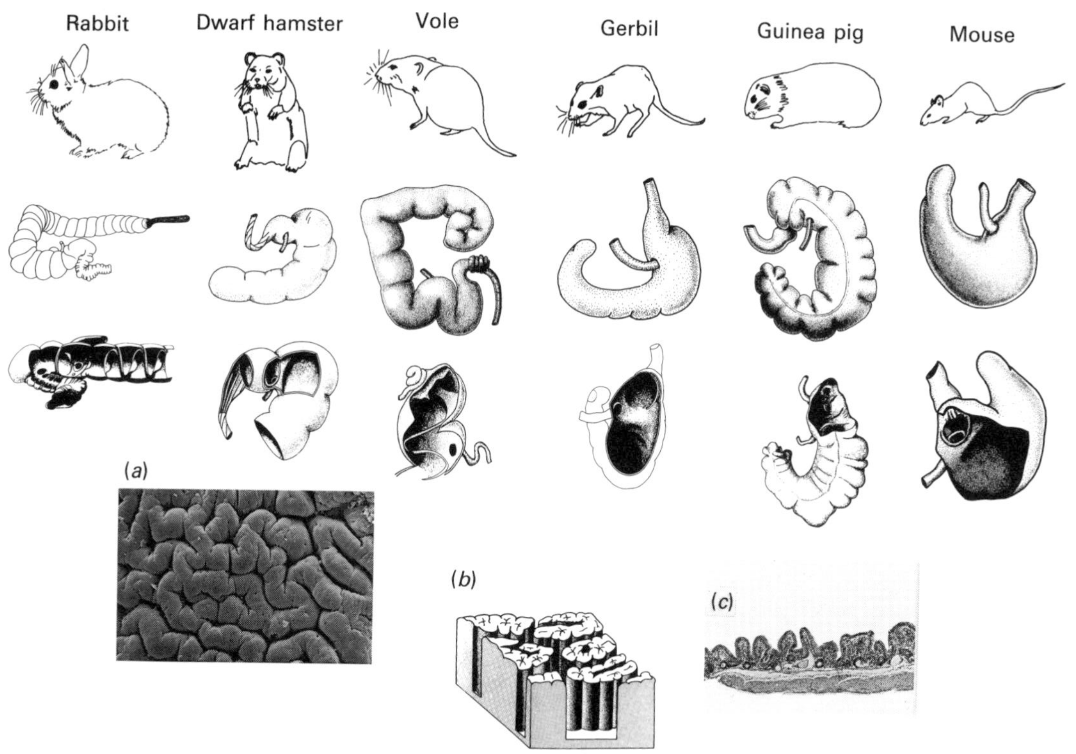

Fig. 15.3. Pictorial presentation of the macroscopic appearance of the caecum in a variety of mammals showing a second-order enlargement of their caecal surface mucosa. The details of the figure are as in Fig. 15.2. See also Table 15.2.

oblique plicae, folds or mounds, which also give a microscopically measurable enlargement to the surface area. As long as these structures can be viewed on a histological section and measured microscopically they are included in the second-order SEF enlargement. At the macroscopic level, additional surface enlargement can result by formation of visible folds (e.g. plicae circulares in the small intestine, semi-lunar folds in the colon: both seen in the human). The existence of second-order enlargements and their possible relationship to dietary strategy is now under investigation.

Morphometry

These visible surface topographic entities can now be quantified by means of these morphometric techniques (see also Snipes, 1991 and Young Owl, Ch. 14 for the determination of basal or ground areas). Previously, most studies dealing with the relationship of areas and volumes to dietary strategies have remained at the basal level (not including microscopic enlargements). Correlations have mostly been made with metabolic body weight and then correlated to dietary strategy (Chivers and Hladik, 1980; Martin *et al.*, 1985).

Ratios

With the morphometrically determined data in addition to the empirical data (see Snipes and Kriete, 1991 for compilation) certain ratios can be instructive in disclosing relationships that may not be immediately apparent. Thus, the ratio large intestine/small intestine for the parameters length, basal area, total area (basal area + SEF from microscopic measurements) and volume can be used to give an indication of the extent of large intestinal involvement in the dietary process. Large ratios indicate greater large intestinal involvement. The analysis included 18 mammalian species for the parameters volume (Fig. 15.4*a*), length (Fig. 15.4*b*), basal area (Fig. 15.4*c*), and 11 for total areas (Fig. 15.4*d*).

The volume ratio has been used as an indication of fermentation capability. The largest volume ratios (Fig. 15.4*a*) are found in rabbits (who indeed have an ideal caecum for such a function), followed by guinea pig, horse and muskrat. These values are all above 1.5. Values between 1.5 and 1 are found in nutria, spalax, golden hamster, rat, gerbil, pig and vole. At least the first two of these species can be considered as species where a fermentation vat function is expedient in the nutritional strategy. Those species with a value under 1 include, in descending order: dwarf hamster, mouse, sheep, goat, dog, cow and human. The last five do not use the caecal/colonic fermentation vat as an integral part of the digestive process: cow, goat and sheep being ruminating forms where the fermentation chamber is set before the small intestine. The value for dog as a strict faunivore is not surprising.

According to Chivers and Hladik (1980), the ratios for volume can be categorized such that values up to 0.7 represent faunivores, whereas values above 2.0 indicate herbivores and between these two values (0.7–2.0) fall the rather larger group of 'intermediate feeders', for want of a better term.

In considering ratios of length (Fig. 15.4*b*), many of the same species appear together as in the volume study although in slightly different order. Here the fact that an animal has a large small intestine (nutria, horse) in addition to a well-differentiated large intestine does not negate the importance of the large intestine in its nutritional strategy but may rather attest to an absorptive advantage of both a copious small intestine as well as a large intestine.

Area ratios (Fig. 15.4*c*, *d*) can be used to give an indication of absorptive capacity (Chivers and Hladik, 1980, 1984). Higher values show a tendency in the large intestine for absorptive or secretory surface to be increased. Here the rabbit, spalax, muskrat, and guinea pig all possess highest values for both basal area and total area ratios (SEF values for horse, pig, sheep, goat and cow were not available). Rabbit, spalax and muskrat had the highest total area ratios (Fig. 15.4*d*). At the other end of the scale with lowest ratios were the human, dog and rat. In the last three histograms the values below 0.3 are regarded as faunivores, intermediate forms are between 0.3 and 0.7 and above 0.7 are herbivores (Chivers and Hladik, 1984; Snipes and Kriete, 1991). All values are correlated to mucosal appearance in Table 15.1.

As can be seen in all four graphs (Fig. 15.4) and in Table 15.1, using different parameters, the same animals can appear in different dietary categories or in different sequences. This fact reflects the large overlap amongst the dietary groups, which has been emphasized by Chivers and Hladik (1980, 1984). It should also be stressed that the arbitrary cut-off points between the three groups were made for simplicity's sake and the real situation is more of a continuum.

Regression curves

Recently, Snipes and Kriete (1991) presented regression curves of area values in correlation to metabolic body weight to discern whether tendencies to dietary habit could be recognised. With the 11 mammalian species where SEF values were available, the following generalizations could be ascertained: the areas of the colon and small intestine scale isometrically to metabolic body

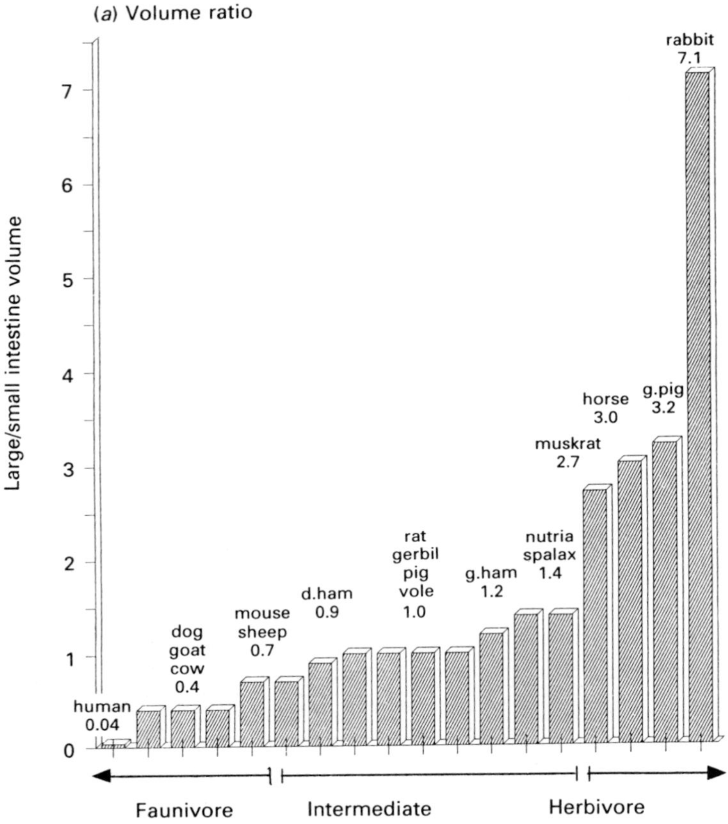

Fig. 15.4. The ratio of large intestinal values divided by small intestinal values for mammals of varying dietary habits gives an indication of the participation of the former in functional aspects of intestinal processes. (*a*) Volume ratios give an indication of fermentation function (fermentation coefficient according to Chivers and Hladik, 1980) in the large intestine. Large values indicate a tendency toward herbivory. Ratios between 0–0.7 indicate faunivory; between 0.7 to 1.5 intermediate feeders and ratios above 1.5 indicate herbivory. (*b*) Length ratios are given to complete the parameters tested here but are the least reliable to compile a categorization. In this graph, values above 0.7 indicate herbivory, between 0.7 and 0.3 intermediate feeders and below 0.3 faunivory. (*c*) Basal area ratio and (*d*) total area ratio give a coefficent of gut differentiation and, therefore, an indication of absorptive function. The dietary categories and the corresponding range of values are the same as in (*b*). d. ham, dwarf hamster; g. ham, golden hamster; g. pig, guinea pig.

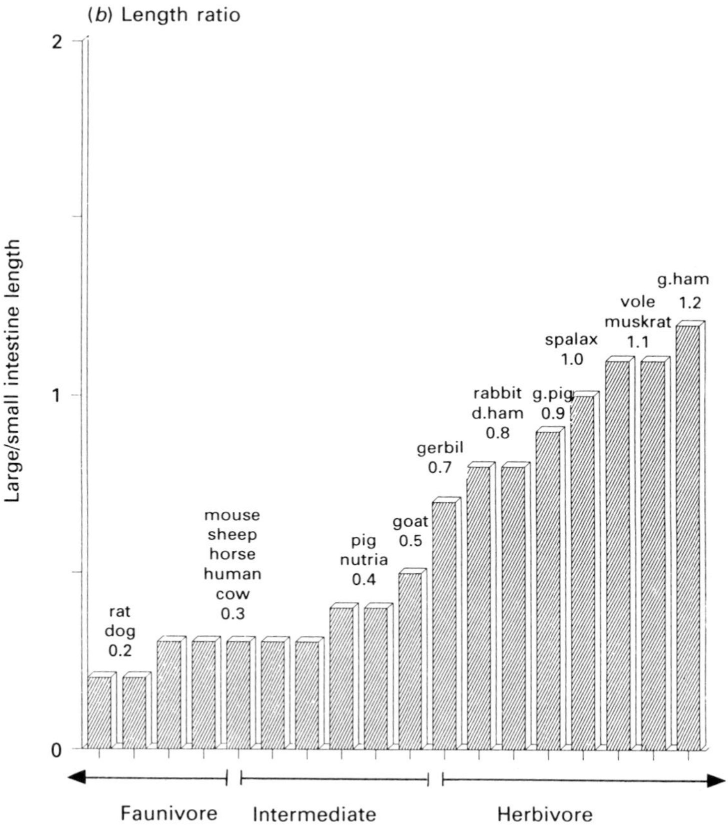

Figure 15.4 *Continued.*

weight, whereas the caecum relates negative allometrically to increasing metabolic body weight. Areas of all intestinal compartments scale negative allometrically to absolute body weight. Volume shows positive allometric scaling to metabolic body weight and approximate isometry to absolute body weight (bar caecum: negative allometry). Table 15.2 shows the total areas (in cm^2) of caecum and colon measured by the present technique for a number of different animals belonging to various orders.

Area/volume

The relationship area/volume gives a physiological factor, i.e. the amount of unit area available per unit volume (cm^2/ml). When this ratio is plotted

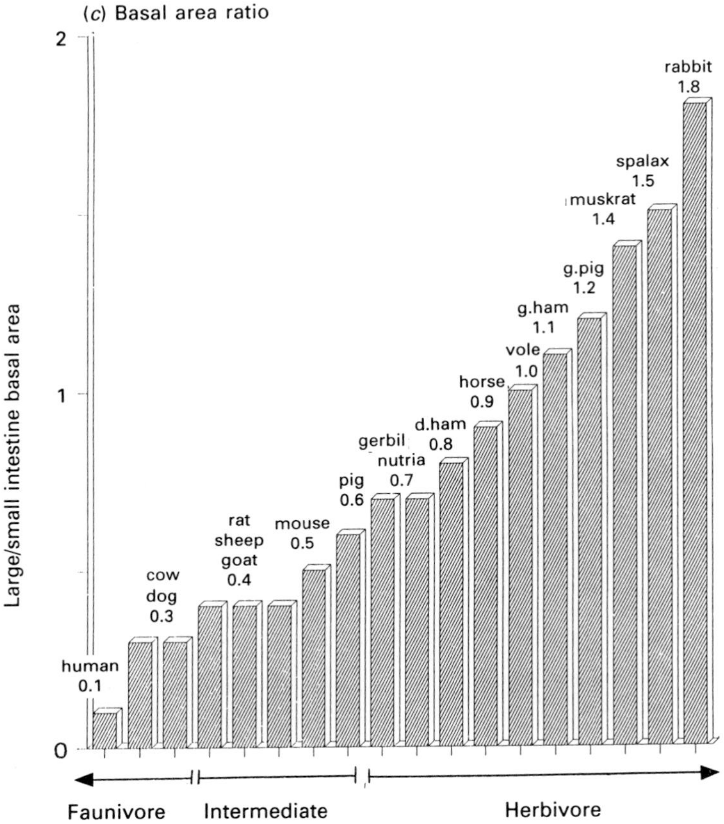

Figure 15.4 *Continued.*

against metabolic body weight it becomes evident that the smaller animals (in weight) have generally more surface area per unit volume available in all areas of their gut, while the larger animals have, respectively, less (Fig. 15.5*a*). From a functional standpoint this could mean that absorption is more efficient in small animals due to more ready access of the potentially absorbable food to the absorbing surfaces. Smaller animals appear to have a more advantageous relationship of area to volume than larger animals. This fact can be interpreted as a correlate to the higher metabolic needs of smaller animals. However, additional factors come into play (the importance of the second-order enlargements was seen in Fig. 15.4) such as passage rates, and

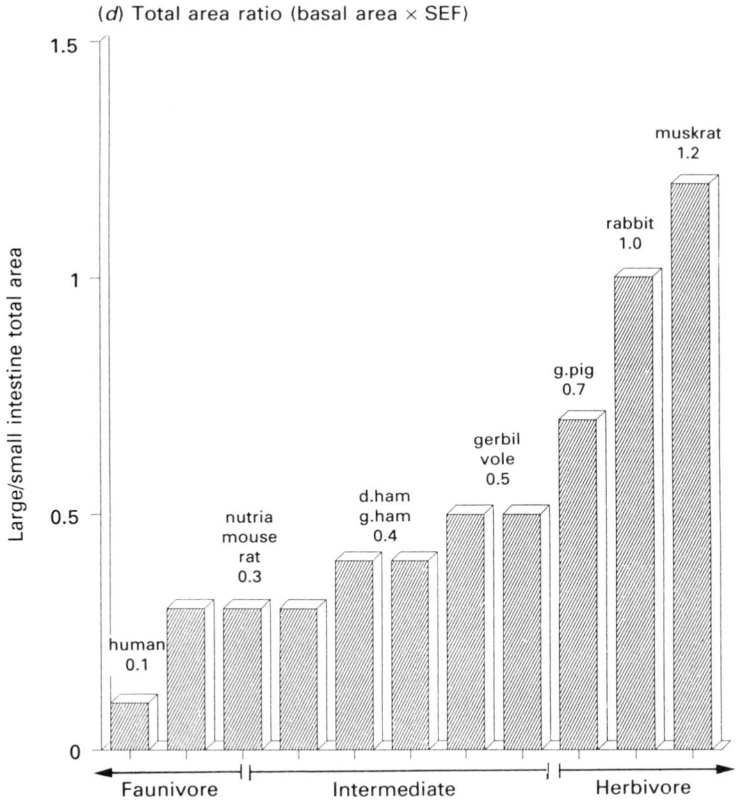

Figure 15.4 *Continued.*

retention times of digesta and their morphological correlates (reviewed recently by Langer and Snipes, 1991).

Another way of representing the increase in surface areas is by comparing the area/volume ratio using the area from basal surface area measurements and that from total measurements (basal + SEF), both versus metabolic body weight (Fig. 15.5*b–d*). The former can be taken as an 'almost' theoretical relationship of area to volume. Here we can see that, as compensation for increased volume in some animals, an increase in SEF occurs thus returning to a more advantageous relationship of area to volume.

The nutritional strategies used by the different species may be only secondary to these geometric configurations; one type uses a fermentation strategy

Table 15.1. *Correlation of the appearance of second-order enlargements of the mucosa and the large/small intestine ratios for volume, length and area with dietary strategy*

(a) The degree of second-order enlargement observed for different species

Second-order enlargements		Flat mucosa (−)
Extensive (*)	Moderate (+)	
		Dog (Cat) (−)
		Human (−)
Rabbit (*)	Mouse (+)	Manatee (−)
Muskrat (*)	Rat (+)	Sheep (−)
Vole (*)	Spalax (+)	Cow (−)
Nutria (*)	Gerbil (+)	Goat (−)
Hamsters (*)	Horse (+)	Kangaroo (−)
		Pig (−)

(Janis, 1976; Bayley, 1978; Hintz *et al.*, 1978; Hume and Warner, 1980) either coupled to rumination (Ulyatt *et al.*, 1975) or caecotrophy (Hörnicke and Björnhag, 1980; Björnhag, 1987; Björnhag, Ch. 17) or a diet of poor quality combined with rapid passage.

Note that rabbit, guinea pig, nutria, vole, gerbil and dwarf hamster, mouse and rat, muskrat, and probably golden hamster and horse (see Wille, 1975) can all be grouped together according to the microscopic appearance of their caecal mucosa (the second-order enlargement component, i.e. the mucosal surface, consists of raised ridges in long looped configurations; Fig. 15.3 and Tables 15.1, 15.2). Another grouping of animals includes those whose caecal mucosa resembles a flatter landscape with pits as depressions into crypts. Animals belonging to this category include cat (dog also), human, cow, pig, manatee, kangaroo, and probably sheep and goat as well (Wille, 1975; for domesticated animals: R. L. Snipes and H. Snipes, unpublished data) (Fig. 15.2 and Tables 15.1, 15.2).

Comparing and correlating these groupings to those attained from ratios of parameters for large intestines and small intestines demonstrates a very definite tendency for animals to stay together in groups that correspond to the morphological groupings based on second-order enlargements. In Table 15.1*a*, *b* these correlations are illustrated. In most cases the animals with highly developed second-order enlargements have the highest ratios for the parameters area, volume and length, the animals with flat mucosas having the lower ratios. Between these two extremes we find a mixture of high and moderate enlargements. The ratings have been divided into three groups, the highest being the herbivores, the middle ratios representing the

Table 15.1 *continued.*

(b) Dietary strategies indicated by large intestine/small intestine ratings using different parameters of intestinal size

	Large/small intestine ratio		
	Herbivore	Intermediate	Faunivore
Volume	(*) Rabbit	(*) Golden hamster	(–) Cow
	(*) Guinea pig	(*) Vole	(–) Goat
	(+) Horse	(–) Pig	(–) Dog
	(*) Muskrat	(+) Gerbil	(–) Human
		(+) Rat	
	← (*) Nutria →		
	← (+) Spalax →		
		← (+) Dwarf hamster →	
		← (+) Mouse →	
		← (+) Sheep →	
Length	(*) Golden hamster	(–) Goat	(–) Dog
	(*) Vole	(–) Pig	(+) Rat
	(*) Muskrat	(*) Nutria	
	(+) Spalax	← (+) Mouse →	
	(*) Guinea pig	← (–) Sheep →	
	(*) Rabbit	← (+) Horse →	
	(*) Dwarf hamster	← (–) Human →	
	← (+) Gerbil →	← (–) Cow →	
Basal areas	(*) Rabbit	(–) Pig	(–) Cow
	(+) Spalax	(+) Mouse	(–) Dog
	(*) Muskrat	(+) Rat	(–) Human
	(*) Guinea pig	(–) Sheep	
	(*) Golden hamster	(–) Goat	
	(*) Vole		
	(+) Horse		
	(*) Dwarf hamster		
	(+) Gerbil		
	(*) Nutria		
Total areas	(*) Muskrat	(+) Gerbil ← (+) Rat →	(–) Human
	(*) Rabbit	(*) Vole ← (+) Mouse →	
		← (*) Nutria →	
	(*) Guinea pig	(*) Golden hamster	
		(*) Dwarf hamster	

The degree of second-order enlargement as shown in (*a*) is given as (*) extensive and (+) moderate second-order enlargement and (–) flat muscoa.
Arrows on each side of a name indicate a borderline case.

Table 15.2. *The total areas of the caecum and colon derived by determining the surface enlargement at the microscopic level for animals from widely varying taxa*

Order/family	Genus, species (common name)	Total area (cm²) Caecum	Colon
Marsupialia			
Phalangeridae	*Phascolarctos cinereus* (koala)	10979.5	9808.5
Rodentia			
Arvicolidae	*Microtus agrestis* (vole)	12.3	15.7
Cricetidae	*Phodopus sungorus* (dwarf hamster)	12.3	16.5
	Meriones unguiculatus (gerbil)	19.7	20.6
	Mesocricetus auratus (golden hamster)	59.7	63.5
	Ondatra zibethicus (muskrat)	417.1	269.1
Spalacidae	*Spalax ehrenbergi* (mole rat)	47.7	29.8
Muridae	*Mus musculus* (mouse)	15.5	20.6
	Rattus norvegicus (rat)	41.2	36.8
Caviidae	*Cavia aperea* (guinea pig)	391.0	303.0
Myocastoridae	*Myocastor coypus* (nutria)	655.9	654.5
Lagomorpha			
Leporidae	*Oryctolagus cuniculus* (rabbit)	1121.1	485.2
Carnivora			
Felidae	*Felis silvestris* (cat)	9.7	124.6
Primates			
Hominidae	*Homo sapiens* (human)	261.7	1764.9
Artiodactyla			
Suidae	*Sus scrofa* (pig)	728.6	6648.5
Bovidae	*Bos primigenius* (cow)	1128.7	11883.0
	Capra aegagrus (goat)	231.1	1582.6
	Ovis ammon (sheep)	397.5	3553.1
Perissodactyla			
Equidae	*Equus przewalskii* (horse)	12137.9	40775.0

intermediate feeders, and the lowest the faunivores. The animals with the extensive enlargement are mostly found in the herbivore grouping, those with moderate second-order enlargement in the intermediate groups, whilst the animals with flat mucosas are spread over the intermediate and faunivore groups. There are a number of animals that are borderline cases. Of the parameters studied, length appears to be the least reliable or consistent.

Thus, a new way of viewing animals in relation to their dietary habits and

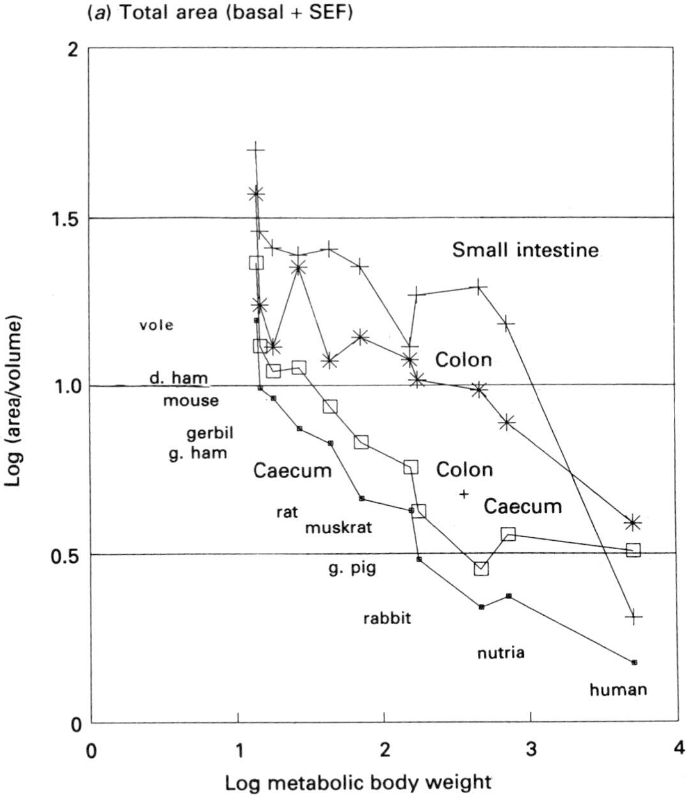

Fig. 15.5. The relationship of area to volume of the various compartments of the intestines. When plotted against log metabolic weight this gives an indication of the ratio of unit area (cm^2) to unit volume (ml) for animals of different sizes. (*a*) Ratios of total area/volume for all compartments of the intestines (small intestine, caecum, colon, colon + caecum). It is evident here that ratios decrease with increasing body weight, such that smaller animals have a more advantageous surface to volume relationship. The ratios of basal area to volume and total area to volume are given for (*b*) small intestine; (*c*) caecum and (*d*) colon. These differences indicate the compensation and adaptations (in the form of surface enlargements, as visualized by total area measurements) to loss of available area due to increasing volumes. d. ham, dwarf hamster; g. ham, golden hamster; g. pig, guinea pig.

strategies has been introduced and an attempt has been made to correlate, at least for the material available in the present study, morphometrically measurable factors with microscopically apparent second-order enlargement structures of the mucosa found in the mammalian caecum and colon.

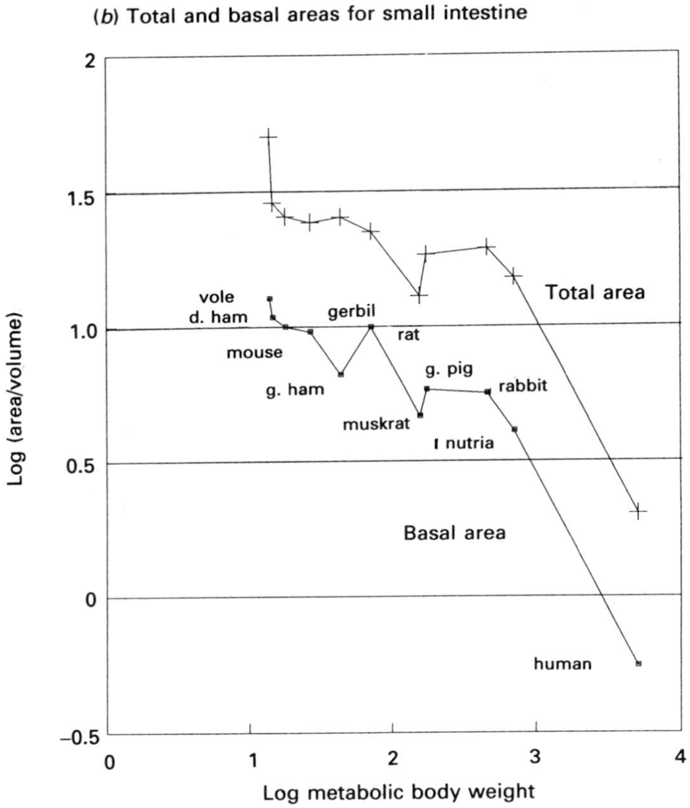

Figure 15.5 *Continued.*

Acknowledgements

The author wishes to take this opportunity to express indebtedness for the technical help of a well-trained and willing staff. Often this short acknowledgement does not appear justified for all the time and effort spent by these people on the various projects. I am truly grateful to Mrs Heidrun Sust and Mrs Heidi Snipes for excellence in all phases of these investigations, especially the morphometric portion, for initiative, patience and perseverance, without which the author could not have carried out these studies. To the other members of the technical staff who generously aided us with their suggestions and time and expert assistance, I am equally grateful, especially to Mrs Andrea Klein, Mrs Magdelena Gottwald and Mrs Corinna Kabot. Mrs

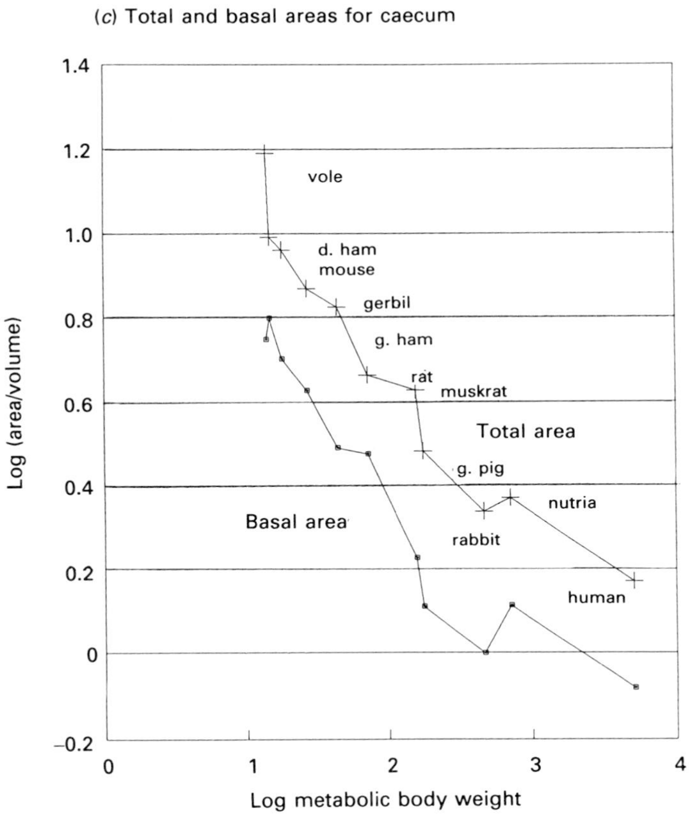

Figure 15.5 *Continued.*

Brigitte Wildner has supported us with her cordial and excellent secretarial help. Dr Peter Langer, colleague and friend, has supported these efforts by sharing interests in this theme and many hours of mutual research, stimulating discussions and friendship. Dr Wieland Stöckmann deserves special merit for his patience with statistical explanations and biological discussions and for his ever present willingness to counsel, advise and edit these studies. Professor Dr Dr H.-R. Duncker guided this work with his constructive criticism and unending reserve of ideas and suggestions.

References

Agnati, L. F. & Fuxe, K. (1985). *Quantitative Neuroanatomy in Transmitter Research*. New York: Plenum Press.

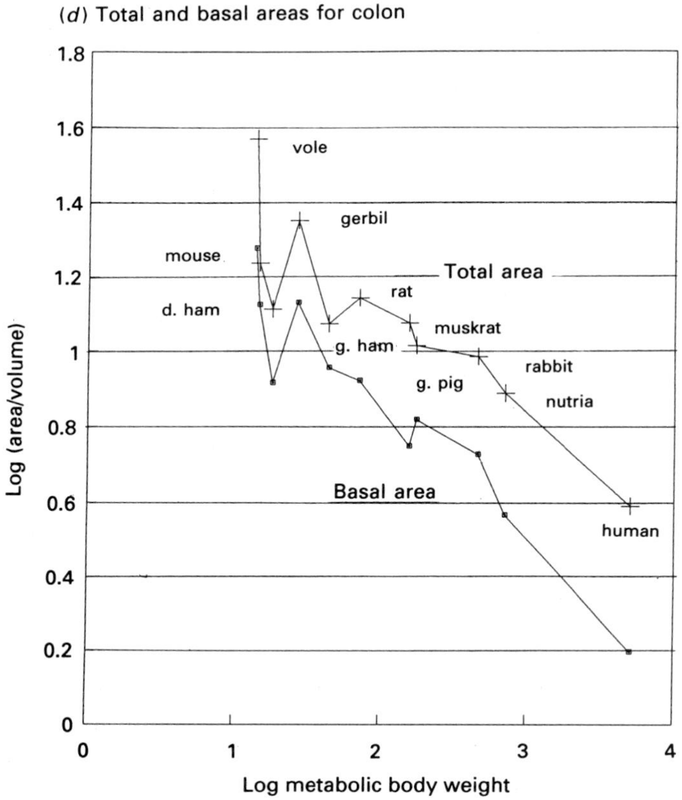

Figure 15.5 *Continued.*

Baddeley, A. J., Gundersen, H. J. G. & Cruz-Orive, L.-M. (1986). Estimation of surface area from vertical sections. *Journal of Microscopy*, **1142**, 259–276.

Baur, R. (1969). Zur Schätzung des kleinsten zulässigen Stichprobenumfanges für stereologische Messungen an histologischen Schnitten. *Experientia*, **25**, 554–555.

Bayley, H. S. (1978). Comparative physiology of the hindgut and its nutritional significance. *Journal of Animal Science* **46**, 1800–1802.

Behmann, H. (1973). Vergleichend- und funktionell-anatomische Untersuchungen am Caecum and Colon myomorpher Nagetiere. *Zeitschrift für wissenschaftliche Zoologie*, **186**, 173–294.

Bell, L. & Williams, L. (1982). A scanning and transmission electron microscopical study of the morphogenesis of human colonic villi. *Anatomy and Embryology*, **165**, 437–458.

Björnhag, G. (1987). Comparative aspects of digestion in the hindgut of mammals. The colonic separation mechanism (CSM) (a review). *Deutsche Tierärztliche Wochenschrift*, **94**, 33–36.

Bruorton, M. R. & Perrin, M. R. (1988). The anatomy of the stomach and caecum of the Samango monkey, *Cercopithecus mitis erythrarchus* Peters, 1852. *Zeitschrift für Säugetierkunde*, **53**, 210–244.

Bucher, O. & Wartenberg, H. (1989). *Cytologie, Histologie und mikroskopische Anatomie des Menschen* 11th edn., p. 402. Stuttgart: Hans Huber Verlag Bern.

Chivers, D. J. (1989). Adaptations of digestive systems in non-ruminant herbivores. *Proceedings of the Nutritional Society*, **48**, 59–67.

Chivers, D. J. & Hladik, C. M. (1980). Morphology of the gastrointestinal tract in primates: comparisons with other mammals in relation to diet. *Journal of Morphology*, **166**, 337–386.

Chivers, D. J. & Hladik, C. M. (1984). Diet and gut morphology in primates. In *Food Acquisition and Processing in Primates*, ed. D. J. Chivers, B. A. Wood & A. Bilsborough, pp. 213–230. New York: Plenum Press.

Debongnie, J. C. & Phillips, S. F. (1978). Capacity of the human colon to absorb fluid. *Gastroenterology*, **74**, 689–703.

Fawcett, D. (1986). *Bloom and Fawcett. A Textbook of Histology*, 11th edn. p. 665. Philadelphia: W. B. Saunders.

Fenwick, B. W. & Kruckenberg, S. (1987). Comparison of methods used to collect canine intestinal tissue for histological examination. *American Journal of Veterinary Research*, **48**, 1276–1281.

Gorgas, M. (1967). Vergleichend-anatomische Untersuchungen am Magen-Darm-Kanal der Sciuromorpha, Hystricomorpha und Caviomorpha (Rodentia). *Zeitschrift für wissenschaftliche Zoologie*, **175**, 331–404.

Gustafsson, B. E. & Maunsbach, A. B. (1971). Ultrastructure of the enlarged cecum in germfree rats. *Zeitschrift für Zellforschung*, **120**, 555–578.

Hendrikson, R. C. (1973). Ultrastructural aspect of mouse cecal epithelium. *Zeitschrift für Zellforschung*, **140**, 445–449.

Hintz, H. F., Schryver, H. F. & Stevens, C. E. (1978). Digestion and absorption in the hindgut of nonruminant herbivores. *Journal of Animal Science*, **46**, 1803–1807.

Hofmann, R. R. (1989). Evolutionary steps of ecophysiological adaptation and diversification of ruminants: a comparative view of their digestive system. *Oecologica*, **78**, 443–457.

Hörnicke, H. & Björnhag, G. (1980). Coprophagy and related strategies for digesta utilization. In *Digestive Physiology and Metabolism in Ruminants*. ed. Y. Ruckebusch & P. Thivend. Lancaster, UK: MTP Press.

Hume, I. D. & Warner, A. C. I. (1980). Evolution of microbial digestion in mammals. In *Digestive Physiology and Metabolism in Ruminants*. ed. Y. Ruckebusch & P. Thivend. Lancaster, UK: MTP Press.

Imoto, S. & Namioka, S. (1978). Volatile fatty acid production in the pig large intestine. *Journal of Animal Science*, **47**, 467–478.

Janis, C. (1976). The evolutionary strategy of the Equidae and the origins of rumen and cecal digestion. *Evolution*, **30**, 757–774.

Karasov, W. H. & Diamond, J. D. (1985). Digestive adaptations for fueling the cost of endothermy. *Science*, **228**, 202–204.

Karasov, W. H. & Diamond, J. D. (1988). Interplay between physiology and ecology in digestion. *Bioscience*, **38**, 602–611.

Kleiber, M. (1961). *The Fire of Life: An Introduction to Animal Energetics*. New York: John Wiley.

Kraus, W. J., Cutts, J. H. & Leeson, C. R. (1977). The postnatal development of the alimentary canal in the opossum. III. Small intestine and colon. *Journal of Anatomy*, **123**, 21–45.

Langer, P. (1987). Der Verdauungstrakt bei pflanzenfressenden Säugetieren. *Biologie in Unserer Zeit*, **17**, 9–14.
Langer, P. & Snipes, R. L. (1991). Adaptations of gut structure to function in herbivores. In *Physiological Aspects of Digestion and Metabolism in Ruminants*, ed. T. Tsuda, Y. Saski & R. Kawashima, pp. 348–384. San Diego: Academic Press.
Leng, E. (1978). Absorption of inorganic ions and volatile fatty acids in the rabbit caecum. *British Journal of Nutrition*, **40**, 509–519.
Lochmiller, L. R., Hellgren, E. C., Gallagher, L. M., Varner, L. M. & Grant, W. E. (1989). Volatile fatty acids in the gastrointestinal tract of the collared peccary (*Tayassu tajacu*). *Journal of Mammology*, **70**, 189–191.
Ludwig, J. (1986). Vergleichend-histologische und morphometrische Untersuchungen am Dickdarm von 30 Wiederkäuer-Arten (Ruminantia Scopli 1777). Inaugural-Dissertation: Veterinärmedizin der Justus-Liebig-Universität Giessen.
Martin, R. D., Chivers, D. J., MacLarnon, A. M. & Hladik, C. M. (1985). Gastrointestinal allometry in primates and other mammals. In *Size and Scaling in Primate Biology*, ed. W. L. Jungers, pp. 61–89. New York: Plenum Press.
McKenzie, R. A. (1978). The caecum of the koala, *Phascolarctus cinereus*: light, scanning and transmission electron microscopic observations on its epithelium and flora. *Australian Journal of Zoology*, **26**, 249–256.
McNeil, N. I., Cummings, J. H. & James, W. P. T. (1978). Short chain fatty acid absorption by the human large intestine. *Gut*, **19**, 819–822.
Mora-Galindo, J. & Martínez-Paloma, A. (1977). Ultrastructural study of the cecal epithelium in normal guinea pig. In *35th Annual Proceedings of the Electron Microscopy Society of America*, ed. G. W. Bailey, pp. 650–651.
Ono, K. (1976). Ultrastructure of the surface principle cells of the large intestine in postnatal developing rat. *Anatomy and Embryology*, **149**, 155–171.
Osawa, R. (1990). Feeding strategies of the swamp wallaby, *Wallabia bicolor*, on the North Stradbroke Island, Queensland. I Composition of diets. *Australian Wildlife Research*, **17**, 615–621.
Osawa, R. & Woodall, P.F. (1990). Feeding strategies of the swamp wallaby, *Wallabia bicolor*, on the North Stradbroke Island, Queensland. II. Effects of seasonal changes in diet quality on intestinal morphology. *Australian Wildlife Research*, **17**, 623–632.
Perrin, M. R. & Curtis, B. A. (1980). Comparative morphology of the digestive system of 19 species of Southern African myomorph rodents in relation to diet and evolution. *South African Journal of Zoology*, **15**, 22–33.
Sakata, T. & von Engelhardt, W. (1981). Luminal mucin in the large intestine of mice, rats and guinea pigs. *Cell and Tissue Research*, **219**, 629–635.
Schmidt-Nielsen, K. (1975). Scaling in biology: the consequences of size. *Journal of Experimental Zoology*, **194**, 287–308.
Snipes, R. L. (1978). Anatomy of the rabbit cecum. *Anatomy and Embryology*, **155**, 57–80.
Snipes, R. L. (1979a). Anatomy of the cecum of the vole, *Microtus agrestis*. *Anatomy and Embryology*, **157**, 181–203.
Snipes, R. L. (1979b). Anatomy of the cecum of the dwarf hamster (*Phodopus sungorus*). *Anatomy and Embryology*, **157**, 329–346.
Snipes, R. L. (1981). Anatomy of the cecum of the laboratory mouse and rat. *Anatomy and Embryology*, **162**, 455–474.
Snipes, R. L. (1982a). Anatomy of the cecum of the gerbil *Meriones unguiculatus (Mammalia, Rodentia, Cricetidae)*. *Zoomorphology*, **100**, 189–202.

Snipes, R. L. (1982b). Anatomy of the guinea-pig cecum. *Anatomy and Embryology*, **165**, 97–111.
Snipes, R. L. (1984a). Anatomy of the cecum of the cat. *Anatomy and Embryology*, **170**, 177–185.
Snipes, R. L. (1984b). Anatomy of the cecum of the West Indian manatee *Trichechus manatus* (Mammalia, Sirenia). *Zoormorphology*, **104**, 76–78.
Snipes, R. L. (1991). Morphology of the mammalian cecum and colon. Microscopic and morphometric considerations. In *Hindgut '91*, ed. T. Sakata & R. L. Snipes, pp. 1–48. Ishinomaki, Japan: Ishinomaki Senshu University Press.
Snipes, R. L., Clauss, W., Weber, A. & Hörnicke, H. (1982). Structural and functional differences in various divisions of the rabbit colon. *Cell and Tissue Research*, **225**, 331–346.
Snipes, R. L., Hörnicke, H., Björnhag, G. & Stahl, W. (1988). Regional differences in the hindgut structure and function in the nutria, *Myocastor coypus. Cell and Tissue Research*, **252**, 435–447.
Snipes, R. L. & Kriete, A. (1991). Quantitative investigation of the area and volume in different compartments of the intestine of 18 mammalian species. *Zeitschrift für Säugetierkunde*, **56**, 225–244.
Snipes, R. L., Nevo, E. & Sust, H. (1990). Anatomy of the caecum of the Israeli mole rat, *Spalax ehrenbergi* (Mammalia). *Zoologischer Anzeiger*, **224**, 307–320.
Snipes, R. L. & Thiele, Ch. (1989). Architecture of the rabbit cecal mucosa. In *Trends in Vertebrate Morphology*, ed. H. Splechtna & H. Hilgers. *Fortschritte der Zoologie*, **35**, 93–97.
Sohma, M. (1983). Ultrastructure of the absorptive cells in the small intestine of the rat during starvation. *Anatomy and Embryology*, **168**, 331–339.
Specht, W. (1977). Morphology of the intestinal wall. In *Intestinal Permeation*, ed. M. Kramer & F. Lauterbach, pp. 4–40. Amsterdam: Excerpta Medica.
Stark, R., Roper, T. J., MacLarnon, A. M. & Chivers, D. J. (1987). Gastrointestinal anatomy of the European badger *Meles meles* L. A comparative study. *Zeitschrift für Säugetierkunde*, **52**, 88–96.
Ulyatt, M. J., Dellow, D. W., Reid, C. S. W. & Bauchop, T. (1975). Structure and function of the large intestine of ruminants. In *Digestion and Metabolism in the Ruminant*, ed. I. W. McDonald & A. C. I. Warner, pp. 119–133, Armidale, Australia: University of New England Publishing Unit.
Vernay, M. (1986). Colonic absorption of inorganic ions and volatile fatty acids in the rabbit. *Comparative Biochemistry and Physiology*, **83A**, 775–784.
Weibel, E. R. (1979). Stereological methods. *Practical Methods for Biological Morphometry*, Vol. 1. London: Academic Press.
Wille, K.-H. (1975). Über die Schleimhautoberfläche des Blinddarmes einiger Haussäuger. Eine raster-elektronenmikroskopische Untersuchung. *Zentralblatt für Veterinärmedizin C*, **4**, 265–273.
Woodall, P. F. & Skinner, J. D. (1993). Dimensions of the intestines, diet and faecal water loss in some African antelope. *Journal of Zoology* (London), **229**, 457–471.
Yamasaki, F. (1971). Comparative anatomical studies on the lymphoid apparata at the ileocaecal region in Lagomorpha. *Okajimas Folica Anatomica Japonica*, **47**, 407–432.
Yamasaki, F. & Komatsu, S. (1971). Peculiar spiral ridges of the caecum of the pika. *Journal of Premedicine Sapporo Medical College*, **12**, 41–54.

16

Weaning time and bypass structures in the forestomachs of Marsupialia and Eutheria

PETER LANGER

At weaning, the herbivorous mammal undergoes a digestive metamorphosis which, although not as obvious to the casual observer, is as significant as the transformation of a tadpole to a frog.

P. A. Janssens and J. H. Ternouth (1987)

Introduction

How do animals digest their food? Considering this question, it is possible to differentiate two groups, namely, those species that are able to digest the ingesta with their own enzymes (**autoenzymatic** digestion) and, secondly, species eating a food that can only be digested with the help of microbial fermentation (**alloenzymatic** digestion). The latter type is necessary when plant material is eaten, as plants contain structural carbohydrates in their cell walls: mainly cellulose with β1-4 linkages between D-glucose monomers. The cell-wall carbohydrates can be encrusted to different extents with lignin and can thus be made highly resistant to microbial degradation. Bacteria and, in many cases, protozoans and sometimes even fungi, are the organisms that have to colonize the gastro-intestinal tract of mammals with alloenzymatic digestion (Savage, 1972, 1983; Church, 1988). Both the large intestine and, in some cases, the forestomach are regions of the tract that have been widened to 'fermentation vats'. Although the main products of microbial activity, namely short-chained fatty acids, can be absorbed in the forestomach and in the large intestine (Stevens, 1988), the efficiency of digestion in both regions of the tract is quite different (Demeyer and de Graeve, 1991). Protein and amino acids are microbially synthesized in the forestomach of ruminants and are digested in the hindstomach and small intestine with subsequent absorption of their end-products (Stevens, 1988). The forestomach lies orad or proximal to those regions where autoenzymatic proteolytic digestion takes place. This location of the fermentation chamber is called 'pre-peptic' (because of the peptic enzymes in the hindstomach or in the unilocular – one-chambered – stomach; Hungate, 1988) or 'pre-acid' (because of acid secretions in the hindstomach or in the unilocular stomach, Hungate, 1976). Fermentation in those parts of the digestive tract that lie aborad or distal to the zone of peptic or acid digestion is less efficient because most of the microbially synthesized proteins and amino acids are voided with the faeces. Stevens (1988, p. 181) stated 'that there is no definitive evidence for active absorption of amino acids by the large intestine of adult mammals'. Microbes in the large intestine transform intralumenal ammonia into bacterial protein and, therefore, this part of the gastro-intestinal tract has a detoxifying function (Kaufmann and Ahrens, 1985). Extensive discussions on the problem of detoxification of chemical plant defences by herbivores are given by Brattsten (1979), van Soest (1982) and Crawley (1983).

Few neonates have the capacity to use the food of their parents readily, especially when these live on plant material (Moir, Ch. 7). Before weaning, all mammals digest autoenzymatically. It is a characteristic feature of mammals that they obtain nutrients in their suckling period from their mother, 'post-paritive matrotrophy' (Blackburn, 1992). The milk they suckle is easily digestible, i.e. its nutrient content is accessible with the animal's own digestive enzymes. Considerable quantities of proteins are available in the milk. The stomach of the newborn young shows little proteolytic activity and intact proteins pass to the duodenal lumen where proteolysis and absorption of amino acids start (Vernier and Sire, 1989). Adult mammals that eat at least some plant material rich in cell-wall constituents have to pass through a transitional period when autoenzymatic digestion of milk in the small intestine becomes less important and alloenzymatic digestion in the forestomach and/or in the large intestine increases in significance. This period of transition, both in nutrition and in digestion, is called the weaning time, i.e. the time when the young mammal is forced to replace mother's milk by other nourishment.

In mammals eating material of animal origin – so-called zoophagous mammals – protein, fat and soluble carbohydrates are digested before, during and after this nutritional and digestive transitional time. On the other hand, in phytophagous (plant-eating) mammals the digestive process in the adult animal has to handle material rich in plant cell wall, which is completely absent from the diet when these animals are still suckling. Microbial colonization of the digestive tract is not just a more or less 'accidental' process taking place because the animal's environment is not sterile, but rather that a well-balanced and complex microbial population has to be established to make the effective fermentation of food material possible. The easily digestible

Table 16.1. *Data on the digestive tract of mammals*

Genus and species	La	We	Di	Ga	Ms	Mc	Gr	Reference
Marsupialia								
Didelphis sp.	101	5	2	1	0	0	0	2
Philander opossum	60	12	2	1	1	1	1	4
Sarcophilus harrisii	140	30	1	1	1	1	1	2
Potorous tridactylus	126	11	3	3	2	2	4	3, 13, 21
Bettongia lesueur	165	50	3	3	2	1	2	9, 10, 13
Thylogale thetis	160	38	3	3	1	1	1	7, 9, 11
Petrogale inornata	286	82	3	3	0	0	0	9, 13
Setonix brachyurus	240	55	3	3	3	3	9	12
Macropus giganteus	550	253	3	3	3	3	9	9, 12, 13
M. rufus	365	130	3	3	3	3	9	9, 13
M. rufogriseus	450	220	3	3	3	3	9	9, 13
M. eugenii	270	56	3	3	3	3	9	9. 19, 23
Eutheria								
Bradypus tridactylus	25	11	3	3	2	2	4	16
Erinaceus europaeus	40	19	1	1	1	1	1	5, 15,17
Talpa europaea	29	3	1	1	1	1	1	5
Homo sapiens	210	45	2	1	1	1	1	18
Canis lupus	49	28	1	1	2	1	2	1, 5
Felis silvestris	60	48	1	1	2	1	2	3
Equus przewalskii	112	76	2	1	2	2	4	14
Sus scrofa	90	73	2	2	3	3	9	6, 19, 24
Catagonus wagneri	75	45	3	3	2	2	4	20
Lama guanicoe	180	120	4	3	4	4	16	25
Hyemoschus aquaticus	90	76	4	3	4	4	16	8
Cervus dama	120	92	4	3	5	5	25	6
Bos primigenius	67	53	4	3	5	5	25	15
Rupicapra rupicapra	147	117	4	3	5	5	25	6
Ovis ammon	168	138	4	3	5	5	25	6
Cricetus cricetus	28	22	2	2	2	2	4	17
Ondatra zibethicus	18	8	2	1	3	3	9	5, 17
Myocastor coypus	53	47	2	1	1	1	1	3
Oryctolagus cuniculus	21	11	2	1	1	1	1	5, 22

Raw data are from the following references: 1. Banfield (1981); 2. Collins (1973); 3. Eisenberg (1981); 4. Eisenberg (1989); 5. Görner and Hackethal (1988); 6. Heptner *et al.* (1966); 7. Johnson (1977); 8. Kingdon (1979); 9. Langer (1980a); 10. Langer (1980b); 11. Langer (1980c); 12. Langer *et al.* (1980); 13. Lee and Cockburn (1985); 14. Meyer (1986); 15. Meyer and Kamphues (1990); 16. Montgomery and Sunquist (1978); 17. Niethammer and Krapp (1982, 1990); 18. Palitzsch (1986); 19. Parker (1977); 20. Sowls (1984); 21. Thompson (1987); 22. Wiesner (1990); 23. Janssens and Ternouth (1987); 24. Kidder and Manners (1978); 25. Cardozo (1954).

milk is also susceptible to microbial fermentation because many bacteria are versatile in their capacity to digest not only plant polysaccharides but also protein (e.g. from milk) (Kotarski *et al.*, 1992). Seen from the point of view of the newborn young, an additional trophic level, the microbial one, is introduced into the digestion of milk. Such an additional trophic level is connected with a loss of energy for the mammalian host (Ryszkowski, 1982). It therefore makes sense that milk in those mammals that digest their particulate food alloenzymatically in a forestomach should bypass the fermentation vat. The bypassing milk is transported in groove-like structures of different complexity along the lesser curvature.

One purpose of this paper is to classify these differentiations of the lesser curvature. They will be discussed in relation to the types of digestion and of differentiation of the entire digestive tract. The length of the lactation period is also taken into consideration. One new aspect of this investigation is to discuss the length of weaning time and its relation to these anatomical and functional differentiations.

Materials and methods

Anatomical data, as well as those referring to lactation and weaning, were compiled for 12 marsupial and 19 eutherian species (Table 16.1). Material

Notes to Table 16.1 (*cont.*)
La, lactation time (days).
We, weaning time (days) is defined here as the time when both milk and particulate food are ingested by the young animal.
Di, types of digestion (terminology after Langer, 1988): (1) autoenzymatic digestion; (2) alloenzymatic digestion in the large intestine; (3) alloenzymatic digestion in the forestomach without rumination; (4) alloenzymatic digestion in the forestomach, with rumination.
Ga, gastric differentiations (terminology after Langer, 1985): (1) unilocular stomach; (2) diverticulated stomach; (3) plurilocular stomasch.
Ms, differentiations of the tunica muscularis[a]: (0) no segmental loop; (1) segmental loop, no folds; (2) two shallow folds; (3) two prominent folds; (4) folds in reticulum and abomasum; (5) folds in reticulum, omasum, and abomasum.
Mc, differentiations in the tunica mucosa[a]: (0) no folds; (1) one, two or more functionally changing folds; (2) shallow and prominent folds; (3) prominent and permanent folds; (4) folds in reticulum and abomasum; (5) folds in reticulum, omasum and abomasum.
Gr, groove differentiations[a]: product of muscular and mucosal differentiations: Gr = Ms × Mc.
[a] The terminology was coined by Waldeyer (1908); Elze (1919); NAV (1983); NA (1989) and Langer and Snipes (1991).
For further details on muscular as well as mucosal characteristics refer to Table 16.3.

Table 16.2. *Sources of material used for anatomical investigations*

Supplied by	Species (common name)
Benirschke, La Jolla, USA	*Catagonus wagneri* (Chacoan peccary)
Hagenbecks Tierpark, Hamburg, Germany	*Lama guanicoe* (guanaco)
Hartwig, Gießen, Germany	*Rupicapra rupicapra* (chamois)
Hendrichs, Bielefeld, Germany	*Potorous tridactylus* (long-nosed potoroo)
Hörnicke, Hohenheim, Germany	*Myocastor coypus* (coypu)
Hume, Armidale, Australia	*Macropus eugenii* (tammar wallaby); *M. rufogriseus* (red-necked wallaby); *M. rufus* (red kangaroo); *Thylogale thetis* (red-necked pademelon)
Meirte, Tervuren, Belgium	*Hyemoschus aquaticus* (water chevrotain)
Moeller, Heidelberg, Germany	*Sarcophilus harrisii* (Tasmanian devil)
Moir, Nedlands, Australia	*Setonix brachyurus* (quokka)
Nelson, Melbourne, Australia	*Petrogale inornata* (unadorned rock-wallaby)
Podloucky, Hannover, Germany	*Didelphis* sp. (opossum); *Philander opossum* (four-eyed opossum)
Römer, Heuchelheim, Germany	*Ondatra zibethicus* (muskrat); *Oryctolagus cuniculus* (rabbit)
Schliemann, Hamburg, Germany	*Bradypus tridactylus* (three-toed sloth)
Thomé, Gießen, Germany	*Equus przewalskii* (horse)
Thorius-Erler, Gießen, Germany	*Bos primigenus* (cattle); *ovis ammon* (sheep); *Sus scrofa* (pig)
Thun-Hohenstein, Salzau, Germany	*Cervus dama* (fallow deer)
Tyndale-Biscoe, Canberra, Australia	*Bettongia lesueur* (burrowing bettong)
Zoo Nürnberg, Germany	*Macropus giganteus* (eastern grey kangaroo)
Own collection	*Canis lupus* (dog); *Cricetus cricetus* (common hamster); *Erinaceus europaeus* (hedgehog); *Felis silvestris* (cat); *Homo sapiens* (human); *Talpa europaea* (vole)

Common names are listed according to van den Brink and Haltenorth (1972); Dorst *et al.* (1973), Strahan (1983) and Redford and Eisenberg (1992).

for anatomical investigations was made available by the persons listed in Table 16.2.

The first column of Table 16.1 gives data from the literature on the length of lactation time in days. It should be kept in mind that these values are not completely unambiguous. For example, in man the data published in the literature vary considerably: Palitzsch (1986) speaks of a lactation time in humans of 7 months, but O'Crohan (1990) who, in the middle of the 19th century, spent his youth on the Blasket Islands off western Ireland, was 4 years old before he was weaned! Other examples of considerable differences in weaning time of young animals can be found in domestic animals, especially in cattle, where lactation and weaning is managed by man. The length of *weaning* is given in days and is defined as *the period when the young animal suckles milk and eats increasing amounts of particulate food that is also ingested by the adult.* Weaning time is the period of overlap between lactation and the early period of particulate food intake.

Four types of digestion, listed in Table 16.1, were differentiated previously by the present author (Langer, 1988): **autoenzymatic digestion** (Di 1) refers to mammals that are able to digest their food exclusively with their own digestive enzymes. Mammals applying the other three types of digestion do not rely exclusively on their own enzymes but also use the enzymatic activity of microbes. We speak of this as **alloenzymatic digestion**. This can take place in two general compartments of the digestive tract, namely, in the large intestine or caecum-colon complex (Di 2) or in forestomachs with expanded volumes (Di 3). In the latter alloenzymatic digestion can be connected with rumination (Di 4) where particle size is degraded and the plant cell wall is opened mechanically after ingestion.

In devising criteria of gastric differentiation it is not advisable to apply the terminology used in the veterinary literature (Nickel *et al.*, 1973; Sisson, 1975): according to these authors a simple stomach consists of a single compartment, a complex stomach has more than one compartment. The latter term is especially very confusing because complexity of the stomach is also established by the types of internal mucosal lining. However, when a stomach is lined with glandular *and* non-glandular mucosa, this is called a composite stomach by the veterinary anatomists (Nickel *et al.*, 1973). Those parts of the organ that are lined with non-glandular mucosa are called forestomach (Sisson, 1975). To avoid these terminological confusions in relation to gastric form, I will speak of the unilocular (Ga 1), diverticulated (Ga 2) and plurilocular (Ga 3) types of stomach. Ga 1 represents a stomach with only one luminal cavity, the so-called unilocular stomach. Under Ga 2 those stomachs are listed that have a small blindsac or diverticulum or a clearly circumscribed region of glandular epithelium within other types of epithelial lining. According to Luppa (1956), this type of diverticulum is called '*Drüsenbeutel*'; Garon and Piérard (1972) call it '*région glandulaire*'. Stomachs with this type of differentiation were termed 'discoglandular stomach' by Carleton (1973) and Langer (1985). The plurilocular group of stomachs (Ga 3) have more than one gastric chamber.

Differentiations of the luminal side of the lesser curvature are characterized in Table 16.1 as Ms, Mc and Gr. To understand these differentiations better, one should bring to mind that – contrary to conditions in the other parts of the gastro-intestinal tract – the **tunica muscularis** of the stomach consists not only of an external longitudinal and an internal circular muscle layer, but also of an internal, most-oblique layer that can be found in the **fornix gastricus** and in the **corpus gastricum**. These oblique fibres are dissected and shown in Fig. 16.1 for the unilocular human stomach.

The region of the lesser curvature of the stomach of the species in Table 16.1 and 16.2 were cross-sectioned with a very sharp knife and then soaked for at least a day with an alcoholic solution of 'boraxkarmin' (Romeis, 1968). After this time, the excess stain was washed out with 70% ethanol until the muscular, mucosal and connective tissue layers could be clearly differentiated. The cross-sectional aspects were drawn to scale; some of the illustrations are given in Fig. 16.2.

In Table 16.3 the tunica muscularis and the tunica mucosa of the gastric wall are considered. Figure 16.2 shows the outlines of the stomach and a semi-schematic cross-section through the region of the lesser curvature. Terms that identify the different types of grooves or groove-like structures parallel to the lesser gastric curvature have been presented in the literature

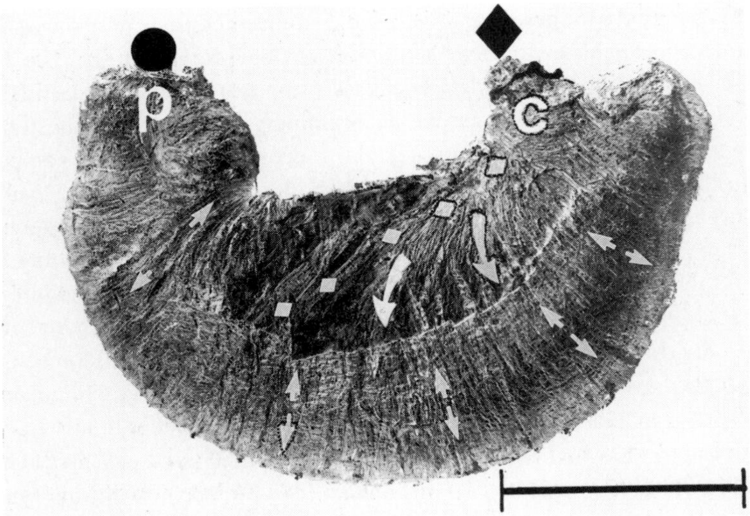

Fig. 16.1. Paries anterior of a human stomach from which the tunica serosa and the stratum longitudinale of the tunica muscularis have been removed. The black diamond indicates the stump of the oesophagus (c, cardia) and the black round dot indicates the stump of the duodenum (p, pylorus). The small white arrows indicate the course of the stratum circulare of the musculature, the two curved white arrows the course of the fibrae obliquae. The row of small white diamonds represent the course of the upper segmental loop (Forssell, 1912, 1913) of the oblique fibres. The length of the scale is 10 cm.

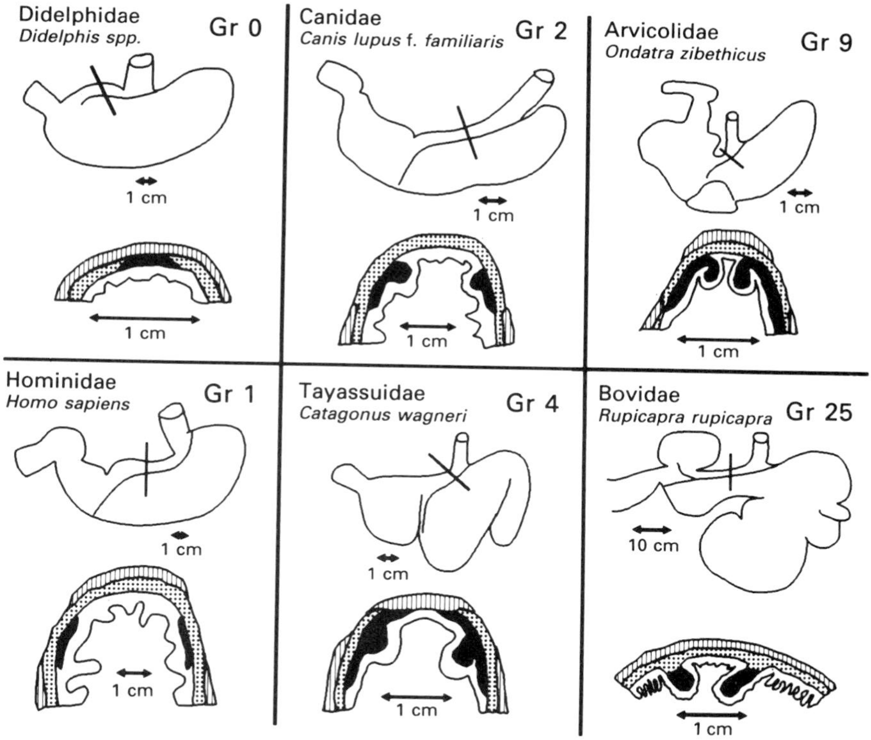

Fig. 16.2. Pictorial representation of the outlines of the stomach of six mammalian species and a cross-section of the lesser curvature. The position of the cross-section is marked on the outline figure. The areas of the cross-sections are: vertical lines, stratum longitudinale; dots, stratum circulare; black, fibrae obliquae; white, tunica mucosa. The differentiation of the groove or groove-like structure (Gr) is given according to Table 16.1.

(Waldeyer, 1908; Elze, 1919; Bauer, 1923; NAV 1983; NA, 1989; Langer and Snipes, 1991). The basis for differentiations of the tunica muscularis is the so-called segmental loop (Forssell, 1912, 1913), running from the cardia towards the greater gastric curvature. This muscular loop forms folds of different prominence, which are classified from Ms 0 (no segmental loop) to Ms 5 with prominent folds in the different compartments of the ruminant forestomach. On the other hand, the differentiation of the tunica mucosa does not always develop in parallel steps to that of the tunica muscularis. The mucosal differentiations are classified from Mc 0 (no folds) to Mc 5, which represents mucosal folds in the forestomach compartments of ruminants. Gr 0 indicates that there is a smooth internal aspect of the lesser curvature with-

Table 16.3. *The muscular and mucosal characteristics along the lesser curvature of mammalian stomachs*

	Ms	Mc	Gr	Species depicted in Fig. 16.2
No groove	0	0	0	*Didelphis* sp.
Canalis ventriculi (Magenstrasse)	1	1	1	*Homo sapiens*
Sulcus salivalis	2	1	2	*Canis lupus*
Sulcus gastricus sive ventriculi				
Shallow	2	2	4	*Catagonus wagneri*
Prominent	3	3	9	*Ondatra zibethicus*
Sulcus reticuli				
No omasum	4	4	16	None
With omasum	5	5	25	*Rupicapra rupicapra*

Ms, tunica muscularis; Mc, tunica mucosa; Gr, groove differentiation. The differentiation of these groups into numbered levels is as given in Table 16.1. The terminology was coined by Waldeyer (1908), Elze (1919), NAV (1983) and NA (1989). The compilation is modified after Langer and Snipes (1991).

out any differentiation of a groove (neither muscular nor mucosal) and the other extreme, Gr 25, represents the complex gastric groove in pecoran ruminants.

To indicate the complexity of the groove-like structures, a factor, groove differentiation, is used which is the product of muscular and mucosal differentiations (Gr = Ms × Mc) (Table 16.1). It is applied here to represent *increasing complexity* by connection of previously separate systems (the musculature and the mucosa), which in turn is an essential part of evolutionary differentiation (Duncker, 1991). Complexity is not simply the sum of differentiations but represents extensive interrelationships. However, precise and detailed information on these interrelationships often are not available (Riedl, 1976).

Results

The data from Table 16.1 are compiled in diagrams (Fig. 16.3) for marsupials or for eutherians showing the length of lactation time versus the types of digestion and the groove differentiations (Table 16.3), which are characterized by the height of the vertical lines. Weaning time is considered in a similar way in Fig. 16.4.

Discussion

While the length of lactation time gives an idea of the period of physiological dependence of the young, weaning time gives information about the period when the young adapts to a nutritionally independent life. Short weaning time means that a relatively long portion of lactation is 'reserved' for exclusive intake of milk without addition of particulate food, a period called 'peak lactation' by Oftedal (1984a) when peak milk yield and peak milk energy output are reached (Oftedal, 1984b).

The information presented in Figs. 16.3 and 16.4 can be 'condensed' in a diagrammatic compilation as in Table 16.4. Groups of mammals can be categorized according to functional and morphological criteria in Marsupialia and Eutheria.

Marsupialia (Table 16.4*a*)

Group M I

Marsupials digesting their food either autoenzymatically or by fermentation in the large intestine form group M I. They always have a unilocular stomach, in which the grooves parallel to the lesser gastric curvature are poorly developed. These mammals have a lactation time that never lies below 60 days (at least in those species considered here) and are characterized by a short weaning time that is less than 30 days. Examples: *Didelphis* spp., *Philander opossum*, *Sarcophilus harrisii.*

Group M II

These are non-ruminating forestomach fermenters with, of course, a plurilocular stomach. Their gastric grooves are either very poorly developed or of intermediate complexity. Their lactation time is generally of intermediate length and the weaning period is short or of intermediate length. Examples: *Potorous tridactylus*, *Bettongia lesueur*, *Thylogale thetis*, *Petrogale inornata*, *Setonix brachyurus*, *Macropus eugenii.*

Group M III

Compared with group M II, the species representing group M III have intermediate differentiation of gastric grooves and long lactation and weaning times. Examples: *Macropus giganteus*, *M. rufus*, *M. rufogriseus.*

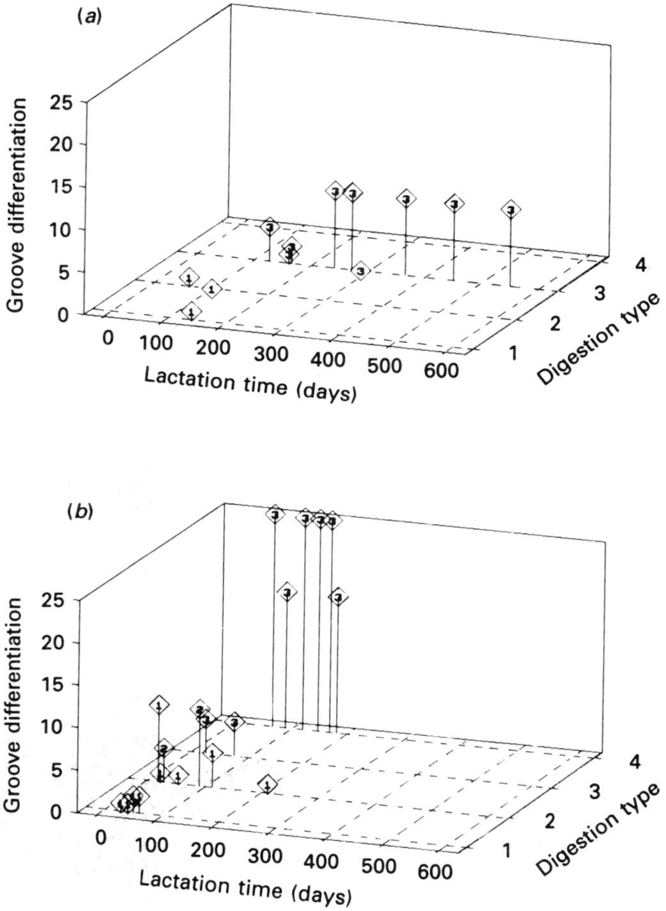

Fig. 16.3. Three-dimensional diagram showing the length of lactation in (*a*) marsupials and (*b*) eutherians versus types of digestion (Di in legend of Table 16.1) and groove differentiations (Gr in Table 16.1 and 16.3). The numbers in the diamonds represent stomach differentiations (Ga), also according to Table 16.1.

Eutheria (Table 16.4*b*)

Group E I

These are mammals that digest autoenzymatically. The stomach is unilocular and – as bypass structures are not necessary – groove differentiations are only minimal. Lactation and weaning times are short. Examples: *Erinaceus europaeus*, *Talpa europaea*, *Canis lupus*, *Felis silvestris*.

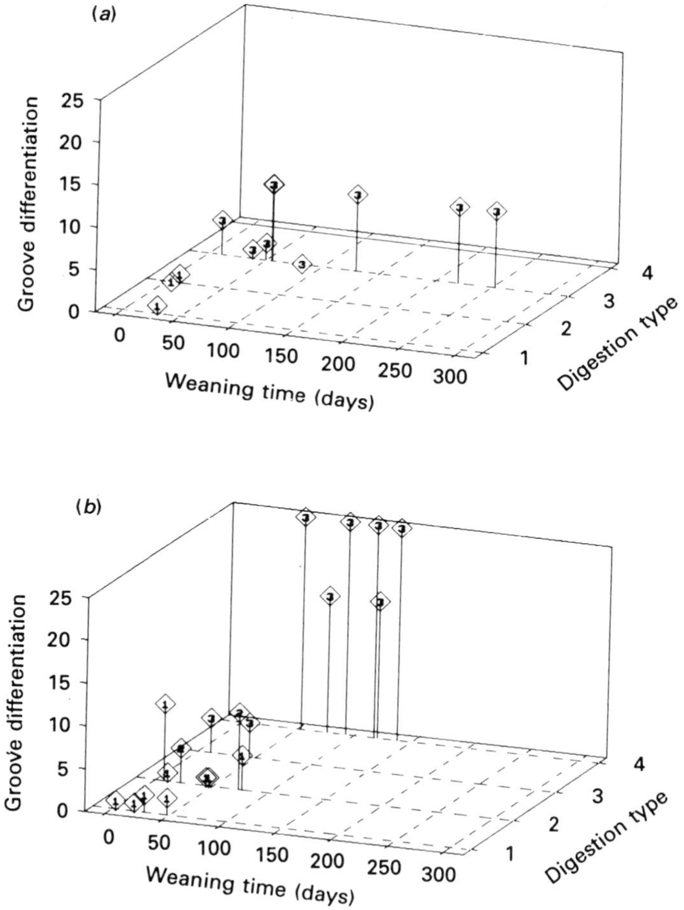

Fig. 16.4. Three-dimensional diagram showing the weaning time in (*a*) marsupials and (*b*) eutherians. Details as in Fig. 16.3.

Group E II

In this group there is alloenzymatic digestion in the large intestine, the stomach is unilocular or diverticulated and gastric grooves show small or intermediate differentiations. Lactation time is either intermediate or – in humans – exceptionally long, but weaning time is short or of intermediate length. Examples: *Homo sapiens*, *Equus przewalskii*, *Sus scrofa*, *Cricetus cricetus*, *Ondatra zibethicus*, *Myocastor coypus*, *Oryctolagus cuniculus*.

Table 16.4. *Anatomical and functional differentiations of the digestive tract*
(a) Marsupialia

	M I	M II	M III
Types of digestion	Autoenzymatic, large intestine	Non-ruminating	Non-ruminating
Gastric differentiations	Unilocular	Plurilocular	Plurilocular
Groove differentiations	Small (Gr ≤ 2)	Small (Gr ≤ 4) Intermediate (Gr ≤ 9)	Intermediate (Gr ≤ 9)
Lactation time (days)	Intermediate (60–140)	Intermediate (126–286)	Long (365–550)
Weaning time (days)	Short (5–30)	Short (11–50) Intermediate (50–82)	Long (130–253)

Group E III

These eutherians are non-ruminating forestomach fermenters, of course with a plurilocular stomach, but with minimal differentiation of gastric grooves. Their lactation time is of intermediate length and their weaning time is short. Examples: *Bradypus tridactylus*, *Catagonus wagneri*.

Group E IV

Eutherian mammals eating low-quality food can digest this with the help of rumination, a process always combined with a plurilocular stomach. Because of microbial fermentation in the forestomach, and because an efficient bypass of this fermentation chamber is necessary to avoid microbial degradation of the milk constituents, complex differentiations of grooves are also necessary. Lactation and weaning times are long. Examples: *Lama guanicoe*, *Hyemoschus aquaticus*, *Cervus dama*, *Bos primigenius*, *Rupicapra rupicapra*, *Ovis ammon*.

Comparison of Marsupialia and Eutheria

From Figs 16.3 and 16.4, as well as from Tables 16.4*a* and 16.4*b*, it is possible to conclude that marsupials are less differentiated: only three groups of species can be found. In the eutherians it is possible to see four distinct group. Ruminating forms with high differentiation of the gastric groove can only be found in the Eutheria.

(b) Eutheria

	E I	E II	E III	E IV
Types of digestion	Auto-enzymatic	Large intestine	Non-ruminating	Ruminating
Gastric differentiations	Unilocular	Unilocular Diverticular	Plurilocular	Plurilocular
Groove differentiations	Small ($Gr \leq 2$)	Small Intermediate ($Gr \leq 9$)	Small ($Gr \leq 4$)	High ($Gr \leq 16$)
Lactation time (days)	Short (28–60)	Intermediate (18–112) Long (210)	Intermediate (25–75)	Long (67–180)
Weaning time (days)	Short (3–48)	Short (8–50) Intermediate (50–76)	Short (11–45)	Long (53–138)

Abbreviations for groove differentiations as in Table 16.1.

Lactation is significantly longer in marsupials than in eutherians, with considerable variation within both groups. The same is also true when only forestomach-fermenting (Ff) marsupials (mean about 300 days) are compared with forestomach-fermenting eutherians (mean about 110 days) ($P < 0.005$) (Fig. 16.5). Lee and Cockburn (1985) showed that marsupial and eutherian young weaned at similar body weights. Because marsupials 'have elaborated lactation instead of extending the period of intrauterine development' (Renfree, 1991, p. 164) a larger proportion of growth is sustained by lactation in marsupials than in eutherians (Cockburn and Johnson, 1988).

It is quite remarkable that weaning time is *not* significantly different between Marsupialia and Eutheria. This holds true for a comparison of all available data from Table 16.1, as well as for a comparison of forestomach fermenters only (Fig. 16.5).

If we compare weaning time of forestomach-fermenting (Ff) Eutheria with those species that are either autoenzymatically digesting or large-intestinal

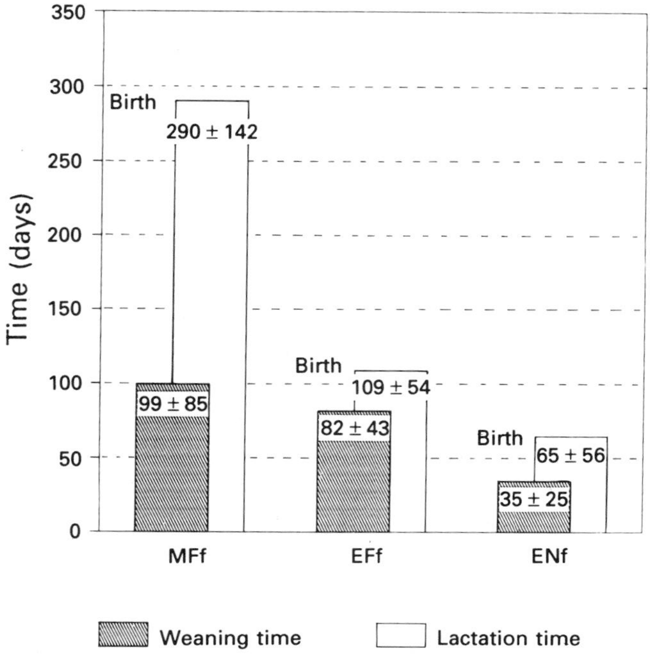

Fig. 16.5. Lactation and weaning times in forestomach-fermenting marsupials (MFf) and eutherians (EFf) and in non-forestomach-fermenting eutherians (ENf). White columns, lactation times (mean ± S.D.); hatched columns, weaning time (mean ± S.D.). The number of observations in each group is MFf (9), EFf (8) and ENf (11).

fermenters, i.e. non-forestomach-fermenting mammals (Nf), there is a significant difference in weaning time (approximately 80 days in forestomach fermenters and approximately 35 days in forms that do not ferment in the forestomach, $P < 0.01$), but *not* in lactation time (Fig. 16.5). How can these findings concerning weaning time be explained?

The fact that weaning is longer in forestomach-fermenting eutherians than in those that do not show this physiological differentiation, as well as the practically identical weaning time in marsupial and eutherian forestomach fermenters, makes it very probable that considerable functional changes take place within the fermentation chamber of forestomach fermenters. These changes are most probably, related to microbial colonization of this chamber. It can be assumed, as well, that this colonization requires time for the establishment of a well-balanced microbial population. In the breast-fed human infant, a species with a unilocular stomach, the ingestion of solids in addition to milk has a major impact on the microbial ecosystem of the gastro-intestinal tract (Perman, 1989). Microbial colonization is influenced by the internal milieu of the tract, e.g. temperature and the epithelial surface that represents the important interface between gastric lumen and wall (Cheng and Costerton, 1980; Czerkawski and Cheng, 1988; Cheng *et al.*, 1991). Salivary and epithelial secretions create a characteristic pH in the digestive tract, which in turn enables colonization by certain microbial species. Microorganisms colonize particular regions of the gastro-intestinal tract early in life, multiply to high population levels soon after colonization, and remain at those levels throughout the lives of healthy well-nourished animals (Savage, 1972). It is interesting that 'some quite distinctive morphological forms of microorganisms are usually found associated with surfaces in only certain areas of the tract' (Savage, 1983, p. 330). According to this author, lactobacilli associate with keratinizing squamous epithelia in the stomach; filamentous bacteria adhere only to small bowel epithelia; fusiform prokaryotes associate only with caecal and colonic epithelia; while yeasts can be found on gastric secreting surfaces.

To set up a microbial population, the establishment of 'late-comers' is influenced by other species that already inhabit the gastro-intestinal tract (Baker, 1984; Leng, 1984; Dehority and Orpin, 1988; Fonty *et al.*, 1987; Fonty and Joblin, 1991; Sonneborn and Greinwald, 1991). For example, Koopman (1984) remarked that the microflora naturally present ensures that pathogenic bacteria will not become established in the intestinal tract; he speaks of 'intestinal colonization resistance' (Koopman *et al.*, 1982). On the other hand, some studies also 'demonstrate the intimate and beneficial interactions that have been established among bacterial species of the gastrointestinal tract . . .' (Bryant, 1972).

In sheep, rumination is established after approximately 80 to 90 days (Fig. 16.6). During this period, mature conditions in relation to bacterial colonization, composition of fermentation products, salivary secretions and forestomach volume have to be developed. The considered physiological and anatomical conditions reach adult proportions within the weaning time. Relatively early, namely about 42 days after birth, i.e. approximately 12 days after the end of peak lactation (Oftedal, 1984a, b) or about 12 days after the beginning of the weaning period, bacterial composition in the rumen of lambs changes to adult conditions (Yokoyama and Johnson, 1988) and only about a week later fermentation and production of short-chain fatty acids reach adult levels (Lyford and Huber, 1988). The same authors showed that about 63 days after birth the salivary gland weight reaches adult proportions in lambs. According to Wallace (1948), Wardrop and Coombe (1960), Church

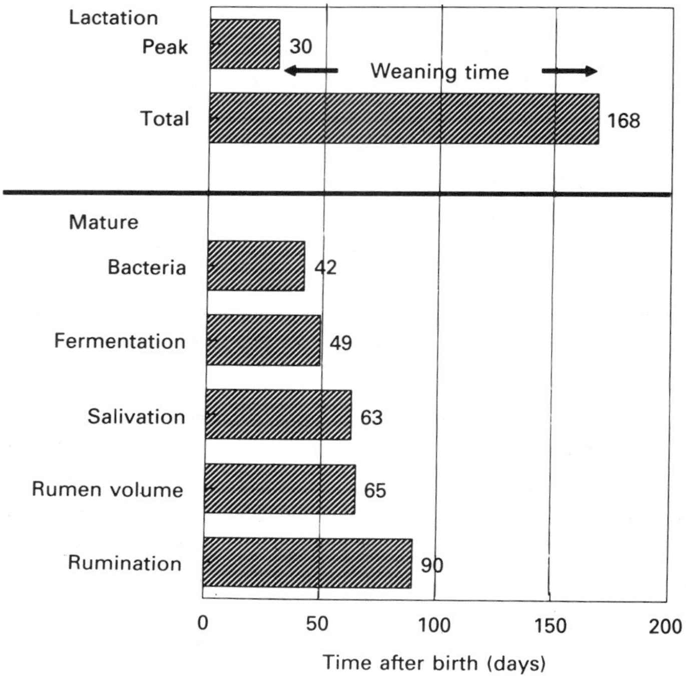

Fig. 16.6. Lactation and weaning times in sheep (*Ovis ammon*) and the time when certain mature conditions of digestion are reached. Peak lactation (Oftedal, 1984a, b) is the period when only milk is ingested. The transitional period to adult food, weaning time, is also marked. From the literature information can be obtained about the time when mature conditions are reached, e.g. for the bacterial flora in the 'fermentation vat' (Yokoyama and Johnson, 1988), microbial fermentation of the food and the intensity of salivation (Lyford and Huber, 1988), adult proportions of the rumen volume (Wallace, 1948; Wardrop and Coombe, 1960; Church *et al.*, 1962; Langer, 1988; Meyer and Kamphues, 1990), as well as the adult intensities of rumination and chewing (Lyford and Huber, 1988).

et al. (1962) and Meyer and Kamphues (1990), the adult proportions of rumen volume in sheep are reached between 60 and 70 days postnatally. After about 90 days, i.e. still within the weaning period, mature patterns of chewing and rumination are established in lambs (Lyford and Huber, 1988). In cattle, a high variability in the time of establishment of adult proportions of the ruminoreticulum has to be related to man's management of bovine lactation. This is the reason why these data are not depicted here.

As it is probable that a link exists between the establishment of a well-balanced microbial forestomach population and the length of weaning time, it is now of interest to examine whether the differentiation of grooves following the lesser gastric curvature is related to the length of weaning time. This question has to be asked because a long period of microbial colonization is combined with the necessity for autoenzymatically digestible substances to bypass the forestomach or pre-acid fermentation vat. Although there is no significant correlation between weaning time and the differentiations of grooves in marsupials, there is a significant correlation in eutherians ($r = 0.7523$, $n = 19$, $P < 0.001$). Highly complex grooves can be found in eutherians with relatively long weaning times (Fig. 16.4*b*). The diagram illustrating the situation in marsupials (Fig. 16.4*a*) at first glance seems to give the same impression, but the high variability in group M II (Table 16.4), both in groove differentiation as well as in weaning time, does not corroborate this statement.

Table 16.5 gives those pairs of functional and anatomical criteria between which a significant correlation can be found. In marsupials and eutherians the type of gastric differentiations (Ga) and types of digestion (Di), as well as the length of lactation (La) and weaning (We) are correlated with each other. In the Eutheria, but not in the Marsupialia, the present material allows the statement that groove differentiations (Gr) are correlated with weaning time (We) in the sense that a functional groove (high value of Gr) can be found in species with long weaning time. It can also be said that the groove differentiation is related to the type of digestion (Di). For example, food of the average zoophagous mammal does not have to bypass a fermentation chamber and therefore a groove is not necessary. This latter differentiation is, however, necessary in forestomach-fermenting phytophages. An extended period in which the fermentation chamber should be bypassed produces evolutionary pressure towards differentiation of the tunica muscularis and the tunica mucosae in the area of the lesser curvature. Investigations with a much

Table 16.5. *Correlation between functional and anatomical criteria in Marsupialia (+) and Eutheria (#) with a significance level $P < 0.001$*

	La	We	Di	Ga
La		+		
We	#			
Di		#		+
Ga			#	
Gr		#	#	#

Abbreviations as in Table 16.1.

wider range of species should be started to corroborate the results of the present pilot study.

References

Baker, S. K. (1984). The rumen as an ecosystem. In *Ruminant Physiology, Concepts and Consequences*, eds. S. K. Baker, J. M. Gawthorne, J. B. Mackintosh & D. B. Purser, pp. 149–160. Nedlands, Western Australia: University of Western Australia.

Banfield, A. W. F. (1981). *The Mammals of Canada*. Toronto: University of Toronto Press.

Bauer, K. H. (1923). Ueber das Wesen der Magenstraße. *Archiv für klinische Chirurgie*, **124**, 565–629.

Blackburn, D. G. (1992). Convergent evolution of vivipary, matrotrophy, and specializations for fetal nutrition in reptiles and other vertebrates. *American Zoologist*, **32**, 313–321.

Brattsten, L. B. (1979). Biochemical defense mechanisms in herbivores against plant allelochemicals. In *Herbivores. Their Interaction with Secondary Plant Metabolites*. ed. G. A. Rosenthal & D. H. Janzen, pp. 199–270. New York: Academic Press.

Bryant, M. P. (1972). Interactions among intestinal microorganisms. *American Journal of Clinical Nutrition*, **25**, 1485–1487.

Cardozo, A. (1954). *Los auquénidos*. La Paz, Bolivia: Editorial Centenario.

Carleton, M. D. (1973). A survey of gross stomach morphology in New World Cricetinae (Rodentia, Musroidea), with comments on functional interpretations. *Miscellaneous Publications of the Museum of Zoology, University of Michigan*, **146**, 1–43.

Cheng, K.-J. & Costerton, J. W. (1980). Adherent rumen bacteria – their role in the digestion of plant material, urea and epithelial cells. In *Digestive Physiology and Metabolism in Ruminants*. ed. Y. Ruckebush & P. Thivend, pp. 227–250. Lancaster, UK: MTP Press.

Cheng, K.-J., Forsberg, C. W., Minato, H. & Costerton, J. W. (1991). Microbial ecology and physiology of feed degradation within the rumen. In *Physiological Aspects of Digestion and Metabolism in Ruminants*. ed. T. Tsuda, Y. Sasaki & R. Kawashima, pp. 595–624. San Diego: Academic Press.

Church, D. C. (1988). *The Ruminant Animal. Digestive Physiology and Nutrition*. Englewood Cliffs, NJ: Prentice Hall.

Church, D. C., Jessup, G. L. & Bogart, R. (1962). Stomach development in the suckling lamb. *American Journal of Veterinary Research*, **23**, 220–225.

Cockburn, A. & Johnson, C. N. (1988). Patterns of growth. In *The Developing Marsupial. Models for Biomedical Research*. ed. C. H. Tyndale-Biscoe & P. A. Janssens, pp. 28–40. Berlin: Springer Verlag.

Collins, L. R. (1973). *Monotremes and Marsupials*. Washington, DC: Smithsonian Institution Press.

Crawley, M. J. (1983). *Herbivory. The Dynamics of Animal–Plant Interactions*. Oxford: Blackwell Scientific.

Czerkawski, J. W. & Cheng, K.-J. (1988). Compartmentation in the rumen. In *The Rumen Microbial Ecosystem*. ed. P. N. Hobson, pp. 361–385. London: Elsevier Applied Science.

Dehority, B. A. & Orpin, C. G. (1988). Development of, and natural fluctuations in, rumen microbial populations. In *The Rumen Microbial Ecosystem*. ed. P. N. Hobson, pp. 151–183. London: Elsevier Applied Science.

Demeyer, D. I. & de Graeve, K. (1991). Differences in stoichiometry between rumen and hindgut fermentation. *Fortschritte der Tierphysiologie und Tierernährung*, **22**, 50–61.

Dorst, J., Dandelot, P., Bohlken, H. & Reichstein, H. (1973). *Säugetiere Afrikas*. Hamburg: Verlag Paul Parey.

Duncker, H.-R. (1991). The evolutionary biology of homoiothermic vertebrates: the analysis of complexity as a specific task of morphology. *Verhandlungen der Deutschen Zoologischen Gesellschaft*, **84**, 39–60.

Eisenberg, J. F. (1981). *The Mammalian Radiations*. Chicago: University of Chicago Press.

Eisenberg, J. F. (1989). *Mammals of the Neotropics. The Northern Neotropics, Vol. I.* Chicago: University of Chicago Press.

Elze, C. (1919). Über Form und Bau des menschlichen Magens. *Sitzungsberichte der Heidelberger Akademie der Wissenschaften, Mathematisch-naturwissenschaftliche Klasse*, **B10**, 1–64.

Fonty, G., Gouet, P., Jouany, J.-P. & Senaud, J. (1987). Establishment of the microflora and anaerobic fungi in the rumen of lambs. *Journal of General Microbiology*, **133**, 1835–1843.

Fonty, G. & Joblin, K. N. (1991). Rumen anaerobic fungi: their role and interactions with other rumen microorganisms in relation to fiber digestion. In *Physiological Aspects of Digestion and Metabolism in Ruminants*. ed. T. Tsuda, Y. Sasaki & R. Kawashima, pp. 655–680. San Diego: Academic Press.

Forssell, G. (1912). Über die Beziehung der auf den Röntgenbildern hervortretenden Formen des menschlichen Magens zur Muskelarchitektur der Magenwand. *Münchner Medizinische Wochenschrift*, **29**, 1588–1592.

Forssell, G. (1913). Über die Beziehung der Röntgenbilder des menschlichen Magens zu seinem anatomischen Bau. *Fortschritte auf dem Gebiete der Röntgenstrahlen*, Ergänzungs-Band 30, 1–265.

Garon, O. & Piérard, J. (1972) Etude morphologique comparée de l'estomac de deux rongeurs Ondatra zibethicus et Marmota monax (Mammalia: Rodentia). *Canadian Journal of Zoology*, **50**, 239–245.

Görner, M. & Hackethal, H. (1988). *Säugetiere Europas*. Stuttgart: Ferdinand Enke Verlag.

Heptner, V. G., Nasimovic, A. A., Bannikov, A. G. (1966). *Die Säugetiere der Sowjetunion I. Paarhufer und Unpaarhufer*. Jena: VEB Gustav Fischer Verlag.

Hungate, R. E. (1976). Microbial activities related to mammalian digestion and absorption of food. In *Fiber in Human Nutrition*. ed. G. A. Spiller & R. J. Amen, pp. 131–149. New York: Plenum Press.

Hungate, R. E. (1988). The ruminant and the rumen. In *The Rumen Microbial Ecosystem*. ed. P. N. Hobson, pp. 1–19. London: Elsevier Applied Science.

Janssens, P. A. & Ternouth, J. H. (1987). The transition from milk to forage diets. In *The Nutrition of Herbivores*. ed. J. B. Hacker & J. H. Ternouth, pp. 281–305. Sydney: Academic Press.

Johnson, K. A. (1977). Ecology and Management of the Red-necked Pademelon, Thylogale thetis, on the Dorrigo Plateau of Northern New South Wales. Armidale, NSW: PhD thesis, University of New England.

Kaufmann, W. & Ahrens, F. (1985). Gärungsabläufe und ihre Bedeutung – eine vergleichende Betrachtung. *Aktuelle Themen der Tierernährung und Veredelungswirtschaft*. Cuxhaven: Lohmann.

Kidder, D. E. & Manners, M. J. (1978). *Digestion in the Pig*. Bristol: Scientechnica.
Kingdon, J. (1979). *East African Mammals. III B (Large Mammals)*. London: Academic Press.
Koopman, J. P. (1984). Kolonisatie resistentie van het maagdarmkanaal. *Tijdschrift Diergeneeskunde*, **109**, 1017–1026.
Koopman, J. P., Kennis, H. M., Lankhorst, A., Prins, R. A., Stadhouders, A. M. & de Boer, H. (1982). The influence of microflora and diet on gastro-intestinal parameters. *Zeitschrift für Versuchstierkunde*, **24**, 184–192.
Kotarski, S. F., Waniska, R. D. & Thurn, K. K. (1992). Starch hydrolysis by the ruminal microflora. *Journal of Nutrition*, **122**, 178–190.
Langer, P. (1980a). Stomach evolution in the Macropodidae, Owen, 1839 (Mammalia: Marsupialia). *Zeitschrift für zoologische Systematik und Evolutionsforschung*, **18**, 211–232.
Langer, P. (1980b). Anatomy of the stomach in three species of Potoroinae (Marsupialia: Macropodidae). *Australian Journal of Zoology*, **28**, 19–31.
Langer, P. (1980c). Functional anatomy and ontogenetic development of the stomach in the macropodine species *Thylogale stigmatica* and *Thylogale thetis* (Mammalia: Marsupialia). *Zoomorphologie*, **93**, 137–151.
Langer, P. (1985). The mammalian stomach: structure, diversity and nomenclature. *Acta Zoologica Fennica*, **170**, 99–102.
Langer, P. (1988). *The Mammalian Herbivore Stomach. Comparative Anatomy, Function and Evolution*. Stuttgart: Gustav Fischer Verlag.
Langer, P., Dellow, D. W. & Hume, I. D. (1980). Stomach structure and function in three species of macropodine marsupials. *Australian Journal of Zoology*, **28**, 1–18.
Langer, P. & Snipes, R. L. (1991). Adaptation of gut structure to function in herbivores. In *Physiological Aspects of Digestion and Metabolism in Ruminants*. ed. T. Tsuda, Y. Sasaki & R. Kawashima, pp. 349–384. San Diego: Academic Press.
Lee, A. K. & Cockburn, A. (1985). *Evolutionary Ecology of Marsupials*. Cambridge: Cambridge University Press.
Leng, R. A. (1984). Microbial interactions in the rumen. In *Ruminant Physiology, Concepts and Consequences*. ed. S. K. Baker, J. M. Gawthorne, J. B. Mackintosh & D. B. Purser, pp. 161–173. Nedlands, Western Australia: University of Western Australia.
Luppa, H.-W. (1956). Zur Morphologie und Histologie des Magens der Bisamratte (Ondatra zibethica). *Wissenschaftliche Zeitschrift der Universität Halle*, **5**, 647–668.
Lyford, S. J. & Huber, J. T. (1988). Digestion, metabolism and nutrient needs in preruminants. In *The Ruminant Animal. Digestive Physiology and Nutrition*. ed. D. C. Church, pp. 401–420. Englewood Cliffs, NJ: Prentice Hall.
Meyer, H. (1986). *Pferdefütterung*. Berlin: Verlag Paul Parey.
Meyer, H. & Kamphues, J. (1990). Grundlagen der Ernährung von Neugeborenen. In *Neugeborenen- und Säuglingskunde der Tiere*. ed. K. Walser, & H. Bostedt, pp. 55–71. Stuttgart: Ferdinand Enke Verlag.
Montgomery, G. G. & Sunquist, M. E. (1978). Habitat selection and use by two-toed and three-toed sloths. In *The Ecology of Arboreal Folivores*. ed. G. G. Montgomery, pp. 329–359. Washington, DC: Smithsonian Institution Press.
NA (1989). *Nomina Anatomica, 6th edn*. Edinburgh: Churchill Livingstone.
NAV (1983). *Nomina Anatomica Veterinaria, 3rd edn*. Ithaca, NY: International Committee on Veterinary Gross Anatomical Nomenclature.

Nickel, R., Schummer, A., Seiferle, E. & Sack, W. O. (1973). *The Viscera of the Domestic Mammals.* Berlin: Verlag Paul Parey.
Niethammer, J. & Krapp, F. (1982). *Handbuch der Säugetiere Europas 2/I. Rodentia II.* Wiesbaden: Akademische Verlagsgesellschaft.
Niethammer, J. & Krapp, F. (1990). *Handbuch der Säugetiere Europas 3/1. Insektenfresser – Insectivora, Herrentiere – Primates.* Wiesbaden: Aula Verlag.
O'Crohan, T. (1990). *The Islandman.* Oxford: Oxford University Press.
Oftedal, O. T. (1984a). Milk composition, milk yield and energy output at peak lactation: A comparative review. *Symposium of the Zoological Society of London*, **51**, 33–85.
Oftedal, O. T. (1984b). Body size and reproductive strategy as correlates of milk energy output in lactating mammals. *Acta Zoologica Fennica*, **171**, 183–186.
Palitzsch, D. (1986). *Pädiatrie.* Stuttgart: Ferdinand Enke Verlag.
Parker, P. (1977). An ecological comparison of marsupial and placental patterns of reproduction. In *The Biology of Marsupials.* ed. D. Gilmore & B. Stonehouse, pp. 273–286. London: Macmillan Press.
Perman, J. A. (1989). Gastrointestinal flora: developmental aspects and effects on nutrients. In *Human Gastrointestinal Development.* ed. E. Lebenthal, pp. 777–786. New York: Raven Press.
Redford, K. H. & Eisenberg, J. F. (1992). *Mammals of the Neotropics. The Southern Cone, Vol. 2.* Chicago: University of Chicago Press.
Renfree, M. B. (1991). Marsupial mammals: enigma variations on a reproductive theme. *Verhandlungen der Deutschen Zoologischen Gesellschaft*, **84**, 153–167.
Riedl, R. (1976). *Die Strategie der Genesis. Naturgeschichte der realen Welt.* München: R. Piper Verlag.
Romeis, B. (1968). *Mikroskopische Technik, 16. Auflage.* München: R. Oldenbourg Verlag.
Ryszkowski, L. (1982). Structure and function of the mammal community in an agricultural landscape. *Acta Zoologica Fennica*, **169**, 45–59.
Savage, D. C. (1972). Associations and physiological interactions of indigenous microorganisms and gastrointestinal epithelia. *American Journal of Clinical Nutrition*, **25**, 1372–1379.
Savage, D. C. (1983). Morphological diversity among members of the gastrointestinal microflora. *International Review of Cytology*, **82**, 305–334.
Sisson, S. (1975). General digestive system. In *Sisson and Grossman's Anatomy of the Domestical Animals.* ed. R. Getty, pp. 104–112. Philadelphia: W. B. Saunders.
Sonnenborn, U. & Greinwald, R. (1991). *Beziehungen zwischen Wirtsorganismus und Darmflora, 2. Aufl.* Stuttgart: Schattauer.
Sowls, L. K. (1984). *The Peccaries.* Tucson, AZ: University of Arizona Press.
Stevens, C. E. (1988). *Comparative Physiology of the Vertebrate Digestive System.* Cambridge: Cambridge University Press.
Strahan, R. (1983). *Complete Book of Australian Mammals.* London: Angus & Robertson.
Thompson, S.D. (1987). Body size, duration of parental care, and the intrinsic rate of natural increase in eutherian and metatherian mammals. *Oecologia*, **71**, 201–209.
van den Brink, F. H. & Haltenorth, T. (1972). *Die Säugetiere Europas.* Hamburg: Verlag Paul Parey.
van Soest, P. J. (1982). *Nutritional Ecology of the Ruminant.* Corvallis, OR: O & B Books.

Vernier, J.-M. & Sire, M.-F. (1989). L'absorption intestinale des protéines sous forme macromoléculaire chez les vertébrés. Implications physiologiques. *Annales de Biologie*, **28**, 255–288.

Waldeyer, W. (1908). Die Magenstrasse. *Sitzungsberichte der Königlich Preussischen Akademie der Wissenschaften* (1908), 595–606.

Wallace, R. (1948). Growth of lambs before and after birth in relation to nutrition. *Journal of Agricultural Sciences*, **38**, 254–302.

Wardrop, I. D. & Coombe, J. P. (1960). The post-natal growth of the visceral organs of the lamb. I. The growth of the visceral organs of the growing lamb from birth to sixteen weeks of age. *Journal of Agricultural Sciences*, **54**, 140–143.

Wiesner, H. (1990). Krankheiten der Neugeborenen und Säuglinge bei Zoo- und Wildtieren. In *Neugeborenen- und Säuglingkunde der Tiere*. ed. K. Walser & H. Bostedt, pp. 508–533. Stuttgart: Ferdinand Enke Verlag.

Yokoyama, M. T. & Johnson, K. A. (1988). Microbiology of the rumen and intestine. In *The Ruminant Animal. Digestive Physiology and Nutrition*. ed. D. C. Church, pp. 125–144. Englewood Cliffs, NJ: Prentice Hall.

17

Adaptations in the large intestine allowing small animals to eat fibrous foods

GÖRAN BJÖRNHAG

Herbivores feeding on coarse food have to rely on microbial digestion to be able to use structural carbohydrates of the food. The fermentation can be located in the stomach prior to the part secreting gastric juice or in the large intestine, or both. The fermentation vat in the large intestine consists of the caecum and/or the proximal colon. The term hindgut is commonly used as equivalent for the entire large intestine although embryologically it refers to colon descendens and rectum only (Hofmann, 1991).

The microorganisms of forestomach fermentation attack all available food components coming into the fermentation chamber. This also includes nutrients that could have been digested by the herbivore's own digestive enzymes. Microbial fermentation always leads to energy losses in the order of 10–20% of the available energy. It also often gives rise to nitrogen losses in the form of ammonia production. Hindgut fermenters have the opportunity to digest the available parts of the food before any microbial fermentation. That means they avoid some of these losses.

The mass-specific energy requirement of homeothermic animals is relatively high and related to body mass$^{-0.25}$, i.e. the smaller the animal the greater its energy need per unit of body mass. Exceptions are found among monotremes, marsupials, hyraxes and sloths, animals having lower metabolic rates than the average for their size. A 25 g vole, for example, needs about 12 times as much energy per kg body mass as a 650 kg horse. This also means that the vole has to eat at least 12 times as much food as the horse when eating the same kind of food. Consequently the passage rate will be 12 times higher in the rodent. The rate of fermentation is the same in a small as in a large animal. Therefore small animals feeding on plant materials with low energy density cannot rely on microbial fermentation entirely. The fermentation is too slow and the food retention time too short for production of sufficient amounts of energy in a form available to the host animal. Small herbivorous animals have to combine autoenzymatic digestion in the foregut with microbial fermentation in the large intestine. They have to be hindgut fermenters.

The retention time of food residues is short in such circumstances, much shorter than the time needed by the microorganisms to reproduce. The retention time of the microorganisms in the gut lumen, therefore, has to be longer than the actual retention time of the food residues; otherwise there will be no microorganisms left except those residing on the gut wall. The microorganisms have to separate from the food residues before the latter are excreted as faeces.

The fraction of digesta having low digestibility is not profitable for small animals to ferment because the required fermentation time is too long. Only the more readily digestible fractions need to be retained along with the microorganisms. This implies a separation of the different fractions of the digesta. The retained fractions of the digesta are brought back towards and mostly into the caecum. The purpose is to concentrate digestive effort on the potentially more fermentable fractions of the digesta. Many small herbivores have developed anatomical and functional adaptations in their large intestine to make separation and retrograde flow possible. Such adaptations are called colonic separation mechanisms, CSM (Sperber, 1985; Björnhag, 1987).

Colonic separation mechanisms (CSM)

CSM in rabbits

The best known colonic separation mechanism is undoubtedly that of the rabbit (*Oryctolagus cuniculus*). The anatomy of the rabbit large intestine is thoroughly described by Snipes (1978) and Snipes *et al.* (1982). It is shown schematically in Fig. 17.1*a*.

All contents from the ileum move into the large caecum. Peristaltic and antiperistaltic movements mix them with the contents of the caecum and the first few centimetres of the proximal colon. The first half of the proximal colon possesses three taeniae and three rows of haustra and the second half has one taenia and one row of haustra. A short, spindle-shaped structure connects the proximal colon with the distal colon. This part is called the **fusus coli**. It has thick muscular layers (Auer, 1925) and acts as a pacemaker for the peristaltic and antiperistaltic activity in the colon (Ruckebusch and Fioramonti, 1976).

Rabbits produce two types of faeces. The normal hard faecal pellets are excreted during most of the time that the animal is active. The other pellets,

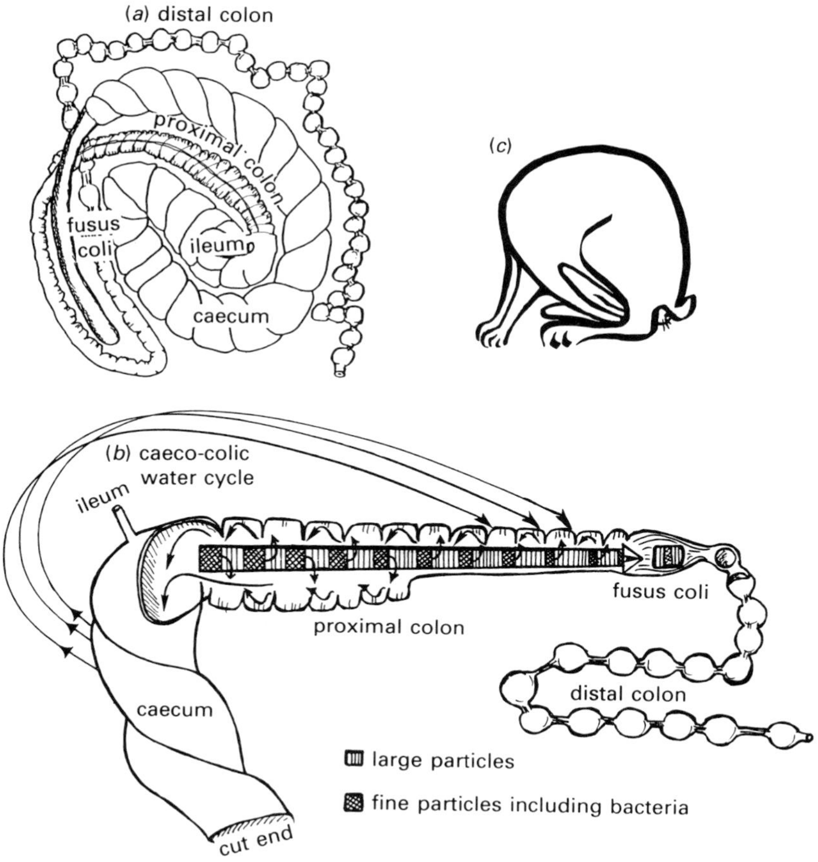

Fig. 17.1. The rabbit large intestine. (*a*) The caecum and colon in a semi-schematic view. (*b*) Illustration of the colonic separation mechanism in the rabbit. (*c*) Caecotrophy in the rabbit.

the soft caecotrophes, are produced during one or two periods per day and are eaten directly from the anus. They consist of almost unaltered caecal contents. The hard faecal pellets are also of caecal origin but their composition is drastically changed during passage through the proximal colon. This change was studied by Björnhag (1972).

A water-soluble, non-absorbable marker (e.g. polyethylene glycol, PEG) given in the drinking water accumulates in the caecum during periods when hard faecal pellets are formed. In the proximal colon the concentrations of the marker and nitrogen successively decrease in the digesta. Simultaneously, the concentration of large particles in the digesta dry matter increases. All

Table 17.1. *Composition of the contents of the large intestine in rabbits killed during the formation of faecal pellets*

	Particles > 0.1 mm in % of total DM	PEG/DM rel. conc.	Nitrogen content of DM (mg/g)
Caecum	37 ± 3	100	38 ± 4
Proximal colon	56 ± 8	73 ± 28	27 ± 7
Distal colon	83 ± 5	12 ± 10	14 ± 3
Number of animals	8	25	19

Standard deviations are given as ±. All differences between values in each column are highly significant ($P < 0.001$). Rel. conc., relative concentration (caecum = 100); DM, dry matter; PEG, polyethylene glycol.
From Björnhag, 1981.

these changes are highly significant, as seen in Table 17.1. Similar results with water-soluble markers were reported by Pickard and Stevens (1972).

These changes in concentration are produced in the following way (see also Fig. 17.1*b*): caecal contents are moved into the proximal colon in small portions. Intense muscular activity of the colonic wall separates water, water-soluble matter and small particles, including microorganisms and endogenous matter, from the coarse matter and moves them towards the haustrated wall. Water secretion from the mucosa facilitates this separation. The pace-maker at the fusus coli initiates antiperistaltic movements of the haustra. Such antiperistalsis may also start as a local response to a free water phase anywhere in the proximal colon (Ehrlein *et al.*, 1983). The contents of the haustra are continuously transported along the colonic wall from haustrum to haustrum towards the caecum at a rate of about 1 mm s^{-1} (Björnhag, 1981). At the same time the fibrous part of the contents moves slowly in the centre of the lumen towards the distal colon. The force for the latter transport probably comes from the caecal outflow. Thus, in the centre of the proximal colon a nearly continuous flow of coarse matter proceeds in the anal direction at the same time as finer materials continuously flow in the opposite direction along the colonic wall.

Water-soluble matter is rinsed from the coarser matter by the secreted water. The surplus water is reabsorbed in the caecum in amounts corresponding to those secreted in the proximal colon (Clauss, 1978). When the fibrous fraction of the food reaches the fusus coli it is formed into faecal pellets from which water and electrolytes are absorbed during passage through the distal colon.

Björnhag and Sperber (1977) discussed the different composition of small and large particles in vegetable food. The contents and the thin cell walls of living plant cells are rapidly fermented. The nutrient reserves of seeds and roots are also rapidly digested. Thick cell walls, especially when lignified, are digested very slowly. In most cases these thick cell walls are protection against mechanical injury or are essential for mechanical stability of the plants. They are usually found as comparatively large and connected structures. Examples of these are the shells and hulls of seeds, the wood and the outermost part of the bark of twigs and branches, the vascular bundles, and the collenchyme bundles of stems and leaves. It may be expected that these mechanically-resistant structures will be less affected by chewing than the parenchymatous parts of the plants. This will lead to a preponderance of relatively indigestible matter in the larger particles. Conversely, the more digestible parts will be divided into smaller particles.

It is advantageous for lagomorphs to pass the larger particles of food as rapidly as possible through the digestive tract and to retain the smaller particles in the caecum for a period sufficiently long for microbial digestion. By this strategy they can eat larger amounts of poorly-digestible food but they utilize only the fraction of higher digestibility.

CSM in rabbits causes accumulation of easily-fermentable food particles, microorganisms and water-soluble matter in the caecum. This supplies the microbes with nutrients and consequently a rapid fermentation takes place. The short-chain fatty acids (SCFA also called VFA, volatile fatty acids) produced are mainly absorbed in the caecum, but the surplus of bacteria and many of the vitamins produced can only be utilized after caecotrophy.

When caecotrophes are formed, no separation mechanism operates. Caecal contents pass through the colon without any marked change in composition. A mucous envelope is laid round each pellet in the second half of the proximal colon. When the pellets reach the anus they are eaten directly as they appear (Fig. 17.1*c*). The caecotrophes are swallowed whole without disrupting the envelope. They are retained in the fundic region of the stomach, and some can remain intact there for up to 7 h. Griffiths and Davies (1963) studied caecotrophes in the stomach. They found bacterial enzymes capable of fermenting starch and sugars to lactic acid inside the mucous envelope. At the same time a phosphate buffer leaks out from the pellet with a strong buffering action in the pH range 4 to 7. The caecotrophes are successively mixed with the food and move into the small intestine where the mucous envelope disappears.

Little is known about the adaptations of CSM and caecotrophy to nutrient requirements and food composition in rabbits. They produce and consume about the same amount of caecotrophes on both balanced diets and high-fibre diets. However, the latter diets seem to give rise to more efficient separation. Food with a low content of valuable nutrients induces stronger separation that makes it possible for the rabbit to save more of these components. This explains why rabbits wearing collars to prevent caecotrophy show a stronger than average separation. They lose considerable amounts of protein, vitamins and electrolytes in faeces compared with non-collared rabbits eating their caecotrophes (Björnhag, 1972).

The structure of the food is important for the function of CSM in rabbits. If they are given finely ground food no separation occurs, and the rabbits produce only one kind of faeces intermediate in composition between caecotrophes and normal hard faecal pellets. These pellets are seldom eaten (G. Björnhag, unpublished data).

CSM in other lagomorphs

Caecotrophy similar to that in the rabbit has been shown to exist in other genera of lagomorphs. The ability to change the production of faeces from a nutrient-rich type to a type poor in nutrients and back again (as in the rabbit) occurs in hares (*Lepus* spp.) (Watson and Taylor, 1955; Lechleitner, 1957; Pehrson, 1983a), cottontails (*Sylvilagus* spp.) (Heisinger, 1962), and pikas (*Ochotona* spp.) (Haga, 1960; Matsuzawa *et al.*, 1981) and suggests that they have a CSM more-or-less similar to the rabbit type. The caecotrophes in the mountain hare are not separated into pellets but appear as 2–3 cm long lumps of a sticky nature. The hare chews the caecotrophes before swallowing them (Pehrson, 1983a).

In caged mountain hares (*Lepus timidus*) fed willow twigs, which are a normal part of their winter food, the caecotrophes had about four times as high a concentration of crude protein as the faeces (Table 17.2). The corresponding value for phosphorus, magnesium and sodium was three-fold. The caecotrophes had eight times as high potassium concentrations as the faeces but the calcium concentration was similar (Pehrson, 1983a).

Caged mountain hares consumed about 12% of their body mass in the form of dry matter of mixed twigs with a digestibility of about 30% and were able to increase their body mass during that time (Pehrson, 1983b).

CSM in marsupials

The common ringtail possum (*Pseudocheirus peregrinus*) shows a separation mechanism coupled to caecotrophy that appears similar to that found in the

Table 17.2. *The concentrations of nitrogen in caecal contents, caecotrophes and faeces in comparison to the food in 12 animal species with CSM*

	Nitrogen content (mg g^{-1} DM)				
	Food	Caecal contents	Caeco-trophes	Faeces	References
Rabbit, *Oryctolagus cuniculus*	24[a]	42	46	15	1, 2
Mountain hare, *Lepus timidus*	10[b]	nm	45	11	3
Rat, *Rattus norvegicus*	21[a]	33	35	20	2, 4
Scandinavian lemming, *Lemmus lemmus*	18[c]	52	nm	17	4
Water vole, *Arvicola terrestris*	nm[d]	60	63	19	5
Kangaroo rat, *Dipodomys microps*	nm[e]	nm	40	27	6
Guinea pig, *Cavia porcellus*	19[a]	31	47	17	7
Chinchilla, *Chinchilla laniger*	20[a]	27	34	18	7
Nutria, *Myocaster coypus*	27[f]	33	34	22	8
Ringtail possum, *Pseudocheirus peregrinus*	13[g]	39	46	12	9
Donkey, horse *Equus asinus, E. caballus*	11[h]	12*	–	8	2, 10
Turkey, *Meleagris gallopavo*	21[i]	55	–	16	11

nm, not measured, DM, dry matter.
Food: *a*, commercial rabbit food; *b*, willow twigs; *c*, mosses and oats; *d*, fresh herbs; *e*, saltbush leaves (*Atriplex confertifolia*); *f*, commercial nutria pellets, oats, beet pulp, hay; *g*, *Eucalyptus* foliae; *h*, straw and oats; *i*, whole oats.
* Nitrogen concentration in the dorsal colon 16 mg g^{-1} DM.
References: 1, Björnhag (1972); G. Björnhag unpublished results; 3, Pehrson (1983a); 4, Sperber *et al.* (1983); 5, I. Sperber and Y. Ridderstråle personal communication; 6, Kenagy and Hoyt (1980); 7, Holtenius and Björnhag (1985); 8, Snipes *et al.* (1988); 9, Chilcott and Hume (1985); 10, Sperber *et al.* (1992); 11, Björnhag and Sperber (1977).

rabbit (Chilcott and Hume, 1985) (Table 17.2). The passage rate of fluid and fine particles through the digestive tract is low compared with the passage rate of coarse matter (Sakaguchi and Hume, 1990), as it is in rabbits and hares. A detailed study of the mechanism in that species has not yet been carried out.

Also the koala (*Phascolarctos cinereus*) (Cork and Warner, 1983) and the greater glider (*Petauroides volans*) (Foley and Hume, 1987) seem to have a similar type of separation mechanism although the adult animals do not practise coprophagy or caecotrophy (Smith, 1979).

CSM in rodents

The CSM in rodents functions in a quite different manner from that in rabbits. Mechanisms have been studied in the Scandinavian lemming and the rat (Sperber *et al.*, 1983), in the guinea pig and the chinchilla (Holtenius and Björnhag, 1985) and in the nutria (Snipes *et al.*, 1988).

Myomorph rodents

The first part of the proximal colon of the Scandinavian lemming (*Lemmus lemmus*), the **ampulla coli**, is as wide as the caecum but very short (Fig. 17.2). The second part is narrow and forms a double spiral. The mucosa in the beginning of the spiral colon is provided with a longitudinal fold, running along the colon for about two circumferences. This fold almost completely divides the lumen of the gut into two separate channels, a narrow channel and a main channel. At its distal end, the narrow channel is connected to a longitudinal groove. Two rows of oblique folds are arranged in a fishbone pattern. The groove is created by means of the oblique folds which do not quite meet (Fig. 17.2). These folds are found from the middle of the inner spiral to the end of the proximal (Fig. 17.3).

The use of the water-soluble marker PEG in the early experiments with the lemming did not indicate any significant retrograde transport of fluid in the proximal colon (Sperber, 1968). However, the concentration of nitrogen decreased rapidly during digesta transport through the proximal colon. Apparently these animals use different mechanisms from those used in rabbits to conserve bacteria and valuable nutrients (Sperber *et al.*, 1983).

All digesta leaving the ileum pass into the caecum and mix with its contents. Caecal contents are in free communication with the first part of the proximal colon, the ampulla coli. The caecum and the main channel of the initial part of the spiral colon are thus filled with a mixture of food residues and bacteria. But the narrow channel contains almost exclusively bacteria and mucus. In the second part of the spiral colon, the bacteria are found mainly between the oblique folds and in the groove. At the end of the spiral colon and in the distal colon, very few bacteria are seen among the food residues. The microorganisms are thus separated from the food residues very efficiently (Sperber *et al.*, 1983).

This may be achieved in the following way: in the spiral part of the proximal colon the contents are mixed with a mucous secretion originating from the thick mucosa in the inner spiral, distal to the narrow channel but outside the groove and oblique folds. The bacteria are trapped in the mucus by means

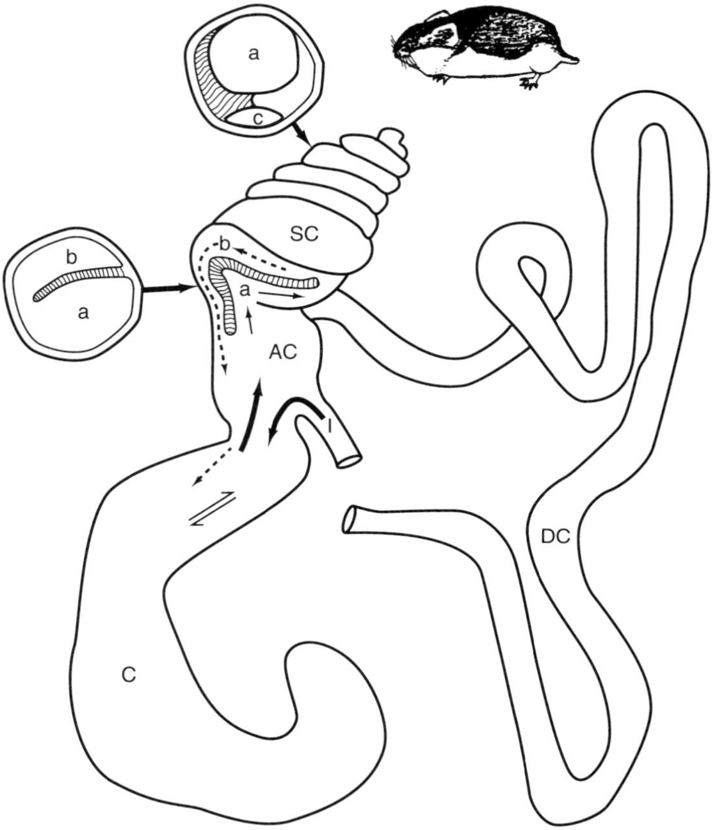

Fig. 17.2. The large intestine of the Scandinavian lemming. AC, ampulla coli; C, caecum; DC, distal colon; I, ileum; SC, spiral colon; a, the main channel in the proximal colon; b, the narrow channel in the first part of the proximal colon; c, the groove in the spiral colon created by pairs of oblique folds, one pair of folds is shown here. The solid arrows illustrate the flow of digesta. The dotted arrows indicate the backflow of bacteria and mucus.

of an aggregating action of the mucus and/or by chemotaxis. The ensuing mixture of mucus and tightly-packed bacteria is brought to the wall of the spiral colon. Most of it is found in the groove and between the oblique folds. Antiperistaltic movements along the base of the groove have been seen in narcotized animals with their intestines immersed in Ringer solution (G. Björnhag and I. Sperber, unpublished data). These movements may be responsible for transporting the mixture into the narrow channel and eventually back to the caecum where it is mixed with caecal contents. Separation

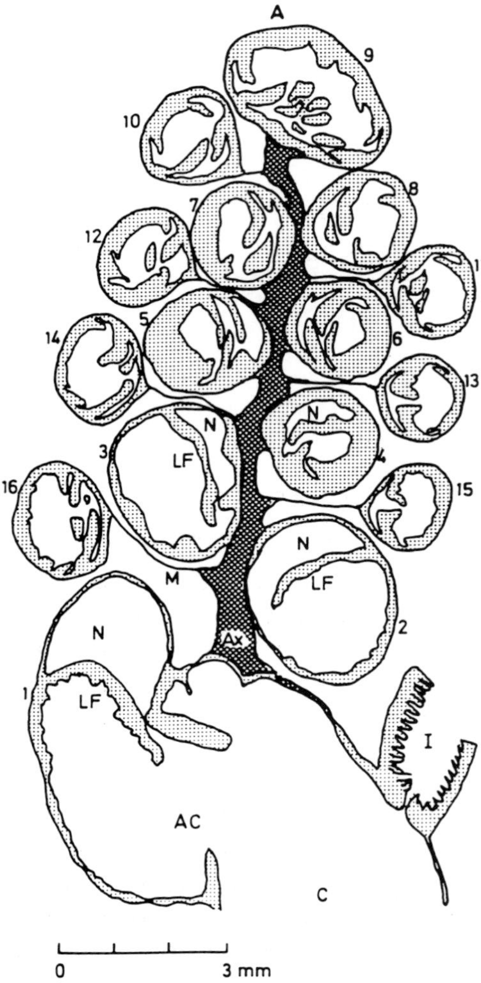

Fig. 17.3. Section through the colonic spiral of the Scandinavian lemming. Composite from sections of a 21-day-old lemming. A, apex; Ax, axial structure; I, ileum; LF, longitudianal fold; M, mesentery; N, narrow channel; AC, ampulla coli; C, caecum. Numbers 1–8 refer to sections through the inner spiral, 10–16 to sections through the outer spiral. From Sperber *et al.*, (1983).

and retrograde transport are indicated by a decrease in nitrogen concentration along the proximal colon, as shown in the wood lemming (*Myopus schisticolor*) in Fig. 17.4 (I. Sperber and Y. Ridderstråle, personal communication). The bacteria are thus retained in the caecum and the proximal colon, while food residues are passed on to the distal colon and are

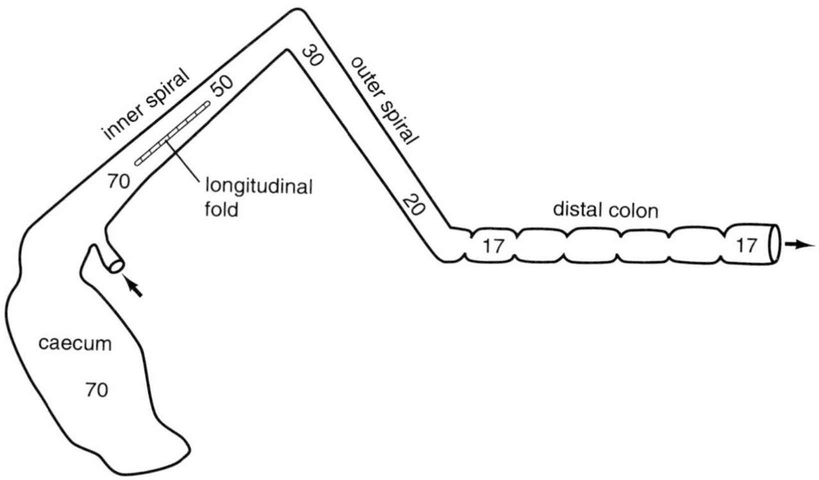

Fig. 17.4. The decrease in nitrogen concentration (mg g^{-1} dry matter) along the colon in the wood lemming (*Myopus schisticolor*).

voided as faecal pellets. During both the day and night, the separation ceases for several short periods. When the colon is empty, caecotrophes may be formed and eventually ingested directly from the anus. This is seen by radioscopy. No analysis of such pellets has been done. The ingested pellets are chewed and mixed into the contents of the stomach (Björnhag and Sjöblom, 1977).

Some data will emphasize the need for a separation mechanism and the backflow of microorganisms. Sperber (1968) reported that a lactating lemming (mass 65 g) with seven young produced faecal dry matter at a rate of 25 g per 24 h or 70 g in fresh weight, which is more than its own body weight. Four other, non-lactating lemmings reared in the laboratory (mass 34–45 g) excreted, on average, 3.0 g dry matter per 24 h in faecal pellets over a 4-day period. The caecum of these animals contained 0.13 g dry matter on average. The mean retention time of dry matter would thus be about 1 h, assuming no addition or loss of dry matter during passage through the colon. When account is taken of the disappearance of bacteria in the colon, the mean retention time for caecal food residues becomes about 40 min (Sperber *et al.*, 1983). The generation interval of the bacteria must on average be 0.69 times their retention time (Hungate, 1966) in order to maintain their number. The caecal bacteria of the lemmings, therefore, have to divide with an average interval of less than 30 min to maintain their number if no other mechanism is involved. Such a high rate of multiplication is quite unlikely for bacteria living without oxygen and easily-digestible food. For example, the bovine rumen microbiota reproduce at an average rate of 1.9 divisions per day (El-Shazly and Hungate, 1965). Most of the bacteria separated in the colon thus have to be returned to the caecum to maintain the bacterial population. Without such a retrograde flow, microbial digestion of food in the caecum would not be possible.

Studies by I. Sperber and Y. Ridderstråle (personal communication) have shown that CSM works in a similar way in another lemming species, the wood lemming (*Myopus schisticolor*) and in three species of voles (*Arvicola terrestris*, *Microtus agrestis* and *Clethrionomys glareolus*).

The anatomical studies of Tullberg (1899) and Behmann (1973) show that practically all myomorph rodents possess in their proximal colon two rows of oblique mucosal folds meeting each other at an angle pointing towards the caecum. Only two carnivorous genera of the myomorph rodents (*Deomys* and *Hydromys*) lack this character. It is tempting to assume that nearly all other members of the myomorph rodent group have a CSM (Sperber *et al.*, 1983).

Rat The proximal colon of rats (*Rattus norvegicus*) also has small oblique furrows that meet at a sharp angle with the apex pointing towards the caecum (Fig. 17.5). A mixture of bacteria and mucus is found in the furrows.

The nitrogen concentration in caecal and colonic contents shows considerable variation (Table 17.3). There are statistically highly significant differences between caecal contents and the contents of the proximal colon ($P < 0.001$), whereas the difference in nitrogen content between the proximal and distal colons is not statistically significant. The nitrogen concentration of the contents of the distal colon is little higher than that of faecal pellets ($P < 0.005$). A more interesting fact is the larger variation that occurs in the nitrogen concentration of the contents of the distal colon than that found in the excreted faecal pellets. By analysis of individual pellets found in the distal colon it can be seen that some rats contained pellets with a much higher nitrogen content than was normally found among excreted faecal pellets. Faecal pellets found in the cage seldom had a nitrogen content higher than 30 mg g^{-1} dry matter, but, in the pellets from the distal colon, nitrogen concentrations of up to 45 mg g^{-1} dry matter were found. In the latter group, 11% of the pellets had a nitrogen content over 30 mg g^{-1} dry matter. This indicates that such pellets are caecotrophes intended for consumption as soon as they appear at the anus and that rats do possess a CSM (Sperber *et al.*, 1983).

The distribution of orally-given PEG in the rat large intestine shows that the CSM is not due primarily to fluid movements. This, as well as the arrangement of furrows and the distribution of crypt mucous cells, are

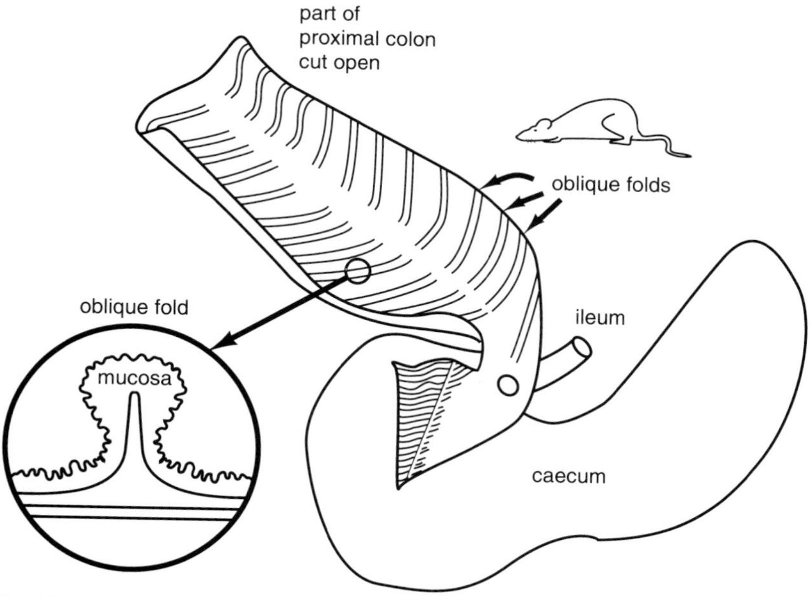

Fig. 17.5. The caecum and the first part of the proximal colon of the rat. The expanded area shows one of the oblique folds in cross section. Modified from Tullberg (1899) and Sperber *et al.* (1983).

Table 17.3. *Nitrogen concentration in the gastro-intestinal contents and in delivered faecal pellets of rats living on commercial rabbit food*

	Nitrogen content (mg g^{-1} DM)
Stomach	26.1 ± 5.5
Small intestine, distal part	19.7 ± 3.4
Caecum	33.2 ± 7.3
Colon, proximal part	23.4 ± 4.9
Colon, distal part	22.0 ± 4.8
Faecal pellets: 240 excreted from 16 rats in 24 h	20.0 ± 3.0

DM, dry matter. Values are the means of 21 animals ± standard deviation; 16 of these rats were used for the faecal pellet collection in the period immediately before assay. Sperber *et al.*, 1983; G. Björnhag, unpublished data.

important points of similarity to the lemming CSM. Retrograde movements in the furrows may thus transport the mixture of bacteria and mucus back into the caecum. The mechanism works more efficiently when the rats eat coarse food. Caecotrophes are formed and consumed during several short periods when the separation and backflow cease. These periods can occur at any time throughout the entire day and night.

Rats reduce caecotrophy when fed a diet high in protein (Araya *et al.*, 1973), but even on a complete and easily-digestible diet some caecotrophy persists (G. Björnhag, unpublished results).

A retrograde movement of the mucus–bacteria mixture may also be indicated by the accumulation of mucus in the caecum of germ-free rats and mice (cf. review by Coates and Fuller, 1977). In the normal animal the mucus may be broken down by bacteria in the caecum.

Caviomorph rodents

Three caviomorph rodent species have been studied, the guinea pig (*Cavia porcellus*), and chinchilla (*Chinchilla laniger*) (Björnhag and Sjöblom, 1977; Holtenius and Björnhag, 1985) and the nutria (*Myocaster coypus*) (Snipes *et al.*, 1988). Like other caviomorph rodents they have two longitudinal folds in the proximal colon (Gorgas, 1967) (Fig. 17.6). In these animals, a mixture rich in bacteria and mucus is found together with food particles in the furrow between the folds. The concentration of nitrogen (and consequently also of microorganisms) outside the furrow decreases along the proximal colon. A CSM similar to the lemming type, but less efficient, may be responsible for this decrease. Antiperistaltic movements caused by the circular muscular layer at the bottom of the furrow are seen also in the guinea pig and the chinchilla (G. Björnhag, unpublished results). These movements may transport the mixture towards and sometimes into the caecum.

The caecotrophes of the guinea pig have a higher nitrogen concentration than the caecal contents (Table 17.2); however, it is at about the same level as that found in the contents of the furrow. The amount of caecotrophes produced at each period of caecotrophy is approximately equal to the estimated amount of contents found inside the furrow. Thus, the caecotrophes do not seem to be made from caecal matter but from contents in the furrow on their way back towards the caecum. This may be explained by a change from antiperistaltic to peristaltic movements over a period only long enough to empty the contents of the furrow into the distal colon.

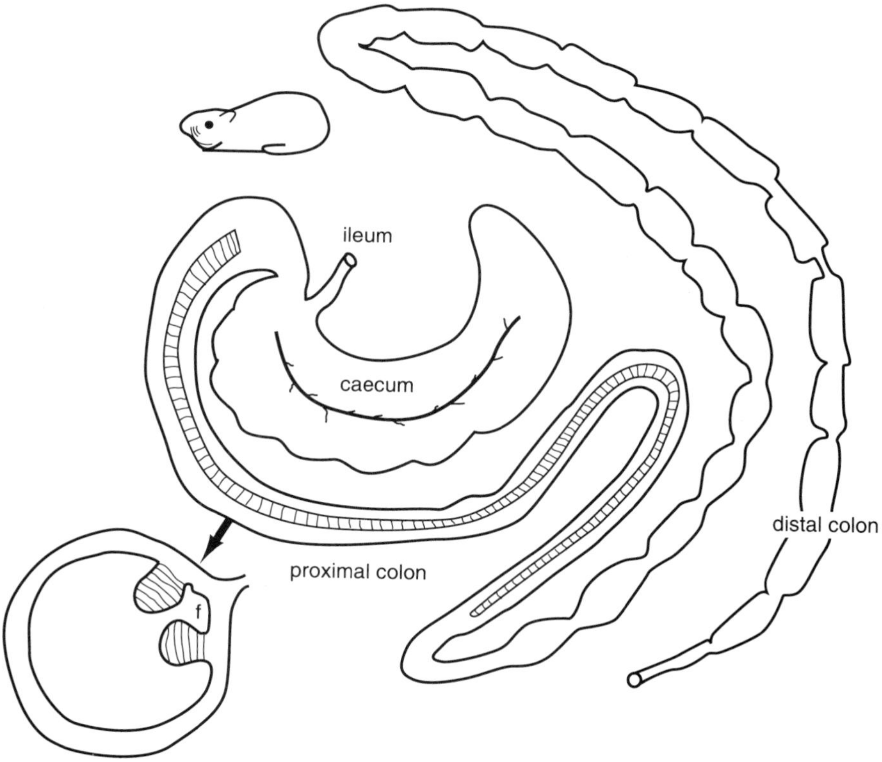

Fig. 17.6. A schematic view of the large intestine of a caviomorph rodent. The expanded area shows a cross section of the proximal colon with the folds hatched; f, furrow. The outline of the folds and the furrow is marked along the proximal colon. Modified from Holtenius and Björnhag (1985).

CSM in other animals

There are also indications that some hystricomorph rodents make use of CSM. Scherman *et al.* (1992) noted that the naked mole rat (*Heterocephalus glaber*) produces two kinds of faecal pellets: 'One is deposited in a communal toilet chamber; the other is reingested'. They also mention that the soft faecal pellets are highly nutritious and laden with microorganisms. The breeding female and the pups are fed such soft, nutrient-rich faecal pellets (Jarvis, 1981).

Birds

Still another kind of separation and retrograde flow occurs in some bird species (see also review by Björnhag, 1989). It has been known for some time that urine can reach the caeca (Browne, 1922). Yasukawa (1959) noted antiperistaltic movements from the junctions of the colon and cloaca travelling up the colon. Skadhauge and Schmidt-Nielsen (1965) found urine in the colon of dehydrated hens. Akester *et al.* (1967) showed by radioscopy that urine in the domestic fowl normally was transported from the cloaca, via the colon into the caeca but never into the ileum. Clemens *et al.* (1975) showed that a water-soluble marker placed in the cloaca of geese was moved into the caeca.

Björnhag and Sperber (1977) studied the effect of this backflow in turkeys, guinea fowl and geese (Fig. 17.7). Contents from the ileum pass through the

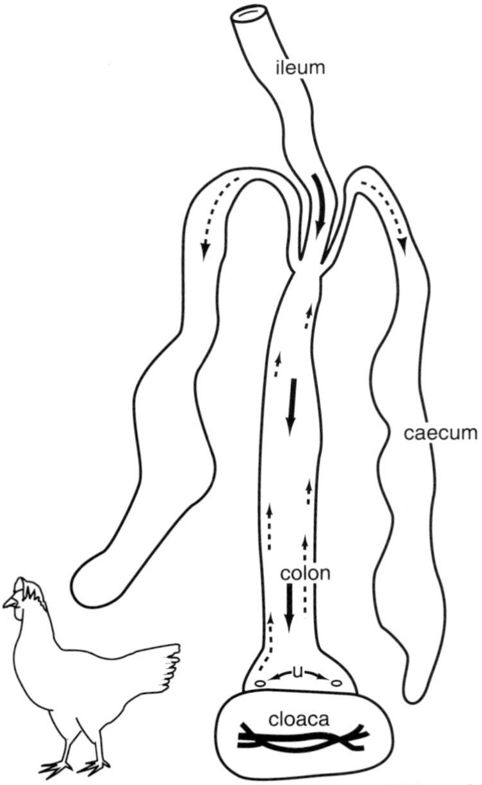

Fig. 17.7. A schematic view of the separation in the galliform hindgut. u, ureters. The solid arrows represent the flow of digesta. The dotted arrows indicate the retrograde flow of urine and small particles.

colon and mix in the cloaca with urine. The water-soluble fraction, including urine, is brought back together with fine particles along the colonic wall towards the ileum. The ileal opening closes and the mixture is forced through the narrow openings into the caeca. Very few large particles are seen in the caecal contents. But an intense fermentation occurs and SCFA are absorbed (Mead, 1989; Goldstein, 1989). Retrograde transport is achieved by means of the circular muscles in the colonic wall. At the same time the contents in the central part of the tube continue towards the cloaca.

Besides moving fermentable matter into the caeca, the separation and retrograde flow also provide an opportunity for reabsorption of water and water-soluble substances during periods when water or electrolytes are scarce. It also furnishes caecal bacteria with nitrogen from uric acid. A low nitrogen content in the food stimulates a larger backflow of urine into the caeca of the chicken (Fig. 17.7) (Björnhag, 1989).

In other bird species a retrograde urine flow into the caeca may occur without any fermentation occurring there. In birds which seldom drink, for example owls, the caeca may serve as an organ to save water (Chaplin, 1989).

Several wild bird species normally eat relatively indigestible food during the winter. Some grouse species adapt to winter food before the time that such food is the only available source. The adaptation has to start early in the autumn and when the food consumption reaches more than a certain volume per day, the caeca start to grow. After several weeks the caeca have reached volumes double the summer size. The digestibility of the coarse food rises as the volume increases (Moss, 1989).

Equines

In the equine hindgut (Fig. 17.8) three fermentation chambers follow each other with anatomical constrictions, bottlenecks, at the end of each (Hofmann, 1991). The first of the bottlenecks is the caeco-colic orifice. The second one is the pelvic flexura and the third one is at the end of the capacious right dorsal colon at the border to the distal colon (*colon transversum*). The three bottlenecks have different effects on digesta transfer.

When contents leave the caecum, large particles move out faster than fluid and small particles (Argenzio *et al.*, 1974). The next bottleneck at the junction between the ventral and the dorsal colon has an opposite effect. Large particles are retained longer than fluid and fine particles (Argenzio *et al.*, 1974). The third bottleneck retains the fluid phase longer than the fraction of large particles. The last separation has been studied by Björnhag *et al.* (1984) in

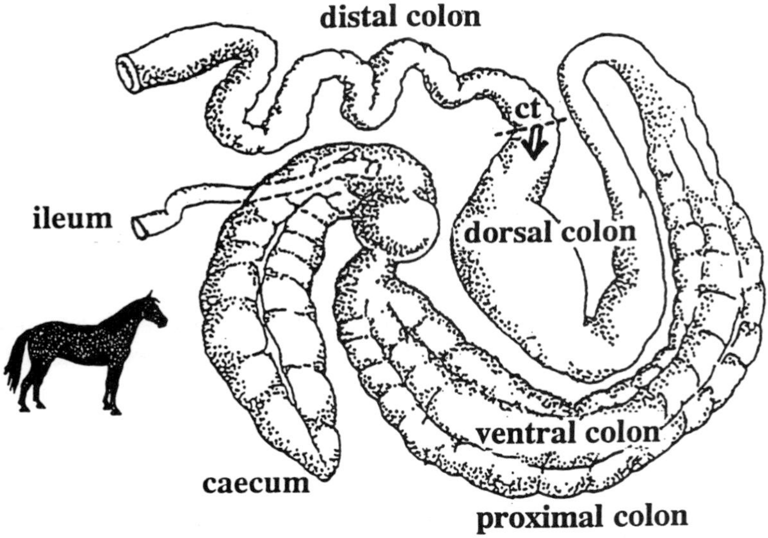

Fig. 17.8. A schematic view of the equine large intestine. ct, colon transversum, border to the small colon. The arrow represents the backflow of the fluid fraction. Modified from Björnhag (1987).

donkeys and ponies given large amounts of coarse food to simulate the natural food during dry or cold periods. Both species behaved in a similar way.

When food contains much coarse matter, the contents of the large colon consist of a fluid and a fibrous fraction that can easily be separated from each other when placed on a wire screen. Contents in the right dorsal colon move dorsally to reach the beginning of the small colon. At this site, much of the fluid phase separates from the coarse fraction, probably due to a muscular contraction of the intestinal wall. The fluid fraction flows back into the large colon bringing also small particles and microorganisms back to maintain an efficient fermentation (Björnhag *et al.*, 1984; Sperber *et al.*, 1992). Clearly the separation mechanism at this third bottleneck allows a high microbial-activity in the dorsal colon without a high nitrogen loss in the faeces when the food consumption and digesta passage rate are high.

Small herbivores without CSM

In two relatively small folivorous hindgut fermenters, the brushtail possum (*Trichosurus vulpecula*) (Wellard and Hume, 1981; Foley and Hume, 1987) and the rock hyrax (*Procavia habessinica*) (Björnhag and Engelhardt, 1987), no sign of a CSM has been detected. Both these species have a slow passage of food residues but also a low basal metabolic rate. In the brushtail possum, both caecum and proximal colon serve as a fermentation chamber. The hyrax has a large caecum and a partly divided, caecum-like colonic enlargement, the colonic sac. The caecum and the colonic sac are separated by a connecting colon with a very high capacity for water absorption. Both the caecum and the colonic sac are adapted for fermentation.

Conclusion

Colonic separation mechanisms are found in species belonging to eight mammalian and two avian families (Table 17.4). The mammalian taxonomy is from Walker (1968). However, very few species outside these families have been studied.

How important are the CSM and caecotrophy for the animals? Two of the caecotrophe's most valuable groups of components are the microbial proteins and vitamins. Both become still more vital for utilization when the food quality is low.

Pehrson (1983a,b) gives data that can illustrate the situation for a mountain hare during wintertime. A 3.5 kg hare can eat about 420 g dry matter as a mixture of twigs (which for free-living hares is their only food for long periods). The nitrogen content is around 1% of the dry matter. The nitrogen digestibility of this food is about 35%. That means 1.5 g nitrogen of food

Table 17.4. *Taxonomic survey of the mammalian and avian families in which species with a CSM are found. Species from very few other families have been studied*

	Families with a colonic separation mechanism	
Class, order, sub-order	Mucus-dependent	Particle-size dependent
Mammalia		
Marsupialia	Phalangeridae	
Rodentia		
Myomorpha	Cricetidae, Muridae	
Caviomorpha	Caviidae, Chinchillidae, Capromydiae	
Lagomorpha		Leporidae
Perissodactyla		Equidae
Aves		
Anseres		Anatidae
Galli		Phasianidae

origin is absorbed per day. The consumption of caecotrophes is not so easy to measure. A reliable estimate is that the caecotrophe production in animals on such food amounts to 12% of the total faecal output (Pehrson, 1983b). Thus, about 40 g dry matter in caecotrophes, with a nitrogen content of about 4.5% (Table 17.2), is consumed per day. A total of 1.8 g nitrogen is consumed per day in caecotrophes. Most of that is available as most of the nitrogen is in microbial protein of high quality. Thus, more than 50% of the daily intake of nitrogen may be supplied with the caecotrophes.

It is not possible to do a similar calculation for the Scandinavian lemming because we do not know the amount of caecotrophes produced per day. The data of separation shows that a 40 g lemming has an outflow of nitrogen from the caecum to the proximal colon that is three times as large as the amount of nitrogen delivered in faeces. Two thirds of this outflow of nitrogen is moved back into the caecum. The nitrogen surplus is consumed by means of caecotrophy. Chilcott and Hume (1985) found that in the ringtail possum caecotrophy contributed twice the maintenance nitrogen requirement.

Nitrogen concentrations in food, caecal contents, caecotrophes and faeces in animals with CSM is shown in Table 17.2.

However, the saving of microorganisms by means of CSM is at least as equally important as the contribution of proteins and vitamins to the food economy. The mean retention time of food in the lemming caecum is a little more than half an hour but the reproduction time of the microorganisms is at least 12 h (p. 298). Without a CSM no microorganism could survive in the caecal contents. The same is true also for larger rodents and for lagomorphs that have much longer retention times that are still shorter than those needed for microbial reproduction.

Without the development of CSM and caecotrophy, it would not have been possible for most small herbivorous mammals and birds to establish on the tundra, in the desert, or in other cold or dry areas. Nor would it have been possible for some larger animals with hindgut fermentation (equines) to remain in their normal habitats during long, dry or cold seasons when only food with low digestibility is found. Development of CSM and caecotrophy has allowed herbivores to survive on poorly digestible food during certain periods or even all the year round. The result is a larger and a more evenly distributed fauna throughout the world.

References

Akester, A. R., Anderson, R. S., Hill, K. H. & Osbaldistone, G. W. (1967). A radiographic study of urine flow in the domestic fowl. *British Poultry Science*, **8**, 209–212.

Araya, H., Araya, J., Negrete, A. & Tagle, M. A. (1973). Coprophagia en ratas alimentadas con dietas de differente valor proteico. *Archivo Latinoamericanos de Nutricion*, **23**, 485–493.

Argenzio, R. A., Lowe, J. E., Pickard, D. W. & Stevens, C. E. (1974). Digesta passage and water exchange in the equine large intestine. *American Journal of Physiology*, **226**, 1035–1042.

Auer, J. (1925). Further note on the fusus coli of the rabbit. *Proceedings of the Society for Experimental Biology and Medicine*, **22**, 301–303.

Behmann, H. (1973). Vergleichend- und funtionell-anatomische Untersuchungen am Caecum und Colon myomorpher Nagetiere. *Zeitschrift für wissenschaftliche Zoologie*, **186**, 173–294.

Björnhag, G. (1972). Separation and delay of contents in the rabbit colon. *Swedish Journal of Agricultural Research*, **2**, 125–136.

Björnhag, G. (1981). The retrograde transport of fluid in the proximal colon of rabbits. *Swedish Journal of Agricultural Research*, **11**, 63–69.

Björnhag, G. (1987). Comparative aspects of digestion in the hindgut of mammals. The colonic separation method (CSM) (a review). *Deutsche tierärztliche Wochenschrift*, **94**, 33–36.

Björnhag, G. (1989). Transport of water and food particles through the avian caeca and colon. *Journal of Experimental Zoology*, Suppl. 3, 32–37.

Björnhag, G. & Engelhardt, W. von (1987). Radioscopy of the alimentary canal of the hyrax. In *Herbivore Nutrition Research*, ed. M. Rose, pp. 65–66. Brisbane: Australian Society of Animal Producers.

Björnhag, G. & Sjöblom, L. (1977). Demonstration of coprophagy in some rodents. *Swedish Journal of Agricultural Research*, **7**, 105–113.

Björnhag, G. & Sperber, I. (1977). Transport of various food components through the digestive tract of turkeys, geese and guinea fowl. *Swedish Journal of Agricultural Research*, **7**, 57–66.

Björnhag, G., Sperber, I. & Holtenius, K. (1984). A separation mechanism in the large intestine of equines. *Canadian Journal of Animal Science*, **64** (Suppl.), 89–90.

Browne, T. G. (1922). Some observations on the digestive system of the fowl. *Journal of Comparative Pathology and Therapeutics*, **35**, 12–32.

Chaplin, S. B. (1989). Effect of cecectomy on water and nutrient absorption of birds. *Journal of Experimental Zoology*, Suppl. **3**, 81–86.

Chilcott, M. J. & Hume, I. D. (1985). Coprophagy and selective retention of fluid digesta: their role in the nutrition of the common ringtail possum, *Pseudocheirus peregrinus*. *Australian Journal of Zoology*, **33**, 1–15.

Clauss, W. (1978). Resorption und Sekretion von Wasser und Elektrolyten im Colon des Kaninchens im Zussamenhang mit der Bildung von Weichkot und Hartkot. Dissertation. Institut für Zoophysiologie der Universität Hohenheim, D-7000 Stuttgart 70, FRG.

Clemens, E. T., Stevens, C. E. & Southworth, M. (1975). Sites of organic acid production and pattern of digesta movement in the gastrointestinal tract of geese. *Journal of Nutrition*, **105**, 1341–1350.

Coates, M. E. & Fuller, R. (1977). The gnotobiotic animal in the study of gut microbiology. In *Microbial Ecology of the Gut*, ed. R. T. J. Clarke & T. Bauchop, pp. 311–346. London: Academic Press.

Cork, S. J. & Warner, A. C. I. (1983). The passage of digesta markers through the gut of a folivorous marsupial, the koala *Phascolarctos cinereus*. *Journal of Comparative Physiology*, **152**, 43–51.

Ehrlein, H.-J., Reich, H. & Schwinger, M. (1983). Colonic motility and transit of digesta during hard and soft faeces formation in rabbits. *Journal of Physiology* (Lond.), **338**, 75–86.

El-Shazly, K. & Hungate, E. (1965). Fermentation capacity as a measure of net growth of rumen microorganisms. *Applied Microbiology*, **13**, 62–69.

Foley, W. J. & Hume, I. D. (1987). Passage of digesta markers in two species of arboreal folivorous marsupials – the greater glider (*Petauroides volans*) and the brushtail possum (*Trichosurus vulpecula*). *Physiological Zoology*, **60**, 103–113.

Goldstein, D. L. (1989). Absorption by the caecum of wild birds: is there interspecific variation? *Journal of Experimental Zoology*, Suppl. 3, 103–110.

Gorgas, M. (1967). Vergleichend-anatomische Untersuchungen am Magen-Darm-Kanal der Sciuromorpha, Hystricomorpha und Caviomorpha (*Rodentia*). *Zeitschrift für wissenschaftliche Zoologie*, **175**, 237–404.

Griffiths, M. & Davies, D. (1963). The role of the soft pellets in the production of lactic acid in the rabbit stomach. *Journal of Nutrition*, **80**, 171–180.

Haga, R. (1960). Observations on the ecology of the Japanese pika. *Journal of Mammalogy*, **41**, 200–212.

Heisinger, J. F. (1962). Periodicity of reingestion in the cottontail. *American Midland Naturalist*, **67**, 441–448.

Hofmann, R. R. (1991). The comparative morphology and functional adaptive differentiation of the large intestine of domesticated mammals. *Fortschritte in der Tierphysiologie und Tierernährung*, **22**, 7–17.

Holtenius, K. & Björnhag, G. (1985). The colonic separation mechanism in the guinea pig (*Cavia porcellus*) and the chinchilla (*Chinchilla laniger*). *Comparative Biochemistry and Physiology*, **82A**, 537–542.

Hungate, R. E. (1966). *The Rumen and its Microbes*. New York: Academic Press.

Jarvis, J. U. M. (1981). Eusociality in a mammal: cooperative breeding in naked mole-rat colonies. *Science*, **212**, 571–573.

Kenagy, G. J. & Hoyt, D. F. (1980). Reingestion of faeces in rodents and its daily rhythmicity. *Oecologia*, **44**, 403–409.

Lechleitner, R. R. (1957). Reingestion in the black-tailed jack rabbit. *Journal of Mammalogy*, **38**, 481–485.

Matsuzawa, T., Nakata, M. & Tshushima, M. (1981). Feeding and excretion in the Afghan pika (*Ochotona rufescens rufescens*), a new laboratory animal. *Laboratory Animals*, **15**, 319–322.

Mead, G. C. (1989). Microbes in the avian cecum: types present and substrates utilized. *Journal of Experimental Zoology*, Suppl. **3**, 48–54.

Moss, R. (1989). Gut size and the digestion of fibrous diets by tetranoid birds. *Journal of Experimental Zoology*, Suppl. **3**, 61–65.

Pehrson, Å. (1983a). Caecotrophy in caged mountain hares (*Lepus timidus*). *Journal of Zoology*, **199**, 563–574.

Pehrson, Å. (1983b). Maximal winter browse intake in captive mountain hares. *Finnish Game Research*, **41**, 45–55.

Pickard, D. W. & Stevens, C. E. (1972). Digesta flow through the rabbit large intestine. *American Journal of Physiology*, **22**, 1161–1166.

Ruckebusch, Y. & Fioramonti, J. (1976). The fusus coli of the rabbit as a pacemaker area. *Experientia*, **32**, 1023–1024.

Sakaguchi, E. & Hume, I. D. (1990). Digesta retention and fibre digestion in brushtail possums, ringtail possums and rabbits. *Comparative Biochemistry and Physiology*, **96A**, 351–354.

Scherman, P. W., Jarvis, J. U. M. & Braude, S. H. (1992). Naked mole rats. *Scientific American*, **267**(2), 42–48.

Skadhauge, E. & Schmidt-Nielsen, B. (1965). Cloacal storage and modification of urine in the fowl. *Federated Proceedings of the American Society of Experimental Biology*, **24**, 634.

Smith, M. (1979). Behaviour of the koala, *Phascolarctos cinereus* Goldfuss, in captivity. II. Parental and infantile behaviour. *Australian Wildlife Research*, **6**, 131–140.

Snipes, R. L. (1978). Anatomy of the rabbit cecum. *Anatomy and Embryology*, **155**, 57–80.

Snipes, R. L., Clauss, W., Weber, A. & Hörnicke, H. (1982). Structural and functional differences in various divisions of the rabbit colon. Cell and Tissue Research **225**, 331–346.

Snipes, R. L., Hörnicke, H., Björnhag, G. & Stahl, W. (1988). Regional differences in hindgut structure and function in the nutria (*Myocastor coypus*). Cell and Tissue Research **252**, 435–447.

Sperber, I. (1968). Physiological mechanisms in herbivores for retention and utilization of nitrogenous compounds. In *Isotope studies on the nitrogen chain*, pp. 209–219. Vienna: IAEA.

Sperber, I. (1985). Colonic separation mechanisms (CSM). A review. *Acta Physiologica Scandinavica*, **124** (Suppl. 542), 87.

Sperber, I., Bjōrnhag, G. & Holtenius, K. (1992). A separation mechanism and fluid flow in the large intestine of the equine. *Pferdeheilkunde*, 1. Europäische Konferenz über die Ernährung des Pferdes, Sonderausgabe, pp. 29–32.

Sperber, I., Björnhag, G. & Ridderstråle, Y. (1983). Function of proximal colon in lemming and rat. *Swedish Journal of Agricultural Research*, **13**, 243–256.

Wellard, G. & Hume, I. D. (1981). Digestion and digesta passage in the brushtail possum, *Trichosurus vulpecula* (Kerr). *Australian Journal of Zoology*, **29**, 157–166.

Tullberg, T. (1899). Über das System der Nagetiere. *Nova Acta Regiae Societatis Scientarum Upsaliensis Seriei tertiae* **18**, 1–511.

Walker, P. (1968). Mammals of the world, 2nd edn. Baltimore: The Johns Hopkins Press.

Watson, J. S. & Taylor, R. H. (1955). Reingestion in the hare *Lepus europaeus* Pal. *Science* **121**, 314.

Yasukawa, M. (1959). Studies on the movements of the large intestine. VII. Movements of the large intestine of fowls. *Japanese Journal of Veterinary Science*, **21**, 1–8.

Part IV

Function

18

Foraging and digestion in herbivores

GEORGE O. BATZLI and IAN D. HUME

The foraging decisions of animals determine which food items enter their digestive tracts, but it is the digestive tract that determines net energy and nutrient gain from ingested food. The details of digestive function depend upon the physiological and morphological characteristics of an organism; these vary with the food habits, body size and phylogeny for each species. As a result, the morphology of vertebrate digestive tracts and the processes of digestion and absorption of nutrients are complex (Johnson, 1987; Stevens, 1988). Our purpose here is not to summarize the voluminous literature on these subjects for mammals but rather to address some recent developments in the field.

This section focuses on herbivores because the diets of these mammals often contain low concentrations of digestible energy or other nutrients (usually associated with high fibre content) that has led to great variety in the form and function of their gastro-intestinal tracts. The nutritional constituents of a herbivore's diet vary with the species and part of plant eaten, the habitat in which the plants are growing, and the time of year that the plants are harvested. Because most herbivorous mammals eat a variety of plant species and parts, and most do so year around, they must deal with the problem of variable diet quality. Of course, a few specialists in aseasonal environments may use a limited range of higher quality resources (such as nectar or fruit), but when they do so they lose the advantage of having more available food (stems and leaves). Apparently, the disadvantage of low nutritional quality for plant shoots compared to plant reproductive parts is balanced by their much greater availability, and most mammalian herbivores include plant shoots in their diet.

Eating plant stems and leaves requires gastro-intestinal adaptations that often seem to work better for larger animals. For instance, whether herbivores use fermentation chambers in their 'foreguts' or in their 'hindguts', as body size decreases it becomes increasingly difficult to rely upon dietary fibre as a source of energy. Therefore, the ways that small mammalian herbivores overcome the constraints of body size so that they can exploit plant shoots has become an important topic for understanding gastro-intestinal function. Hume (Ch. 19) considers the effects of gastro-intestinal morphology and body size on the digestive performance of rodents; Batzli *et al* (Ch. 20) discuss the ways that microtine rodents compensate for higher dietary fibre and Cork (Ch. 21) reviews the status of current ideas regarding body size constraints on mammalian diets.

In addition to high levels of dietary fibre, plants that herbivores eat often contain secondary compounds that act as allelochemicals. Because the kind and amount of secondary compounds often determines which parts and how much of a plant a herbivore can eat, interest in the role of these compounds in herbivore nutrition has grown in recent years. Gut microbes aid both in the digestion of fibre and the detoxication of plant secondary compounds, and interest in the contribution of microbes to the digestive physiology of mammals has also continued to grow. The role of plant allelochemicals in herbivore nutrition is discussed in Ch. 22 and the effects of microbial fermentation products (short-chain fatty acids) on the gastro-intestinal tract are reviewed in Ch. 23.

Though far from a complete list of current topics in herbivore nutrition, the chapters in this section provide a sense of the divergent topics currently being pursued. We hope they also convey the sense of excitement and rapid progress that pervades the field.

References

Johnson, L. R. (1987). *Physiology of the Gastrointestinal Tract.* New York: Raven Press.

Stevens, C. E. (1988). *Comparative Physiology of the Vertebrate Digestive System.* Cambridge: Cambridge University Press.

19

Gut morphology, body size and digestive performance in rodents

IAN D. HUME

Amongst the Mammalia, the rodents are the most diversified order and include the greatest number of species and individuals (Perrin and Curtis, 1980). Although it is often assumed that rodents are basically herbivorous, Landry (1970) has argued that 'the primitive adaptation of the rodent mandibulo-dental apparatus was for an omnivorous diet' and 'their extraordinary success as an order can be attributed primarily to the flexibility of their dietary adaptations'. He listed four basic anatomical adaptations of rodent teeth:

1. Rootless, continuously-growing incisors with enamel restricted to the anterior face
2. A large diastema between the incisors and the cheek teeth
3. A longitudinal sulcus, the glenoid fossa, along which the articular process of the mandible slides, so that the jaw functions at either the forward position, when the incisors occlude but the cheek teeth do not, or the rear position, when the incisors do not occlude but the cheek teeth do
4. Folds of skin of the upper lip that push inward through the diastema and meet behind the upper incisor and above the tongue to close off the oral cavity, so that the incisors occlude outside the mouth.

Because of these dental adaptations, rodents can efficiently clip off and grind plant stems, leaves and buds; stab and seize invertebrate prey; shear vertebrate flesh; remove bark; and manipulate small seeds (Landry, 1970). The earliest-known rodents, the Paramyidae, have cheek teeth that are remarkably similar to those of present-day sciurids (squirrels) (Landry, 1970). Many sciurid rodents include animal material in their diets. For example, Tevis (1953) found that yellow-pine chipmunks (*Eutamias amoenus*), which feed heavily on the seeds of ponderosa pine (*Pinus ponderosa*), also use invertebrates; stomach contents consisted of 27% arthropod remains in spring, 22% in summer and 13% in autumn.

Other rodent groups are more herbivorous. These include some sciuromorphs, particularly larger members such as beaver (*Castor canadensis*) and marmots (*Marmota* spp.), and many caviomorph rodents such as porcupines (e.g. *Erethizon dorsatum, Hystrix cristata*), the capybara (*Hydrochoerus hydrochaeris*) and the mara (*Dolichotis patagonum*), all of which are large, with body masses greater than 1 kg. In contrast, the suborder Myomorpha, with 1137 species in 264 genera, contains only one species larger than 1 kg; this is the muskrat (*Ondatra zibethicus*) in the cricetid subfamily Arvicolinae (Microtinae). The Arvicolinae, with 110 species in 18 genera, consists mainly of lemmings and voles and is probably the most herbivorous group of myomorph rodents (Batzli 1985) despite their small size (all except the muskrat weigh less than 300 g).

Evolutionary trends among the Rodentia as adaptations for herbivory include modifications not only to the teeth (Vorontsov, 1962) but also to the stomach (Vorontsov, 1962; Carleton 1973, 1981; Perrin and Curtis, 1980) and the hindgut (i.e. the caecum and colon) (Tullberg, 1899; Gorgas, 1967; Vorontsov, 1962; Behmann, 1973). Vorontsov (1962) listed some of these trends:

1. Increase in the number of transverse ridges on the cheek teeth, hypsodonty, and appearance of continuously growing cheek teeth that develop roots only late in life or not at all
2. Strengthening of the masticatory musculature
3. Increase in total digestive tract capacity
4. Replacement of unilocular (i.e. undivided) stomachs with bilocular and perhaps even trilocular stomachs
5. Reduction in the proportion of the stomach lined by glandular mucosa and an increase in the area of cornified, non-glandular epithelium
6. Reduction in the relative length of the small intestine
7. Increase in relative size of the caecum and colon
8. Increase in complexity of the caecum by the formation of sacculations, haustrations and spiral folds
9. Increase in the complexity of the proximal colon by ampullae, oblique folds and an increase in the number of turns of the double spiral.

Fig. 19.1 illustrates these trends in the stomach, and Fig. 19.2 in the caecum and colon.

Vorontsov (1962) argued that the above trends indicated a change from feeding on energy-dense but hard-to-get food, such as invertebrates and seeds, to less energy-dense but easier-to-get food in the form of the vegetative parts of plants. In particular, division of the stomach into two partially separ-

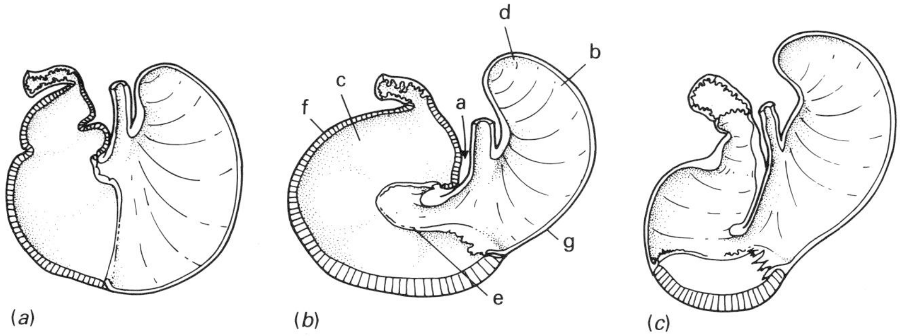

Fig. 19.1. Variations in the gross stomach anatomy of microtine rodents. (*a*) Unilocular – hemiglandular – stomach of *Synaptomys cooperi*; (*b*) intermediate stomach of *Alticola roylei*; (*c*) bilocular – discoglandular – stomach of *Phenacomys longicaudis*. a, incisura angularis; b, corpus; c. antrum; d, fornix ventricularis; e, bordering fold; f, glandular mucosa; g, cornified squamous (non-glandular) epithelium. Redrawn from Carleton (1981).

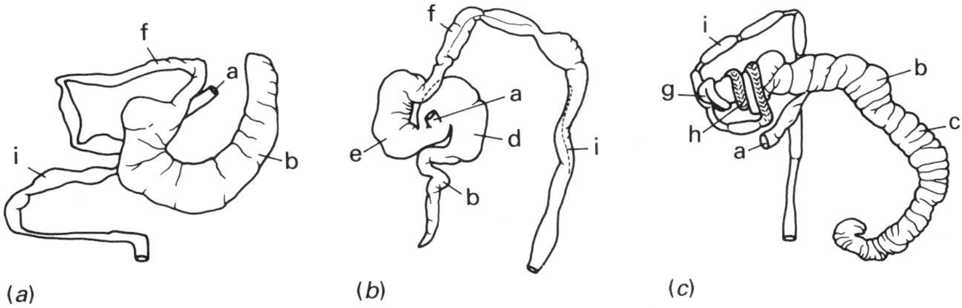

Fig. 19.2. Variations in the gross morphology of the caecum and colon of myomorph rodents. (*a*) Simple caecum and colon of *Rattus norvegicus*; (*b*) intermediate caecum and colon of *Sicista subtilis*; (*c*) large, haustrated caecum and complex colon of *Arvicola terrestris*. a, ileum; b, caecum; c, haustration; d, ampulla caecum; e, ampulla colon; f, proximal colon; g, colonic spiral; h, oblique folds of proximal colon mucosa; i, distal colon. Adapted from Behmann (1973).

ated chambers and reduction in the amount of glandular mucosa would be expected to maintain a higher pH in the fornix ventricularis or forestomach, which Vorontsov (1962) thought would allow a cellulolytic fermentation to proceed.

However, Carleton (1973, 1981) proposed an alternative explanation for the dental and gastric differentiations reported by Vorontsov (1962) and shown in his own work on New World cricetine and microtine rodents. Although the open-rooted, complex molars and capacious and complex caecum clearly support selection for using structural plant parts, seeds, once husked, would also be efficiently masticated by such cheek teeth and the relatively high pH in the forestomach would allow prolonged action of salivary amylase on starches. In Carleton's view (1981), the elaborations listed above mean that rodents with these features are able to use a broad spectrum of foods, including both vegetative and reproductive parts of plants, rather than being highly specialized on only certain classes of food.

A modification to Carleton's gastric amylolytic reservoir theory (1981) was proposed by Perrin and Kokkin (1986). They suggested that the forestomach of two African myomorph rodents, *Mystromys albicaudatus* (white-tailed rat) and *Cricetomys gambianus* (African giant rat) is an amylolytic reservoir where prolonged salivary amylase digestion of starches and glycogen is supplemented by amylase produced by a dense population of bacilli attached to numerous filiform papillae lining the forestomach (fornix ventricularis and corpus). This implies increased digestive efficiency of diets high in seeds. The mean retention time (MRT) of stained particles of unspecified size(s) in the whole digestive tract of *M. albicaudatus* (80–110 g body mass) was only 12.1 h (Perrin and Maddock, 1983), too short for significant cellulose digestion to occur, although there could be some digestion of hemicellulose (Keys *et al.*, 1970). Longer MRTs (90 h for large (2 mm × 2 mm) plastic particles) in the larger (1.0–1.5 kg) *C. gambianus* (Knight and Knight-Eloff, 1987) suggest greater capacity for cellulose digestion. These MRTs may be overestimates, however, as the animals were fasted for 24 h prior to dosing with marked food. *M. albicaudatus* shows a clear preference for insects, fruits and seeds, while *C. gambianus* feeds on seeds, fruits and tubers. Little in the way of vegetative plant parts is included in the diet of either species.

In contrast, microtine rodents take significant amounts of the leaves, stems and roots of monocotyledons and dicotyledons (60–80% in *M. ochrogaster*, the prairie vole; Batzli, 1985). This is surprising for such small animals (most weigh less than 100 g). For this reason, the nutritional ecology of microtines has been the most researched of all the wild rodents (e.g. Batzli, 1985), as have aspects of their digestive physiology, particularly the structure and function of the caecum and colon (e.g. Golley, 1960; Björnhag and Sjöblom, 1977; Snipes, 1979). The remarkable phenomenon of separation of fine digesta particles and microbes in the proximal colon and their selective retention in the caecum of myomorph and caviomorph rodents is described in Ch. 17.

Comparison of the digestive performance of microtines with that of sciurid rodents provides an opportunity to examine the nutritional consequences of some of the evolutionary trends listed by Vorontsov (1962). The family Sciuridae (chipmunks, squirrels and marmots) feeds on a variety of seeds, nuts, fruits, inflorescences, fungi and some herbs, but does not feed on grasses to any great extent (Ingles, 1965). The sciurid digestive system is also much simpler than that of microtines (Fig. 19.3). On the basis of these clear dietary and digestive tract differences between sciurids and microtines, it can be predicted that, on a common high-fibre diet, microtines should digest more than sciurids of similar body size. Hume *et al.*, (1993) recently tested this prediction by measuring digestive performance of Townsend voles (*M. townsendii*) and yellow-pine chipmunks (*Eutamias amoenus*) (both 50–60 g body mass) fed commercial (Purina lab chow) diets of low- and high-fibre content (17% and 35% neutral-detergent fibre). Mean retention times (MRTs) of fluid digesta (marked with Co-EDTA) (Udén *et al.*, 1980) and large (0.5–1.0 mm) particles (mordanted with Cr) (Ellis and Beaver, 1984) given as a pulse dose with the low-fibre diet were also measured in order to compare digestive tract function between the two rodent groups. Sciurids range in body size from 50 g chipmunks to 5 kg marmots. The increase in the ratio of absolute gut capacity to metabolic requirements that accompanies increasing body size (Demment and van Soest, 1985) can be expected to increase digestive performance, particularly on fibrous diets, through longer retention of digesta within the gut of the larger animals. Hume *et al.* (1993) included two additional species of sciurids in their study, Columbian ground-squirrels (*Spermophilus columbianus*) (600–700 g body mass) and hoary mar-

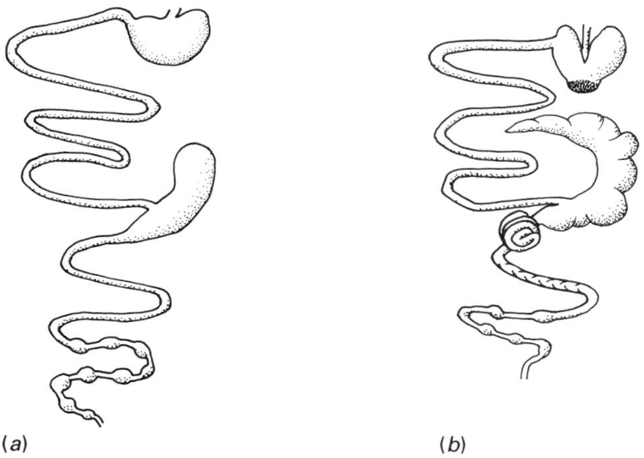

Fig. 19.3. The digestive tracts of (*a*) sciurid and (*b*) microtine rodents.

mots (2–3 kg), to test the prediction that the disadvantage within the Sciuridae of a relatively simple gut (Fig. 19.3) should be offset by increasing body size.

In general, the results of Hume *et al.* (1993) support the two predictions (Table 19.1). First, dry matter digestibility was greater ($P < 0.05$) in the voles (microtine) than in the chipmunks (sciurid), and the difference was greater on the higher fibre diet. Second, digestibility was similar in the chipmunks and ground-squirrels on both diets, but in the largest of the sciurids, the marmots, digestibility was higher ($P < 0.05$), similar to the voles on the high-fibre diet and even higher ($P < 0.05$) than the voles on the low-fibre diet.

The superior digestive performance of the voles relative to the chipmunks was correlated with selective retention of the fluid marker in the voles, presumably in the complex and voluminous caecum (Fig. 19.3; Snipes, 1979), but not in the chipmunks. The significance of this is that in the voles fine particles are swept along with and are retained with the fluid digesta (Sakaguchi and Hume, 1990). Because of their relatively high ratio of surface area to volume, fine particles are potentially more digestible than large par-

Table 19.1. *Comparison of the digestive performance of microtine and sciurid rodents. Body mass, digestive performance and mean retention times of fluid and large particle markers in four rodents.*

	Microtine	Sciurid		
	Vole	Chipmunk	Squirrel	Marmot
Body mass (g)				
Low fibre	55 ± 4	62 ± 2	663 ± 21	2308 ± 134
High fibre	61 ± 4	55 ± 3	629 ± 17	2522 ± 131
Apparent digestibility of dry matter (%)				
Low fibre	77.4 ± 1.4^{b}	74.6 ± 0.4^{a}	75.4 ± 0.9^{ab}	81.0 ± 0.5^{c}
High fibre	50.9 ± 1.0^{b}	40.6 ± 3.0^{a}	40.9 ± 1.1^{a}	50.3 ± 1.2^{b}
Mean retention time (h)				
Co (fluid digesta)	14.8 ± 1.0	12.7 ± 1.0	22.5 ± 2.2	24.8 ± 3.2
Cr (large particles)	13.1 ± 2.4^{d}	14.1 ± 1.2^{e}	22.1 ± 1.1^{f}	28.9 ± 3.4^{d}

Values are given as means ± S.E., $n = 6$.
^{abc} Within a diet, mean digestibilities bearing different superscripts differ significantly ($P < 0.5$) by one-way analysis of variance.
Mean retention time for large particles was significantly different from that for fluid digesta by paired t-test: ^{d} $P < 0.05$; ^{e} $P < 0.01$; ^{f} not significant.
From Hume *et al.* (1993).

ticles. Thus in voles, digestive effort is concentrated in the caecum on the food residues of highest quality. The fine particles also include bacteria. Thus another advantage of selective retention of fine particles is that a high concentration of bacteria is maintained in the voles' caecum and this must also help to maximize digestion of the fine particles there (Björnhag, Ch. 12). Coprophagy would be expected to increase MRTs, especially of the fluid marker, in the voles (Björnhag and Sjöblom, 1977), but this was not obvious in the results (Table 19.1).

In the three sciurids, there was either no difference in MRT between the two markers (ground-squirrels) or there was selective retention of the large-particle marker (chipmunks, marmots). This is probably the net result of slower flow of particles than of fluid out of the stomach and the absence of separation of the two markers in the hindgut. Thus retention times of large and fine particles in the caecum were probably similar. However, MRTs increased with body size so that, in the largest sciurids the marmots, the disadvantage of having no selective retention of fine particles in the hindgut was overcome by greater total gut capacity.

Among those rodents in which there is selective retention of fluid and fine particulate digesta in the hindgut, there is also a correlation between digestive performance and body size, and thus with MRT. This is illustrated by the work of Sakaguchi *et al.* (1987) in which digestibility of fibre by rabbits and five rodents fed a common diet was related to the MRT of large particles in the whole digestive tract (Fig. 19.4a). The relationship is even closer when fibre digestibility is regressed on turnover time of particles in the caecum rather than on whole-tract MRT (Fig. 19.4b).

Such relationships need to be explored further by including a wider range of body sizes and taxonomic groups and feeding more natural diets before the flexibility of the dietary adaptations of the Rodentia can be fully explained in functional terms.

References

Batzli, G. O. (1985). Nutrition. In *Biology of New World Microtus*, ed. R. M. Tamarin, Special Publication of the American Society of Mammalogists, **8**, 779–811.

Behmann, H. (1973). Vergleichend– und funktionell–anatomische Untersuchungen am Caecum und Colon myomopher Nagetiere. *Zeitschrift für wissenschaftliche Zoologie*, **186**, 173–294.

Björnhag, G. & Sjöblom, L. (1977). Demonstration of coprophagy in some rodents. *Swedish Journal of Agricultural Research*, **7**, 105–114.

Carleton, M. D. (1973). A survey of gross stomach morphology in New World Cricetinae (Rodentia: Muroidea) with comments on functional interpretations.

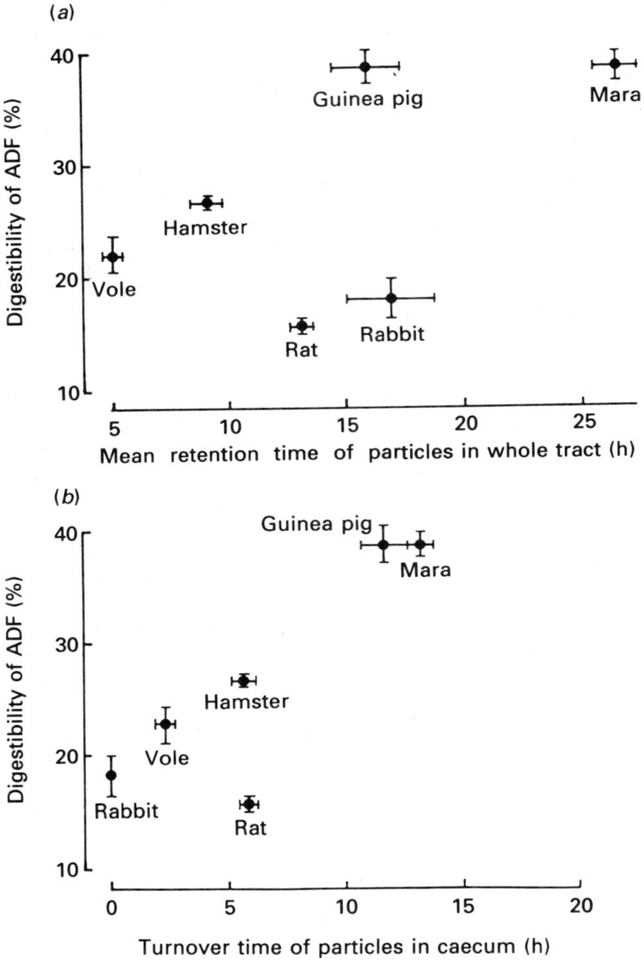

Fig. 19.4. Digestibility of acid-detergent fibre (ADF) in relation to (*a*) mean retention time of particles in the whole digestive tract; (*b*) turnover time (or residence time) in the caecum of rabbits, voles, guinea pigs, rats, hamsters and maras. Points are means from five to seven animals, with standard error bars. From Sakaguchi *et al.* (1987).

Miscellaneous Publication, Museum of Zoology, No. 146, pp. 1–43, Michigan: University of Michigan.

Carleton, M. D. (1981). A survey of gross stomach morphology in Microtinae (Rodentia: Muriodea). *Zeitschrift für Säugetierkunde*, **46**, 93–108.

Demment, M. W. & van Soest, P. J. (1985). A nutritional explanation for body-size patterns of ruminant and non-ruminant herbivores. *American Naturalist*, **125**, 641–672.

Ellis, W. C. & Beaver, D. E. (1984). Methods for binding rare earths to specific feed particles. In *Techniques in Particle Size Analysis of Feeds and Digesta in Ruminants*, ed. P. M. Kennedy, pp. 154–165. Edmonton, Alberta: Canadian Society of Animal Science Occasional Publication No. 1.

Golley, F. B. (1960). Anatomy of the digestive tract of *Microtus*. *Journal of Mammalogy*, **41**, 89–99.

Gorgas, M. (1967). Vergleichend-anatomische Untersuchungen am Magen-Darm-Kanal der Sciuromorpha, Hystricomorpha und Caviomorpha (Rodentia). *Zeitschrift für wissenschaftliche Zoologie*, **175**, 237–404.

Hume, I. D., Morgan, K. R. & Kenagy, G. J. (1993). Digesta retention and digestive performance in sciurid and microtine rodents: effects of hindgut morphology and body size. *Physiological Zoology*, **66**, 396–411.

Inglis, L. G. (1965). *Mammals of the Pacific States*. Stanford: Stanford University Press.

Keys, J. E. Jr, van Soest, P. J. & Young, E. P. (1970). Effect of increasing dietary cell wall content on the digestibility of hemicellulose and cellulose in swine and rats. *Journal of Animal Science*, **31**, 1172–1177.

Knight, M. H. & Knight-Eloff, A. K. (1987). Digestive tract of the African giant rat, *Cricetomys gambianus*. *Journal of Zoology* (Lond.), **213**, 7–22.

Landry, S. O. (1970). The Rodentia as ominivores. *Quarterly Review of Biology*, **45**, 351–372.

Perrin, M. R. & Curtis, B. A. (1980). Comparative morphology of the digestive system of 19 southern African myomorph rodents in relation to diet and evolution. *South African Journal of Zoology*, **15**, 22–33.

Perrin, M. R. & Kokkin, M. J. (1986). Comparative gastric anatomy of *Cricetomys gambianus* and *Saccostomus campostris* (Cricetomyinae) in relation to *Mystromys albicaudatus* (Cricetinae). *South African Journal of Zoology*, **21**, 202–210.

Perrin, M. R., & Maddock, A. H. (1983). Preliminary investigations of the digestive processes of the white-tailed rat *Mystromys albicaudatus* (Smith 1984). *South African Journal of Zoology*, **18**, 128–133.

Sakaguchi, E. & Hume, I. D. (1990). Digesta retention and fibre digestion in brushtail possums, ringtail possums, and rabbits. *Comparative Biochemistry and Physiology*, **96A**, 351–354.

Sakaguchi, E., Itoh, H., Uchida, S. & Horigome, T. (1987). Comparison of fibre digestion and digesta retention time between rabbits, guinea-pigs and hamsters. *British Journal of Nutrition*, **58**, 149–158.

Snipes, R. L. (1979). Anatomy of the cecum of the vole, *Microtus agrestis*. *Anatomy and Embryology*, **157**, 181–203.

Tevis, L. J. (1953). Stomach contents of chipmunks and mantled squirrels in northeastern California. *Journal of Mammalogy*, **34**, 316–324.

Tullberg, T. (1899). Ueber das System der Nagetiere. *Nova Acta Regiae, Societatis Scientarum Upsaliensis,* **18**, 1–511.

Udén, P., Colucci, P. E. & van Soest, P. J. (1980). Investigation of chromium, cerium and cobalt as markers in digesta rate of passage studies. *Journal of the Science of Food and Agriculture*, **31**, 625–632.

Vorontsov, N. N. (1962). The ways of food specialisation and evolution of the alimentary tract in Muroidea. In *Symposium Theriologicum, 1960*, ed. J. Kratochvil & J. Pelikan, pp. 360–377. Praha: Czechoslovak Academy of Science.

20

The integrated processing response in herbivorous small mammals

GEORGE O. BATZLI, ALLEN D. BROUSSARD† and ROBERT J. OLIVER

The amount of energy and nutrients that herbivorous mammals extract from a diet depends upon the amount of food eaten (intake) and its digestibility, but intake and digestibility do not vary independently. Apparent digestibility for any particular diet depends upon its chemical constituents, its retention time in the gastro-intestinal (GI) tract and the absorptive capacity of the GI tract. The last is a function of epithelial surface area and nutrient transport rates across the epithelium (Karasov and Diamond, 1988). Within limits, forage intake increases as the digestibility of the food decreases; retention time of food within the GI tract decreases as intake increases; and digestibility decreases as retention time decreases (Sibley, 1981; van Soest, 1982; Robbins, 1983). Retention time also decreases as the volume of the GI tract decreases. All of these variables (intake, retention time, GI size and digestibility) can change in response to changes in energy demand or fibre content of the diet (Hammond and Wunder, 1991). We call the relationships among these changes, and their effect on digestible energy (or nutrient) intake, the **integrated processing response** (**IPR**).

A wide variety of birds and mammals show an IPR, as indicated by changes in size, morphology and nutrient absorption rates of their GI tract with changes in diet quality (Gross *et al.*, 1985; Karasov and Diamond, 1988; Brugger, 1991 and references therein). The IPR of small mammals is particularly interesting because a considerable body of theory, primarily based on ruminants, argues that small mammals (adult mass less than 1 kg) should not be able to meet their energy requirements on high fibre diets, such as mature grass (Demment and van Soest, 1985). Nevertheless, several species of herbivorous arvicoline (microtine) rodents routinely do so (Batzli, 1985).

† Deceased.

All arvicoline rodents grind their food to fine particles and some (lemmings) pass large amounts of food rapidly through their GI tract, apparently living mostly on highly digestible cell contents (Batzli *et al.*, 1980; Batzli, 1985). However, other arvicolines (voles) digest more of their diet (Batzli and Cole, 1979) and laboratory experiments with prairie voles (*Microtus ochrogaster*) indicate that both the content and size of the GI tract increase in response either to increased energy demand for thermoregulation or to increased fibre content (lower digestibility) in their diet (Gross *et al.*, 1985). In the most comprehensive experiment to date, Hammond and Wunder (1991) measured intake, digestibility and size of the GI tract for prairie voles on low-fibre and high-fibre laboratory rations. They found that voles on high-fibre diets showed lower digestibility but higher intake and larger gut capacity, so that digestible energy intake was as high or higher than on low-fibre diets. The ability of arvicoline rodents to compensate for low quality of diet by increasing intake and thereby maintaining a nearly constant digestible energy intake has been reported previously (Batzli and Cole, 1979; Batzli, 1986). However, these earlier studies did not measure changes in size of the GI tract, which provides a mechanism to explain this remarkable compensatory capability. Increased GI capacity slows passage rates so that digestive and absorptive processes have longer to operate on a given amount of intake, and increased GI tissue mass can increase the area and absorptive capacity of GI epithelium (Karasov and Diamond, 1988).

Though the laboratory evidence suggests that the IPR may be an adaptive response to the changes in diet quality that herbivores face in natural settings, few small mammal studies have been done using natural diets. Sibley *et al.* (1990) reported increased GI size for rabbits fed clover and ryegrass pellets (21 and 24% fibre, respectively) compared to rabbits fed commercial pellets (13% fibre). They also found larger GI sizes for rabbits collected from areas with chalk soils, which are thought to produce lower quality forage than clay soils, but the actual diets of these rabbits were not examined.

The purpose of our work was to test the IPR hypothesis, which states that the IPR improves the ability of herbivorous mammals to utilize natural diets with different fibre content. From this hypothesis we predicted that: (a) voles from natural populations that consume diets with higher fibre content should have greater GI capacity and (b) voles should improve their ability to utilize those natural diets to which they have been acclimated. To determine if the GI tracts of voles in natural populations do change in response to food quality we collected meadow voles (*M. pennsylvanicus*) from two habitats in which they have different diets. Voles in alfalfa oldfields primarily eat high-quality (low-fibre) legumes, voles in bluegrass oldfields primarily eat moderately high quality grasses and forbs, and voles in tallgrass prairie primarily eat low quality (high-fibre) grasses and forbs (Cole and Batzli, 1979; Lindroth and Batzli, 1984). We expected voles in tallgrass prairie to have larger GI tracts than voles in bluegrass oldfields. To determine if acclimation improved the voles' ability to utilize a diet, we fed high quality natural food (alfalfa) and low quality natural food (mature grass) to meadow voles and prairie voles in the laboratory. We expected voles acclimated to grass diets to have larger GI tracts and to digest grasses better than voles acclimated to alfalfa diets.

Methods

Field samples

From 17 June to 4 August 1988 we set snap traps in two habitats, bluegrass oldfield and tallgrass prairie, at the University of Illinois Ecological Research Area located 5 km north-east of Urbana, Illinois. Museum Special traps were set near signs of activity and checked once a day. We hoped to capture at least ten prairie voles (*M. ochrogaster*) and ten meadow voles (*M. pennsylvanicus*) in each habitat. After intensive trapping (1044 trap-nights in bluegrass and 840 trap-nights in prairie) we captured 12 meadow voles in bluegrass and 11 meadow voles in prairie but only one prairie vole in each habitat. We ceased trapping at that time to avoid seasonal changes in diet. Therefore, all data reported from the field refer only to meadow voles.

We took captured animals directly into the laboratory and noted their sex, reproductive condition, mass and length (head and body). We then removed their gastro-intestinal tract and divided it into four sections: stomach, small intestine, caecum and colon. The mass of each component was weighed while wet, the contents removed by rinsing with distilled water into a pre-weighed aluminium pan, and the section reweighed. Each section was then extended along a ruler to measure its length to the nearest millimetre (except for the first six animals caught, all of which were from bluegrass), taking care not to stretch the intestines. The tip of the spiral portion of the colon was not extended to avoid damaging it. To determine volume of each section we tied one end with thread, filled the section with water using a wash bottle, then emptied the water into a small graduated cylinder and determined its volume to 0.5 ml. This proved to be difficult because of the small size and delicate nature of the GI tracts of voles and the volume of some organs needed to be measured piecemeal. Because no pressure was applied when filling the sections, these results represent minimal volume. We then dried each section and its contents to a constant weight at 60 °C and determined their mass with an analytical balance to 0.1 mg.

To compare the quality of food eaten in the two habitats, we measured the acid-detergent fibre (cellulose, lignin and silica) content of the stomach and caecal contents (Goering and van Soest, 1970). We used contents from both the stomach and from the caecum to see if the fibre of caecal contents would differ from that of the food originally consumed. To obtain sample sizes of 0.5 g or more, we combined the stomach or caecal contents of some of the voles captured in the same habitat, which reduced sample sizes to seven for each habitat and may have reduced the variability of our samples.

Laboratory feeding trials

We conducted laboratory feeding trials to determine the effect of acclimation to different natural diets on the digestibility of the diets and on the size of the GI tracts. Two types of food were collected from the field in autumn (September and October) of 1990 and oven dried at 60 °C for the trial diet. The low quality (high-fibre) diet consisted of a mixture of grass shoots (*Dactylis glomerata* and *Bromus inermis*) that contained 27.2 ± 1.1% ($\bar{x} \pm 1$ S.E.) fibre ($n = 4$), whereas the high quality diet consisted mostly of alfalfa branchlets and leaves (*Medicago sativa*) that contained 12.5 ± 1.1% fibre ($n = 2$). Few stems of the alfalfa, which had 34.8 ± 0.3% fibre ($n = 2$), were eaten by the voles. Most voles maintained their body mass (<10% change in mass) when fed either diet and only such voles were used in the experiment.

The experimental design included four different groups: prairie voles acclimated to grass, prairie voles acclimated to alfalfa, meadow voles acclimated to grass and meadow voles acclimated to alfalfa. Our goal was to have six voles in each group but only seven meadow voles were available, so we fed four grass and three alfalfa. The acclimation period in each case lasted 21 days, after which we conducted two 3-day digestibility trials on each experimental group. We monitored dry mass of intake and dry mass of faecal production for each animal, then calculated apparent dry matter digestibility. During the first digestibility trial, each group received its acclimation diet; during the second digestibility trial each group received the other diet. All trials began and ended at the same time of day. At the end of the digestibility trials, all animals were sacrificed and measurements of GI tracts were made as described for the field samples, except that no attempt was made to measure GI volume for these voles.

Results

Field samples

The stomach and caecal contents of voles caught in the prairie contained more acid-detergent fibre than did those of voles caught in bluegrass, mean values of 24% versus 19% (Fig. 20.1). The differences in dietary fibre between habitats were highly significant but those between the stomachs and caeca were not (2-way ANOVA: $F_{1,24} = 14.6$, $P < 0.001$ for habitats and $F_{1,24} = 0.2$, $P = 0.672$ for GI compartments).

Meadow voles trapped in prairie appeared to be slightly larger than those trapped in bluegrass. Differences in head and body lengths were significant ($P = 0.05$), but differences in body masses were not (Table 20.1). Mass to length ratios, often used as an index of condition, were very similar (Table 20.1), though live body masses for some individuals (one in bluegrass and three in prairie) were underestimated because they had been partially chewed (probably by other voles) before their removal from the traps.

Because the IPR hypothesis predicts that GI tracts for voles collected in prairie will be larger than those for voles collected in bluegrass, we compared the size of GI compartments using one-tailed *t*-tests. We ran *t*-tests without

Table 20.1. *Body size and size of GI compartments of meadow voles collected from two habitat types: bluegrass oldfield and tallgrass prairie*

	Habitat type		
	Bluegrass	Prairie	P (adj)[a]
Body size			
Head and body length (mm)*	104 ± 2	110 ± 1	0.052
Body mass (g)	28.0 ± 1.2	29.9 ± 1.5	0.345
Mass/length (g cm^{-1})	2.69 ± 0.11	2.71 ± 0.13	0.930
Total GI tract			
Tissue dry mass (mg)	198 ± 13	194 ± 14	0.831 (0.267)
Contents' dry mass (mg)**	719 ± 63	982 ± 73	0.006 (0.021)
Length (mm)	532 ± 17	599 ± 17	0.011 (0.192)
Volume (ml)***	11.2 ± 0.6	15.9 ± 1.0	< 0.001 (0.003)
Tissue mass/length (mg cm^{-1})	3.18 ± 0.24	3.23 ± 0.20	0.443

Values are given as means ± S.E. for 12 meadow voles from bluegrass habitat and 11 from prairie.

* $P < 0.10$; ** $P < 0.05$; *** $P < 0.001$ for probabilities for comparison of habitats by *t*-tests (two-tailed for body size and one-tailed for GI tract).

[a] Adjusted P values calculated for GI data as described in the text.

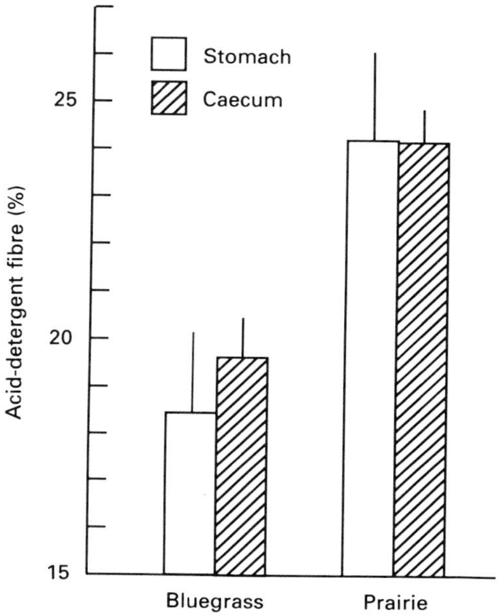

Fig. 20.1. Fibre content as mean percentage of acid-detergent fibre in the contents of GI tracts of meadow voles collected in bluegrass oldfields or tallgrass prairie. Vertical lines give ± 1 S.E. (n = 7 for each mean).

and with adjustments for body size (GI length/head and body length; GI dry mass or volume/live body mass). As predicted, all significant differences between habitats indicated that voles collected from prairie had larger GI tracts (Table 20.1). Both dry mass of contents and volume were significantly larger for voles from the prairie (37% increase for dry mass and 42% increase for volume). Total length was 13% greater in the prairie animals, but the difference was not significant when adjusted for body size. Tissue dry mass and tissue mass per unit length of GI tract showed only small, insignificant differences. The same general pattern held for all the GI compartments and will not be considered in detail here.

Laboratory feeding trials

Meadow voles ate more than prairie voles no matter what the diet or acclimation regimen, and, because digestibilities did not differ for the two species, meadow voles also had greater digestible energy intake (Fig. 20.2). The same pattern occurred whether intake was adjusted by dividing by live body mass (gbm) or with metabolic body mass (gbm$^{0.75}$), although the differences

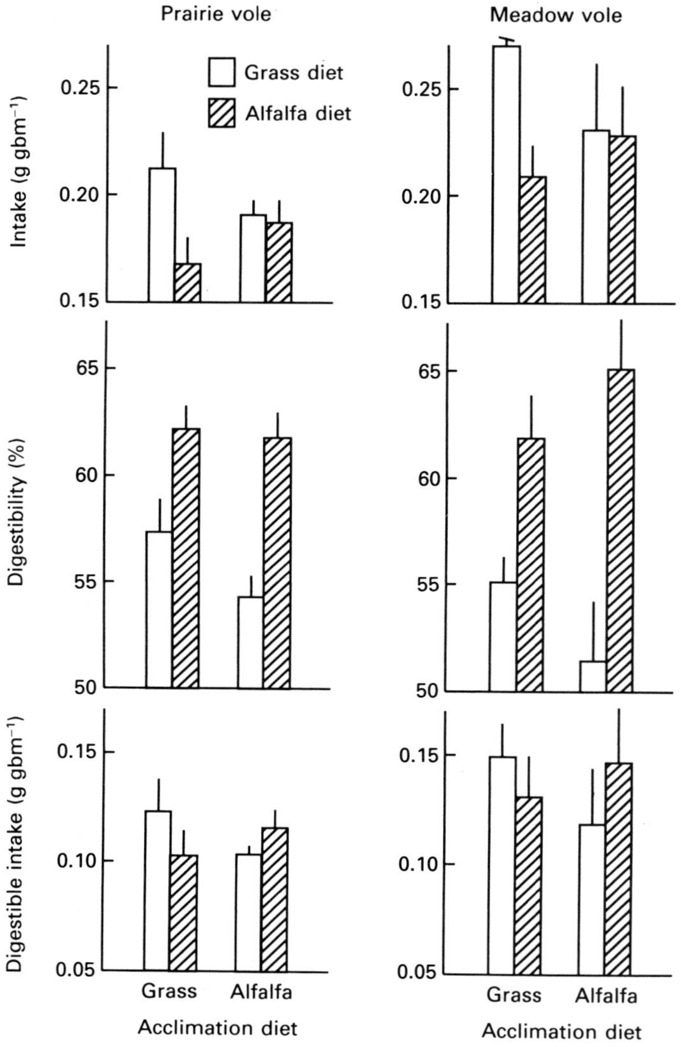

Fig. 20.2. Mean intake, digestibility and digestible intake of grass and alfalfa diets fed to prairie voles and meadow voles after acclimation to grass or alfalfa diets for 3 weeks. Vertical lines give ± 1 S.E. (n = 6 for all prairie vole means, n = 4 for meadow voles acclimated to grass and n = 3 for meadow voles acclimated to alfalfa). gbm, g body mass.

between treatments were slightly more significant using metabolic body mass (Table 20.2). Both species of vole ate significantly more grass than alfalfa, but there was a significant interaction between test diet and acclimation diet (Table 20.2); voles ate more of the test diet if they had been acclimated to it (Fig. 20.2).

Neither the species of vole nor the acclimation regimen affected overall digestibilities, but alfalfa was always more digestible than grass (Figure 20.2,

Table 20.2. *Intake, digestibility and digestible intake of diets for voles acclimated to high quality (alfalfa) and low quality (grass) diets shown as three-way ANOVA tables*

Source	Degrees of freedom	Sum of squares	*F*	*P*
Intake (g gbm$^{-0.75}$)				
Vole species (A)***	1	0.2081	39.3	< 0.001
Acclimation diet (B)	1	0.0017	0.316	0.578
Diet fed (C)**	1	0.0388	7.32	0.011
AB	1	0.0036	0.679	0.417
AC	1	0.0015	0.283	0.598
BC*	1	0.0301	5.68	0.024
ABC	1	0.0018	0.332	0.569
Error	30	0.1590		
Digestibility (%)				
Vole species (A)	1	0.0003	0.253	0.618
Acclimation diet (B)	1	0.0009	0.731	0.399
Diet fed (C)***	1	0.0572	46.7	< 0.001
AB	1	0.0005	0.397	0.534
AC	1	0.0038	3.11	0.088
BC*	1	0.0052	4.24	0.048
ABC	1	0.0010	0.780	0.384
Error	30	0.0367		
Digestible intake (g gbm^{-075})				
Vole species (A)***	1	0.0653	28.8	< 0.001
Acclimation diet (B)	1	0.0015	0.677	0.417
Diet fed (C)	1	0.0001	0.050	0.824
AB	1	0.0006	0.278	0.602
AC	1	0.0008	0.374	0.546
BC**	1	0.0197	8.70	0.006
ABC	1	0.0017	0.736	0.398
Error	30	0.0679		

Means and S.E. are shown in Fig. 20.2. The data for intake and digestible intake were corrected for metabolic body size before analysis.
* $P < 0.05$; ** P 0.01; *** $P < 0.001$.

Table 20.2). A significant interaction again occurred between test and acclimation diets; test diets (particularly grass) had higher digestibilities in animals that had been acclimated to them. Though meadow voles extracted more digestible dry matter whatever the diet, voles within each species received the same amount of digestible energy from grass as from alfalfa (Fig. 20.2, Table 20.2), which indicates their ability to regulate digestible energy intake. However, a highly significant interaction between test and acclimation diets reflected the ability of voles to extract more digestible dry matter from a diet if they had been acclimated to it (20% more from grass for prairie voles and 27% more from grass for meadow voles when acclimated to it; 10% more from alfalfa for prairie voles and 13% more from alfalfa for meadow voles when acclimated to it).

Meadow voles acclimated to grass diets had significantly longer (14%) whole GI tracts but not a greater mass of GI contents (Table 20.3). Dry tissue mass was 12% greater (non-significant), so tissue mass per unit length did not differ with diet (Table 20.3). Trends were generally consistent for all GI compartments, but the trends for dry mass of contents (non-significant) were contrary to those expected (greater in animals acclimated to alfalfa), except for the small intestine.

Prairie voles showed trends that were generally inconsistent with our expectations. All the trends (none significant) for the whole GI tract, except

Table 20.3. *Body size and size of GI compartments of meadow voles fed high quality (alfalfa) and low quality (grass) diets for 3 weeks*

	Acclimation diet		
	Alfalfa	Grass	*P* (adj)[a]
Body size			
Head and body length (mm)	120 ± 3	120 ± 2	0.984
Live body mass (g)	41.4 ± 8.3	43.7 ± 4.8	0.757
Total GI tract			
Tissue dry mass (mg)	312 ± 49	350 ± 66	0.343 (0.441)
Contents dry mass (mg)	797 ± 49	784 ± 104	0.460 (0.187)
Length (mm)**	709 ± 10	806 ± 24	0.012 (0.022)
Tissue mass/length (mg cm^{-1})	4.42 ± 0.76	4.40 ± 0.92	0.493

Values are given as means ± 1 S.E. for four meadow voles acclimated to alfalfa and three acclimated to grass.
** $P < 0.05$ for probabilities (P) for comparison of diets by t-tests (two-tailed for body size and one-tailed for GI tract).
[a] Adjusted P values calculated for GI data as described in the text.

Table 20.4. *Body size and size of GI compartments of prairie voles fed high quality (alfalfa) and low quality (grass) diets for 3 weeks*

	Acclimation diet		
	Alfalfa	Grass	P (adj)[a]
Body size			
Head and body length (mm)	106 ± 3	102 ± 2	0.299
Live body mass (g)	30.7 ± 2.0	28.6 ± 2.7	0.451
Total GI tract			
Tissue dry mass (mg)	317 ± 39	304 ± 43	0.421 (0.480)
Contents dry mass (mg)	719 ± 110	648 ± 72	0.301 (0.466)
Length (mm)	607 ± 28	564 ± 33	0.175 (0.256)
Tissue mass/length (mg cm^{-1})	5.27 ± 0.65	5.44 ± 0.73	0.432

Values are given as means ± 1 S.E. for six prairie voles acclimated to alfalfa and six acclimated to grass.
Probabilities (P) given for comparison of diet types based on t-tests (two-tailed for body size and one-tailed for GI tract).
[a] Adjusted P values calculated for GI data as described in the text.

that for tissue mass per unit length, indicated larger GI capacity for voles acclimated to alfalfa (Table 20.4). Individual GI compartments showed inconsistent trends, only two of which were significant: a 20% longer caecum ($P < 0.05$) and 42% greater dry mass of colonic contents ($P < 0.1$) for voles acclimated to alfalfa.

Discussion

Data from the field clearly supported the prediction of the IPR hypothesis that the GI capacity of voles increases as poorly digestible components of the diet (fibre) increase. The fibre content of the diet, the dry mass content of the GI tract and the GI capacity of meadow voles were all greater in tallgrass prairie habitats than in bluegrass oldfields (Fig. 20.1, Table 20.1). The lack of significant change in total length or tissue mass per unit length of the whole GI tract suggests that increased volume was primarily achieved by radial distention of the GI rather than by the addition of new tissue. If distention accounts for a major part of acclimation to a fibrous diet, such acclimation could occur very rapidly.

Sperber *et al.* (1983) reported that the colonic spiral found in arvicoline rodents functions to separate fluids and fine particles, including bacteria, from food residues and to return them to the caecum (also see Ch. 17). Therefore, we might expect a lower concentration of fibre in the caecal contents. However, because the more digestible fractions of the diet will have been absorbed before the food reaches the ileo-caecal junction, material entering the caecum from the small intestine must be more fibrous than that found in the stomach. Apparently, these two processes (absorption in the small intestine and return of small particles from the colon) balanced one another so that stomach and caecal contents of meadow voles from the field had the same concentrations of fibre (Fig. 20.1).

Feeding trials in the laboratory also supported the IPR hypothesis. By measuring digestibilities on both high- and low-fibre diets after acclimation for 3 weeks to one diet type, we found that acclimation to a diet enhanced voles' ability to extract digestible energy by 10–13% for alfalfa and 20–27% for grass (Fig. 20.2, Table 20.2). Though some previous work indicates that caecal fermentation and reingestion of faeces provides nutritional benefits to voles (Cranford and Johnson, 1989), so far as we are aware, our results provide the first measurement of the nutritional benefit of the IPR. The substantial effect of acclimation on low quality diets was expected, but the acclimation to higher quality diets, even though smaller, was somewhat surprising. Taken as a whole, our results clearly suggest that the IPR is indeed an adaptive response to changing food quality in natural settings.

Our results also confirm previous results indicating that voles can extract the same amount of digestible material from natural diets even though they differ in digestibility (Batzli, 1985; Hammond and Wunder, 1991). This is achieved by compensatory changes of intake in relation to digestibility of the diets, though voles acclimated to grass showed greater differences for intake of the two diets than did voles acclimated to alfalfa (Fig. 20.2). Our digestibility trials were only for 3 days, and it may take longer for voles acclimated to a low-fibre diet to adjust their intake of a high-fibre diet than it does for voles acclimated to a high-fibre diet to adjust their intake of a low-fibre diet.

Our laboratory results for GI capacity were not entirely as expected. Most measures did not differ between voles acclimated to grass and voles acclimated to alfalfa (Tables 20.3 and 20.4). Only the lengths of the GI tracts of meadow voles fed grass were significantly greater. The dry mass of GI contents tended to be greater in meadow voles acclimated to alfalfa, and all measurements of GI capacity for prairie voles tended to be greater when acclimated to alfalfa. These results contrast strongly with those of Gross *et al.* (1985) and Hammond and Wunder (1991) for prairie voles. They found increased length, contents and tissue mass for several compartments and for the total GI tract of voles fed higher-fibre diets. Though the difference in our results could reflect differences in the diets used (natural versus prepared), recall that the voles we acclimated to alfalfa had been fed grass for 3 days (digestibility trials) just before they were sacrificed. As a result their GI contents consisted of grass. For similar reasons, the GI contents of voles acclimated to grass contained alfalfa when they were sacrificed. It seems likely that the trends that we expected from acclimation had already begun to be reversed by 3 days on the opposite diet. Changes in amount of intake, as reflected in mass of GI contents and in GI capacity owing to stretching, seem particularly likely to respond rapidly to dietary changes. Additional experiments that include variable periods of time on acclimation diets and reversals of acclimation diets will be required to determine how quickly changes in intake and GI capacity do occur.

Acknowledgements

We are grateful to Leah Freeman, Lynn Culhane and Christopher Stanczyk for their help in conducting the laboratory experiment and to Drs Steven Cork and Ian Hume for reviewing the paper. Financial support for this work was provided by a grant from the University of Illinois Research Board and the National Science Foundation (Grant No. BSR 90–06825) to the senior author.

References

Batzli, G. O. (1985). Nutrition. In *Biology of New World Microtus*, ed. R. Tamarin, pp. 779–811. American Society of Mammalogists Special Publication No. 8.

Batzli, G. O. (1986). Nutritional ecology of the California vole: effects of food quality on reproduction. *Ecology*, **67**, 406–412.

Batzli, G. O. & Cole, F. R. (1979). Nutritional ecology of microtine rodents: digestibility of forage. *Journal of Mammalogy*, **60**, 740–750.

Batzli, G. O., White, R. G., MacLean, S. F., Jr, Pitelka, F. A. & Collier, B. (1980). The herbivore-based trophic system. In *An Arctic Ecosystem: the Coastal Plain of Northern Alaska*, ed. J. Brown *et al.*, pp. 335–410. Stroudsburg, PA: Dowden, Hutchinson and Ross.

Brugger, K. E. (1991). Anatomical adaptation of the gut to diet in red-winged blackbirds (*Agelaius phoeniceus*). *Auk*, **108**, 562–567.

Cole, F. R. & Batzli, G. O. (1979). Nutrition and population dynamics of the prairie vole, *Microtus ochrogaster*, in Central Illinois. *Journal of Animal Ecology*, **48**, 455–470.

Cranford, J. A. & Johnson, E. O. (1989). Effects of coprophagy and diet quality on two microtine rodents (*Microtus pennsylvanicus* and *Microtus pinetorum*). *Journal of Mammalogy*, **70**, 494–502.

Demment, M. W. & van Soest, P. J. (1985). A nutritional explanation for body-size patterns of ruminant and nonruminant herbivores. *American Naturalist*, **125**, 641–672.

Goering, H. K. & van Soest, P. J. (1970). *Forage fiber analysis*. USDA Agricultural Handbook No. 379.
Gross, J. E., Wang, Z. & Wunder, B. A. (1985). Effects of food quality and energy needs: changes in gut morphology and capacity of *Microtus ochrogaster*. *Journal of Mammalogy*, **66**, 661–667.
Hammond, K. A. & Wunder, B. A. (1991). The role of diet quality and energy need in the nutritional ecology of a small herbivore, *Microtus ochrogaster*. *Physiological Zoology*, **64**, 541–567.
Karasov, W. H. & Diamond, J. M. (1988). Interplay between physiology and ecology in digestion: intestinal transporters vary within and between species. *Bioscience*, **38**, 602–611.
Lindroth, R. L. & Batzli, G. O. (1984). Food habits of the meadow vole (*Microtus pennsylvanicus*) in bluegrass and prairie habitats. *Journal of Mammalogy*, **65**, 600–606.
Robbins, C. T. (1983). *Wildlife Feeding and Nutrition*. New York: Academic Press.
Sperber, I., Björnhag, G. & Ridderstråle, Y. (1983). Function of the proximal colon in lemming and rat. *Swedish Journal of Agricultural Research*, **13**, 243–256.
Sibley, R. M. (1981). Strategies of digestion and defecation. In *Physiological Ecology: An Evolutionary Approach to Resource Use*, ed. C. R. Townsend & P. Calow, pp. 109–139. Oxford: Blackwell Scientific.
Sibley, R. M., Monk, K. A., Johnson, I. K. & Trout, R. C. (1990). Seasonal variation in gut morphology in wild rabbits (*Orycotolagus cuniculus*). *Journal of Zoology* (Lond.), **221**, 605–619.
van Soest, P. J. (1982). *Nutritional ecology of the ruminant*. Corvallis, OR: O&B Books.

21

Digestive constraints on dietary scope in small and moderately-small mammals: how much do we really understand?

STEVEN J. CORK

With decreasing body mass among mammals, the proportion of species eating fibrous plant tissues declines and the proportion selecting low-fibre plant and animal tissues increases (Bell, 1969; Jarman, 1974; Clutton-Brock and Harvey, 1977; Hume, 1984; Hofmann, 1989). The trend is clear among all groups of mammals (see discussion below), but the most comprehensive evidence for it comes from studies of wild ruminants (Bell, 1969; Jarman, 1974; Hofmann, 1989). This is because ruminants are among the most visible and abundant of mammalian herbivores and have been intensively studied due to their potential for domestication. Explanations for the decline in fibre-feeding with decreasing body mass among ruminants have been postulated in recent years, based on empirical and mechanistic models of energy yield from fermentative digestion of fibre and the unequal allometries of energy requirements and fermentation capacity (Parra, 1978; Demment and van Soest, 1985). These models are based on data from medium and large species and predict strong constraints on fermentative use of fibrous diets at body masses less than about 10–20 kg. However, this concentration on ruminants, the smallest of which is about 3 kg body mass, limits extrapolation of the models to predict or explain feeding constraints in other mammals, especially small species and those with digestive strategies not based on foregut fermentation. I show in this chapter that most current ruminant-based models make few or no useful predictions about constraints on fibre-utilisation in small non-ruminant herbivores.

Data on feeding styles among non-ruminant herbivores, including foregut fermenters and 'hindgut' fermenters, indicate few species smaller than about 15 kg body mass that primarily eat high-fibre plant parts such as mature foliage of grasses, shrubs or trees (Landry, 1970; Baker, 1971; Clutton-Brock and Harvey, 1977; Chivers and Hladik, 1980, 1984; Eisenberg, 1981; Hume, 1982, 1984; Cork and Foley, 1991; see also p. 339). In this chapter, I suggest that 15 kg is a convenient, if arbitrary, point at which to distinguish between 'fibre-tolerant' and 'fibre-intolerant' mammals generally. Above 15–20 kg, use of fibrous diets is seen among a wide range of mammals and the evolution of feeding styles seems to have been determined more by ecological pressures than size-dependent constraints on digestive function (Janis, 1976; Hofmann, 1989). Recent research has identified several extreme metabolic and digestive adaptations in mammals ranging from a few hundred grams to about 13 kg body mass that are not found in larger species, that are found only in the relatively few species that use fibrous diets to a substantial degree and that appear to partially offset the expected physiological disadvantages for small mammals of eating high-fibre food (Björnhag, 1972, 1987; Cork and Foley, 1991; Foley and Cork, 1992). These trends suggest that physiological constraints and limitations on the use of high-fibre diets become critical for not just ruminants but all mammals at small body masses and that investigation of extreme digestive adaptations of species less than about 15 kg will give insights into what those constraints are and how they might be modelled.

In this chapter I discuss the interrelationships between body size, digestive strategy and dietary scope among mammals, focussing on digestive and metabolic characteristics of mammals smaller than 15 kg. I aim to assess what we really understand and what we still need to find out about constraints and limitations on evolutionary adaptation for utilising fibrous plant tissues among mammals. I discuss evidence for fibre-intolerance in species less than about 15 kg in body weight, examine current models relating body size to digestive strategy among mammals, show that these models fail to predict or explain many of the adaptations seen among mammals less than 15 kg, and I discuss in detail differences and similarities in digestive and metabolic function among members of two key groups of mammals in this mass-range that use extreme diets: the so-called 'arboreal folivores', spread across a wide range of mammalian taxa, and the terrestrially foraging herbivorous rodents, especially microtines.

Some notes about terminology

This meeting revealed disagreement about the use of terminology for describing the proximal colon and caecum in mammals. This region is called 'hindgut' by many physiologists but anatomists argue that, in terms of blood supply, innervation and embryology, it is 'midgut'. In this chapter, the term 'hindgut fermenters' refers to mammals that ferment plant material primarily in the caecum, the proximal colon, or both.

Conventionally, 'small' mammals have been defined as being less than about 5 kg body mass (Bourlière, 1975). However, as I will show here, mammals in the range 1–15 kg are physiologically small compared with medium-sized and large species of 50–100 kg and more. It is to emphasise this point that I use here the term 'small and moderately small' to describe mammals of 1–15 kg.

Evidence for fibre-intolerance below 15 kg body mass

Various authors have categorised mammalian herbivores into feeding types based on content of dietary fibre (i.e. plant constituents resistant to digestion by mammalian enzymes) in their natural diet (e.g. Vorontsov, 1962; Hofmann and Stewart, 1972; Jarman, 1974; Chivers and Hladik, 1980, 1984; this volume, Ch. 5). Dietary fibre is recognised as a major nutritional challenge for mammals because it is only partially digestible (by microbial symbionts in the gut), yields energy slowly compared with other dietary constituents and limits intake of other, more available, nutrients (van Soest, 1982).

Hofmann and Stewart (1972) classified ruminants into bulk-roughage feeders (which feed unselectively on high-fibre grasses), concentrate-selectors (which feed selectively on fruits, seeds, young leaves of shrubs and other moderate- and low-fibre plant parts) and intermediate feeders. Above about 20 kg body mass among ruminants, the full range of feeding styles are found, suggesting that ecological pressures rather than mass-dependent constraints on diet utilisation have guided the evolution of feeding styles in this mass range (Hofmann, 1989). However, most ruminants less than about 15 kg are concentrate-selectors and there are no small unselective bulk-roughage feeders in this mass range (Hofmann and Stewart, 1972; Demment and van Soest, 1985), suggesting that physical and physiological constraints on efficient digestive utilisation of roughage diets become critical at low body masses.

Terrestrially-foraging macropodoid marsupials (kangaroos, wallabies and rat-kangaroos) are foregut-fermenting herbivores that occupy similar feeding niches in Australia and Papua/New Guinea to those occupied by ruminants elsewhere. They also show relationships between dietary strategies and body mass similar to those seen among ruminants, except that there are macropodoids as small as a few hundred grams and there are no extant macropodoids to compare with cattle, buffalo and larger ruminants. The largest macropodoids (20–80 kg) are bulk-roughage feeders and intermediate feeders (Hume, 1984), but the vast majority of species less than 15 kg are concentrate-selectors eating a mixed diet of low-fibre shrub leaves (and/or tips of young grasses in a few species) and fruit or seeds (Jarman, 1984; Hume, 1982, 1984). Most macropodoids less than 10 kg eat only moderate amounts of high-fibre food and all species less than 1 kg are highly selective for low-fibre items such as flowers, twig tips, fruits, seed pods, some fungi and succulent roots (Hume, 1982, 1984).

Among arboreal primates and marsupials, the proportion of mature leaf in the diet falls dramatically between body masses of 15 kg and 1 kg (Clutton-Brock and Harvey, 1977; Hume, 1982; Cork and Foley, 1991). The vast majority of species in both groups in the range 1–15 kg eat fruit and/or seeds primarily, supplemented in many cases with animal matter. In both groups, there are a small number of species of mass 700–1000 g that include moderate amounts of foliage in their diet but only with the help of extreme digestive adaptations (Cork and Foley, 1991; see p. 351). However, leaf-eating is extremely rare below 700 g body mass: apart from two species of New Guinean ringtail possums of 150 and 450 g body mass that are suspected from their stomach contents to practise significant folivory (I. D. Hume, personal communication), all other species so far studied select their diet from invertebrates, fruits, seeds, flowers, nectar, fungi and plant exudates (Chivers and Hladik, 1980, 1984; Hume, 1982, 1984; Henry *et al.*, 1989).

Most rodents are plant-eaters, but the majority include animal matter in their diet on a regular basis (Landry, 1970). The majority of species less than 10 kg primarily eat seeds, fruits and/or insects (Landry, 1970; Baker, 1971; Eisenberg, 1981). Within the vast majority of genera, all species less than 1 kg eat little or no leaf or other high-fibre items. The exceptions are some species of microtines and cricetids that, like the exceptional primates and marsupials, have extreme digestive adaptations (see p. 354).

Therefore, in all groups of mammals for which data are available, there is a decrease in the proportion of species using fibrous plant parts with decreasing body mass and this decrease appears particularly pronounced below 15 kg. In the two following sections, I discuss the ability of current theoretical models to predict and explain the fact that this trend is so dramatic and widespread among mammals. In later sections, I use data from arboreal folivores and terrestrial rodents to suggest improvements to these models.

Current models relating body size to digestive strategy in mammals (allometry-based models)

There have been several very useful attempts since the early 1970s to explain differences in feeding choices and digestive strategies amongst herbivorous mammals in relation to underlying constraints associated with body size (e.g. Bell, 1969; Jarman, 1974; Janis, 1976; Clutton-Brock and Harvey, 1977;

Parra, 1978; Demment and van Soest, 1985; Baker and Hobbs, 1987; Hobbs, 1990; Illius and Gordon, 1992; Justice and Smith, 1992). I will refer to these as **allometry-based** models, to distinguish them from **reactor-based** models that are discussed in the following section.

Most of the allometry-based models identify constraints on foregut fermenters, especially medium-sized and large ruminants, with little or no discussion of constraints on alternative digestive strategies. A recurring theme in these models is that fermentative use of fibrous plant tissues becomes decreasingly feasible with decreasing body mass among mammals because, approximately, mass-specific energy requirements scale with body $mass^{0.75}$ whereas the maximum rate that energy can be obtained via fermentative digestion scales to body $mass^1$ (because gut capacity scales with body $mass^1$ and the fermentation rate for a given diet is not expected to vary significantly between species of mammal). Taking empirically-derived values for fermentation rates and basal metabolic requirements, it has been calculated that a (ruminant) digestive strategy which relies entirely on fermentation to release digestible energy is not able to supply the energy requirements for basal metabolism for an animal below a body mass of about 9 kg (Demment and van Soest, 1985). However, since Demment and van Soest's (1985) model was published, data on field (or average daily) metabolic rates (FMR) in a wide range of mammals have become available (Nagy, 1987). Applying these estimates of FMR to Demment and van Soest's model (Fig. 21.1) indicates that a strategy of primary reliance on fermentation of high-fibre diets becomes unviable at much higher body masses than 9 kg.

Models based on these principles give insights into the limits on use of fibrous diets in ruminants and other mammals whose digestive strategies maximise retention and fermentation of the fibre (Parra, 1978; Demment and van Soest, 1985) but make few predictions about limits on digestive strategies that do not necessarily maximise retention and fermentation. For example, the models acknowledge that the constraints associated with maximising fibre retention could be avoided in small foregut fermenters by digestive mechanisms that allow the rate of passage of fibre through the foregut to be determined by intake rate rather than rate of breakdown of particles (such mechanisms are found among small ruminants (Hofmann, 1989), tragulids (Langer, 1974) and macropodoid marsupials (Hume and Carlisle, 1985; Hume, 1982, 1984, 1989; Freudenberger *et al.*, 1989)) but no predictions are made about limits on such alternative digestive strategies. Similarly, large hindgut fermenters are able to utilise fibrous diets by maintaining high intakes and passage rates at the expense of lower fibre digestibilities than those found in

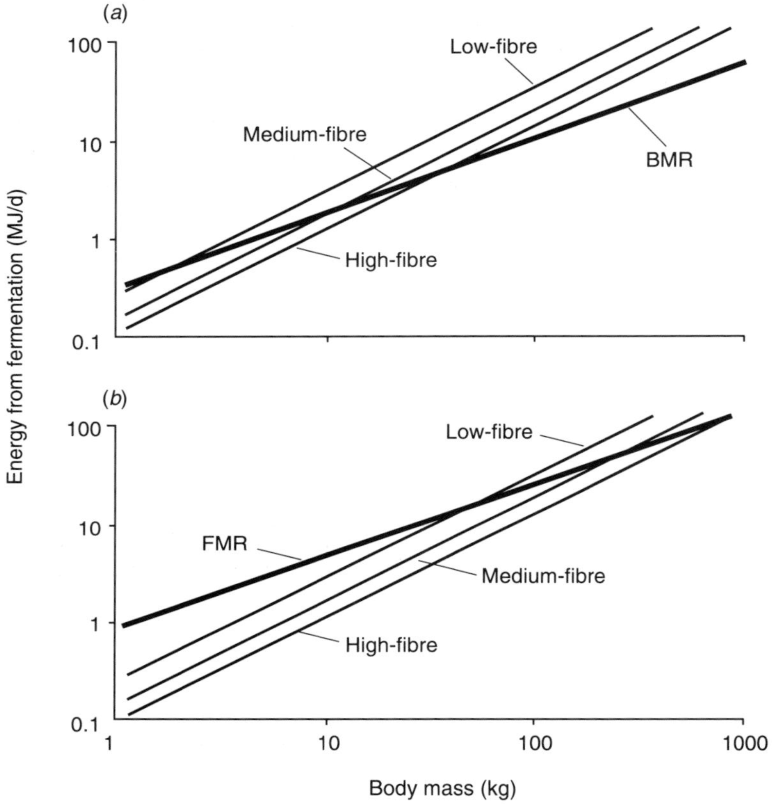

Fig. 21.1. Predicted relationship between body mass and energy obtained from microbial fermentation (MJ per day) of three different plant-based diets (using data and assumptions given by Demment and Van Soest, 1985). Predictions are based on measured fermentation rates in the rumen of antelopes of 3.6, 49 and 200 kg body mass assumed to be eating low-, medium-, and high-fibre diets respectively. Energy from fermentation is compared with: (*a*) basal metabolic rate (BMR) (after Kleiber, 1975) and (*b*) field (average daily) metabolic rate (FMR) (after Nagy, 1987).

ruminants (Janis, 1976; Duncan *et al.*, 1990) but, until recently, there has been no consideration of whether this strategy could be adopted by mammals of all sizes.

Two recent models (Illius and Gordon, 1992; Justice and Smith, 1992) have approached predicting and explaining dietary differences in relation to body size in a more mechanistic way and have greatly advanced understanding of digestive constraints and limitations in small herbivores. Both models use rates of food intake and passage, gut capacities and metabolic requirements (predicted mostly from data on medium to large mammals) to simulate energy gain from different diets in relation to body size among mammals. Neither model assumes primary reliance on fermentative digestion to meet energy requirements.

The model by Justice and Smith (1992) focusses specifically on small hindgut fermenters and illustrates well the problems facing small herbivores but does not ultimately predict limits to utilisation of fibrous diets. The model

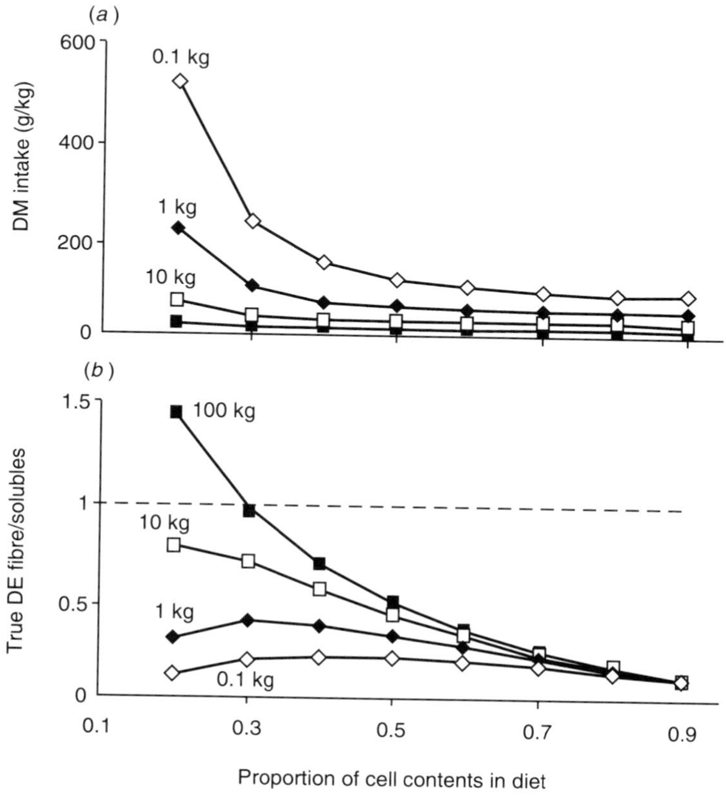

Fig. 21.2. Simulation (using the model of Justice and Smith, 1992) for hindgut-fermenting mammals of various body masses to show the relationship between the proportion of cell contents in the diet (1 − fibre proportion) and (*a*) the gross intake of dry matter (DM) and (*b*) the proportion of true digestible energy (DE) (i.e. energy that is released from the diet, digested and absorbed from the digestive tract) from fibre (cell walls) versus solubles (cell contents). Values used in the simulation were those typical of medium-quality alfalfa (Justice and Smith, 1992): *U* (the proportion of fibre not digestible) = 0.51; *k* (the fractional fermentation rate of fibre) = 0.114/h (= 2.736 per day). All other parameters as given by Justice and Smith (1992).

predicts that mass-specific food intakes for maintaining energy balance increase exponentially with decreasing body mass (Fig. 21.2*a*) and, hence, that the relative energy yield from fibre compared with cell contents is much less in small than in large species (Fig. 21.2*b*). Justice and Smith (1992) concluded that small mammals can derive a substantial proportion of their digestible energy (DE) from fibre, but this overestimates the net value of fibre to the animal, as illustrated in Table 21.1. Although energy released from fermentation of fibre accounts for 31% of apparent DE in a 0.15 kg species eating an 80% fibre diet and 46% in a 1 kg species (Table 21.1), this is far exceeded by the loss of endogenous energy in faeces which, although included as cell contents in the determination of apparent DE, is a consequence of both the high intake of dry matter required when high-fibre diets are eaten (Justice and Smith, 1992) and the microbial biomass produced by fermentation of the fibre itself (Mason and Palmer, 1973). Hence, although not recognised by the authors, the model by Justice and Smith (1992) indicates that achieving energy equilibrium for small mammals eating fibrous diets depends mainly on maintaining a sufficiently high overall intake of food for energy yield from digestion of cell contents to balance endogenous losses due to fibre intake.

The model does not, however, predict where limits to this strategy occur.

Table 21.1. *Prediction of energy intake and excretion at energy equilibrium for 0.15 kg and 1.0 kg hindgut-fermenting mammals eating an 80% fibre diet*

Model variable	Body mass (kg)	
	0.15	1.0
Dry matter intake (g per day)	68.4	221.8
Gross energy intake (kJ per day)	1259.22	4083.25
True digestibility of fibre (%)	4.39	8.31
True digestibility of cell contents (%)	102	102
True digestible energy from fibre (kJ per day)	44.22	271.37
True digestible energy from cell contents (kJ per day)	256.86	832.95
Endogenous energy in faeces (kJ per day)	159.91	518.56
Apparent digestible energy (kJ per day)	141.17	585.76

From Justice and Smith, 1992.
It is assumed that the energy density of the diet was uniformly 18.4 kJ/g, the fractional rate of fibre digestion was 2.736 per day (0.114 h), and 0.51 of the fibre was indigestible.

It allows food intake and digesta passage to increase unconstrained until predicted requirements for digestible energy are met. Hence, it predicts that mammals of 0.1, 1 and 10 kg can achieve energy equilibrium on an 80% fibre diet if dry matter intakes of 52, 22 and 8% of body mass respectively are maintained (Fig. 21.2a), but it cannot predict whether or not these intakes are attainable. (The authors acknowledge that even the most highly specialised herbivores among the small rodents are apparently unable to survive on a diet of 60% fibre, but the model does not explain why.) Nevertheless, the model raises challenging questions by showing that, within a realistic range of diets and intakes, several small species ferment dietary fibre markedly more completely than predicted, suggesting compensatory mechanisms of the sort discussed later in this chapter.

The model by Illius and Gordon (1992) specifically compares the ruminant digestive strategy with hindgut fermentation and it is the only model that makes predictions of limits to food intake in small herbivores. It predicts that, for both foregut and hindgut fermentation strategies, small species require higher-quality (lower-fibre) diets than large species and that, especially in small species, hindgut fermentation allows higher intake of digestible energy than ruminant-style foregut fermentation. It differs from previous models (e.g. Janis, 1976; Demment and van Soest, 1985) in that it does not predict medium-sized ruminants to be necessarily superior to nonruminants at utilising moderately fibrous diets. The model predicts the upper limits on digesta load and passage rate empirically from data on mass of gut contents in a range of species and it allows food intake to be determined by the ability of digestion and passage to keep the digesta load below the upper limits. This model makes realistic predictions of food intake and digestibility for a wide range of herbivores but, currently, it does not make realistic predictions of limits to diet choice among small species (see p. 357).

Models relating gut function and digestive strategy to chemical reactor theory (reactor-based models)

Recently, mechanistic models based on chemical reactor theory have been applied to digestive systems (Penry and Jumas, 1986, 1987; Alexander, 1991; Martínez del Rio, Cork and Karasov, Ch. 3). These reactor-based models provide a useful framework for considering the optimal balance between enzymic (catalytic) and fermentative (autocatalytic) digestion in mammals. Most importantly, they have provided an explanation for why catalytic reactions are more efficiently carried out in unmixed, tubular portions of the digestive tract (e.g. small intestine) whereas autocatalytic reactions are best performed in uniformly mixed compartments (e.g. rumen, caecum).

Reactor-based models make realistic predictions about the order and number of compartments and linking tubes in mammalian herbivores that either maximise retention and digestion of fibre or that select low-fibre diets (Alexander, 1991), but they neither predict nor explain rapid-throughput strategies for using fibrous diets in mammals of any size. Because reactor-based models seek to define optimal gut configurations for maximising digestion of different diets, they predict that diets high in fibre are best used by slow passage of food through guts containing two mixing/retention chambers (i.e. both forestomach and hindgut fermentation) separated by a tubular plug-flow reactor (the small intestine) (Penry and Jumas, 1986, 1987; Alexander, 1991). Only for diets moderately high in cell contents is a single fermentation chamber (expanded hindgut) preceded by a plug-flow reactor predicted (Alexander, 1991). Hence, the use of high-fibre pastures by large equids that optimise intake and throughput rather than the extent of digestion (Janis, 1976; Duncan *et al.*, 1990) is neither anticipated nor explained (Alexander, 1991). Furthermore, if the models are constrained to high intakes of high-fibre food, which is the situation facing small mammals, physically impossible gut configurations (two fermentation chambers with no intervening plug-flow reactor) are predicted (Alexander, 1991).

Digestive and metabolic strategies

In the preceding two sections, I have drawn attention to the deficiencies of current models, not to denigrate the use of these techniques, but to illustrate that too few constraints have been taken into account as yet to make the models useful for understanding digestive characteristics and strategies among small herbivores. Below, these problems are brought into focus by consideration of digestive and metabolic strategies in two groups of mammals: the so-called 'mammalian arboreal folivores' (among primates, edentates, marsupials and rodents) and the terrestrial rodents that use a wide range of plant tissues as food.

Arboreal mammalian folivores

Feeding styles

The upper body-mass limit to arboreality in forests seems to be imposed by what tree branches can support and is around 15 kg (Eisenberg, 1978). Consistent with this restriction on body mass, the vast majority of arboreal mammals select low-fibre, high-energy foods (e.g. fruits, flowers and invertebrates). In the whole of the Mammalia, only about ten species – principally among a few genera of Australian marsupials, New and Old World primates, edentates and rodents – are both highly arboreal and principally folivorous (Eisenberg, 1978). All of these species are larger than 700 g and most are in the body-mass range 5–15 kg (Clutton-Brock and Harvey, 1977; Eisenberg, 1978; Cork and Foley, 1991). Below, I summarise feeding styles among arboreal folivores (from a review by Cork and Foley, 1991) to assist in the interpretation of their digestive and metabolic adaptations that are discussed in the next section.

The most folivorous primates are in the African/Asian sub-family Colobinae (*Colobus, Procolobus, Presbytis, Nasalis*) and range from about 9 to 21 kg in body mass. However, even the most folivorous of these species spend less than 40% of their feeding time on mature foliage and, often, over 50% on seeds, fruits and flowers (see also DaSilva, 1992). Most members of the Madagascan families Lemuridae (*Hapalemur, Lepilemur, Lemur*) and Indriidae (*Avahi, Indri, Propithecus*) are less folivorous than colobids but the sportive lemurs, *Lepilemur* spp., are notable exceptions being not only highly folivorous but also having one of the lowest body masses (approximately 700 g) of all arboreal folivores. The howler monkeys, *Alouatta* spp., are the most significant leaf-eaters among the primates of Central and South America, but are much less folivorous than the folivorous colobids.

Among the edentates, *Bradypus tridactylus*, the three-toed sloth (2–5 kg body mass), is highly folivorous, although few quantitative estimates of diet composition are available (see review by Cork and Foley, 1991).

The most folivorous arboreal mammals are marsupials. The most comprehensive data on feeding habits are on the four species inhabiting *Eucalyptus* forests of south-eastern Australia. The koala *Phascolarctos cinereus* (5–13 kg) probably is the most folivorous of all arboreal mammals, eating *Eucalyptus* spp. foliage almost exclusively and including a large proportion of mature foliage when young foliage is scarce. The greater glider, *Petauroides volans* (0.8–1.7 kg), also is highly folivorous but strongly selects young foliage. The common ringtail possum, *Pseudocheirus peregrinus* (0.7–1 kg), like *Lepilemur* spp. among the primates, is much more folivorous than predicted from its small mass, eating mature and young eucalypt foliage and buds, fruits, and flowers of understorey species. The common brushtail possum, *Trichosurus vulpecula* (2–3 kg), is much less folivorous than the other three, its diet including ground-storey plants, fruits, occasional insects and a variable amount of foliage. Smaller marsupials inhabiting Australian eucalypt forests, like small primates, avoid foliage completely and feed largely on sap, nectar and/or insects and invertebrates (see review by Cork and Foley, 1991).

In wet tropical rain-forests, arboreal folivory is practised by other related species of possums (in the genera *Trichosurus, Pseudocheirus, Pseudochirops* and *Hemibelideus*) and tree kangaroos (*Dendrolagus* spp.). The few published data on tree kangaroos suggest only moderate folivory and their dentition also suggests utilisation of foliage with low lignin concentration and/or fruits (see review by Cork and Foley, 1991).

Only three rodent species (two flying squirrels, *Petaurista* spp. and one *Anomalurus* sp.) of the many in forests of south-east Asia are considered to be at least moderately folivorous. These are among the largest of the arboreal rodents (800–1400 g) (Eisenberg, 1978; Muul and Liat, 1978).

Members of two other genera of small mammals (the Dermopteran genus *Cynocephalus* and the Hyracoidean genus *Dendrohyrax*) are credited with moderate folivory but there are too few data to draw firm conclusions (Eisenberg, 1978).

Digestive adaptations

Among all major groups of arboreal mammals, length, capacity and surface area of the digestive tract increase with degree of folivory (Muul and Liat, 1978; Hume, 1982; Chivers and Hladik, 1980, 1984; Martin *et al.*, 1985). Three aspects of gut morphology and function are particularly relevant to differences in dietary scope among arboreal mammals: primary site of fermentation (foregut versus hindgut); presence/absence of digesta separation in the hindgut; and the practice of caecotrophy by some species (Table 21.2).

Hindgut fermenters are distributed across the full body-size range of arboreal mammals, including the most highly folivorous species (koalas, greater gliders, *Lepilemur*) and the least folivorous species (all insectivores and insectivore/frugivores). In contrast, foregut fermentation is found only among the largest arboreal mammals (colobids, sloths, tree kangaroos) and all but the three-toed sloth (an oddity for reasons discussed below) are at least partially frugivorous or granivorous. Not surprisingly, foregut-fermenting arboreal mammals are found only in wet tropical forests (where fruits are usually abundant) whereas hindgut-fermenting species are found in temperate and both wet and dry tropical forests (Cork and Foley, 1991).

Colobid primates, three-toed sloths and tree kangaroos differ from ruminants in at least one important respect: in none is there an equivalent of the reticulo-omasal orifice which restricts passage of large particles of fibre from the rumen (Dellow, 1982; Hume, 1982; Langer, 1988). There appears to be

Table 21.2. *Digestive and metabolic adaptations of small mammalian herbivores*

Mammal	Body mass (kg)	Fibre in diet	Main fermentation site	Caeco/colonic digesta separation/ caecotrophy	Bypass of forestomach	Field metabolic rate[a]
Koala (marsupial)	5–13	High	Caecum + proximal colon	Yes/no		Very low
Greater glider (marsupial)	0.9–1.2	Moderate	Caecum	Yes/no		Low
Ringtail possum (marsupial)	0.7–0.9	Moderate	Caecum	Yes/yes		Low
Sportive lemur (primate)	0.7–0.8	Moderate	Caecum (+ proximal colon?)	Yes/yes		?
Fibre-feeding rodents	< 0.1	Moderate	Caecum	Yes/some		Often high
Brushtail possum (marsupial)	2–3	Low–moderate	Caecum + proximal colon	No/no		Low
Most non-colobid arboreal primates	0.05–15	Low	Caecum (often small)	No/no[b]		Varies
Howler monkey	7	Low–moderate	Caecum	?/?		Low
Granivorous, frugivorous, faunivorous rodents	< 0.5	Low	Caecum (often small)	No/no		Often high
Tree kangaroos (marsupials)	4–13	Low–moderate	Forestomach	No/no[b]	Possibly	Low
Rat kangaroo	0.3–3.5	Low	Forestomach	No/no	Yes	Low
Colobid primates	4–21	Low–moderate	Forestomach	No/no[b]	Possibly	?
Three-toed sloth (edentate)	2–3	Moderate–high	Forestomach	No/no[b]	Apparently not	Very low
Small antelope and tragulids (ruminants)	3–40	Low–moderate	Forestomach	No/no	Yes	Average

[a] Relative to predicted FMR for comparably sized eutherians (Nagy, 1987).
[b] Conclusions not confirmed by experimentation but highly likely based on similar species.
Further details in the text

strong convergence of forestomach anatomy between arboreal colobids and kangaroos (Fig. 21.3). Unlike the sacciform forestomach of ruminants, the forestomach of macropodoid marsupials (kangaroos and wallabies) and colobids consists of a sacciform region followed by a tubiform region into which the oesophagus opens (Langer *et al.*, 1980; Langer, 1988). The relative proportions of sacciform and tubiform forestomachs in colobids, the entry of the oesophagus into the latter and the offset of the former from the main flow of digesta are similar to those in the smaller macropodoid marsupials (wallabies and rat kangaroos) which, like colobids, eat mixed diets of both high- and low-fibre items (Langer, 1988; Hume, 1989). In the small macropodoids, significant amounts of ingesta can bypass the main fermentative (i.e. sacciform) region of the forestomach (Hume *et al.*, 1988). This allows at least some cell contents to be digested in the acid-enzymic hindstomach, although there is as yet no convincing evidence of selective direction of cell walls and cell contents into different regions of the stomach as has been postulated (Kinnear *et al.*, 1979; Hume and Carlisle, 1985). Adaptations of this type should permit tree kangaroos and colobid primates to maintain high food intakes when eating foliage and might allow them to minimise fermentation of fruits and seeds (which is less efficient than enzymic digestion of the sugars and starches of these diets; van Soest, 1982), although the latter remains to be demonstrated in either tree kangaroos or colobids.

The forestomach in three-toed sloths resembles that of macropodoids and colobids in some, but not all, respects. The central portion of the stomach is not tubiform but consists of two pouches, the central pouch and the connecting pouch, which account for about 60% of the stomach's volume (Langer, 1988; W. J. Foley *et al.*, unpublished data). There is also an anterior fundus (Figure 21.3*b*). The arrangement of pouches suggests that solute digesta could be routed past the fundus, but the limited data available indicate that digesta generally pass very slowly through the stomach of sloths (Montgomery and Sunquist, 1978; W. J. Foley, unpublished data) and that there is little or no difference in passage rate between digesta components (W. J. Foley, unpublished data). In contrast to other arboreal mammals, sloths may be able to exploit a strategy of long retention and slow fermentation of fibre because evolution of a very low metabolic rate (20% of that predicted from other mammals; Nagy and Montgomery, 1980) has freed them from the high rates of energy demand from digestion faced by other similarly-sized mammals.

Among hindgut-fermenting primates, marsupials, rodents, dermopterans and hyraxes, the caecum and/or proximal colon are expanded (Chivers and Hladik, 1980, 1984; Hume, 1982; Cork and Foley, 1991)(Fig. 21.4) and their

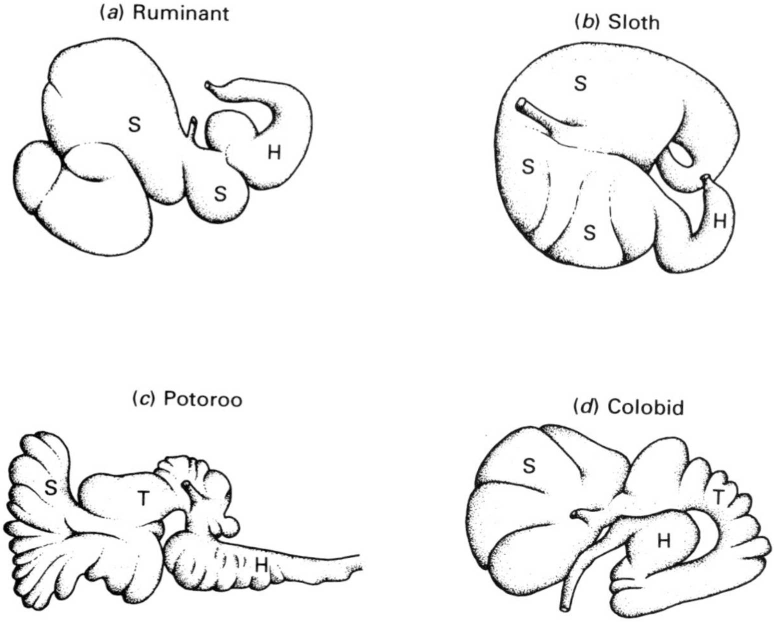

Fig. 21.3. External morphology of the stomach in (*a*) a medium-sized ruminant and (*b–d*) three small foregut-fermenting non-ruminant mammals. Species are: (*a*) Roe deer *Capreolus capreolus* (Langer, 1988); (*b*) sloth *Bradypus tridactylus* (Langer, 1988); (*c*) potoroo *Potorous tridactylus*, a small wallaby (Hume, 1982); (*d*) Langur monkey *Presbytis cristatus*, a colobid primate (Bauchop, 1978). S, sacciform forestomach; T, tubiform forestomach; H, hindstomach.

length and surface area are proportional to the degree of folivory practised (Bauchop, 1978; Eisenberg, 1978; Muul and Liat, 1978; Chivers and Hladik, 1980, 1984; Hume, 1982). Among these hindgut fermenters, a distinction can be made between **separators** and **non-separators**. In the hindgut of separators, particulate digesta are separated by a retrograde movement of solutes, so that large particles are excreted relatively rapidly while small particles are selectively retained (Björnhag, 1972, 1987; Cork and Warner, 1983; Chilcott and Hume, 1985; Foley and Hume, 1987a; Cork and Foley, 1991). The most highly folivorous hindgut-fermenting arboreal mammals (koalas, greater gliders, ringtail possums, and *Lepilemur* spp.) are all separators (Cork and Warner, 1983; Chilcott and Hume, 1985; Foley and Hume, 1987a) and all non-separators are either non-folivorous or are much less folivorous than the separators. Among the separators, two genera, *Lepilemur* among the primates and *Pseudocheirus* among the marsupials, are known or thought to be caecotrophic (i.e. they ingest, periodically, specially produced faeces that are made directly from caecal contents)(Hladik *et al.*, 1971; Chilcott and Hume, 1985).

Very long retention times of digesta markers have been reported in koalas and other folivorous hindgut-fermenting marsupials (Cork and Warner, 1983; Chilcott and Hume, 1985; Foley and Hume, 1987a; Cork and Foley, 1991), but these seem to reflect retention of solutes and fine particles that move with the solutes rather than retention of large particles of dietary fibre. In all of the folivorous hindgut-fermenting marsupials so far examined in detail (koalas, greater gliders and ringtail possums), fermentation provides a minor

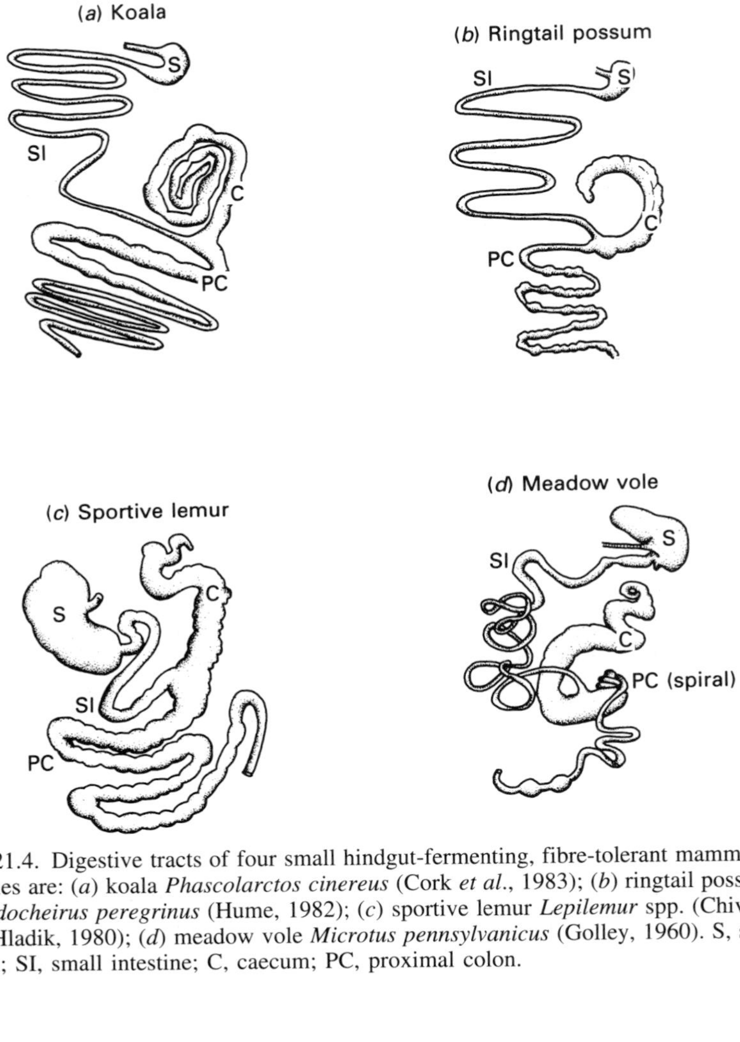

Fig. 21.4. Digestive tracts of four small hindgut-fermenting, fibre-tolerant mammals. Species are: (*a*) koala *Phascolarctos cinereus* (Cork *et al.*, 1983); (*b*) ringtail possum *Pseudocheirus peregrinus* (Hume, 1982); (*c*) sportive lemur *Lepilemur* spp. (Chivers and Hladik, 1980); (*d*) meadow vole *Microtus pennsylvanicus* (Golley, 1960). S, stomach; SI, small intestine; C, caecum; PC, proximal colon.

proportion of digestible energy, the main source being cell solubles (Cork *et al.*, 1983; Cork and Hume, 1983; Chilcott and Hume, 1984a; Foley, 1987; Foley *et al.*, 1989).

Metabolic rate

Many arboreal mammals have rates of basal metabolism less than the average for other mammals of similar body mass. These low rates of metabolism are found in folivorous, frugivorous and insectivorous species among the arboreal primates, edentates and marsupials (see review by Cork and Foley, 1991). Although there is some dispute about whether low metabolic rates are adaptive or pre-adaptive in relation to food types among mammals (Elgar and Harvey, 1987; McNab, 1987), it is clear that low energy requirements can benefit frugivores and folivores because of the potential rarity of fruits and the low availability of nutrients and high concentrations of defensive compounds in foliage (McNab, 1978; Nagy and Montgomery, 1980; Nagy and Martin, 1985; Cork *et al.*, 1983). On the other hand, small species appear to require higher than average metabolic rates if they are to remain continuously endothermic rather than become periodically torpid (McNab, 1986) and this might help to explain the absence of folivory among species lighter than about 700 g.

The only two mammals that maintain almost exclusive use of mature tree foliage, three-toed sloths and koalas, have very low rates of basal metabolism and low total energy expenditure in the field (21% and 36%, respectively, of those predicted for placental mammals of the same body mass)(Nagy and Montgomery, 1980; Nagy and Martin, 1985; Nagy, 1987). Many marsupials have metabolic rates lower than comparably-sized placentals, but the field metabolic rate in koalas is 89% of that predicted for a marsupial (Nagy, 1987). The two other highly folivorous marsupials, the ringtail possum and greater glider, have basal metabolic rates about 70% of the mean for placental mammals (both are close to the mean for marsupials) and field metabolic rates 79% and 69%, respectively, of those predicted for eutherians (107% and 101% of predicted rates for marsupials) (Nagy, 1987; Cork and Foley, 1991). Tree kangaroos and colobid monkeys, both folivore/frugivores, also are reported to have low rates of basal metabolism but further studies are required because the measurements on *Colobus* (Müller *et al.*, 1983) were not made under strictly basal conditions. No data on field metabolic rate are available for these species. Another folivore/frugivore, the howler monkey, has a field metabolic rate that is 56% of the predicted rate (Nagy and Milton,

1979; Nagy, 1987) despite having a reported basal metabolic rate close to the predicted rate.

Terrestrially foraging rodents

Little is known in detail about digestive function or dietary habits in arboreal rodents but some parallels with arboreal folivores can be drawn for terrestrially foraging herbivorous rodents.

Most small rodents are seed-eaters, fruit-eaters or insect-eaters (Landry, 1970; Baker, 1971; Eisenberg, 1981). However, some species, notably microtines including voles and lemmings, eat herbage primarily and often include fibrous tissues as major dietary items (Baker, 1971; Batzli and Cole, 1979; Batzli, 1985; Hansson, 1985). Several rodents are coprophagous (Kenagy and Hoyt, 1980; Cranford and Johnson, 1989) but caecotrophy, the periodic ingestion of specialised faeces formed as the result of the concentration of solutes and bacteria in the caecum, occurs in relatively few species (Hörnicke and Björnhag, 1980; Björnhag, 1987). There is considerable evidence that caecotrophy in any mammal requires digesta separation similar to that described in the highly folivorous arboreal marsupials but digesta separation is found without caecotrophy in several species (Björnhag, 1972, 1987; Cork and Warner, 1983; Foley and Hume, 1987a). Among rodents, increasing use of fibrous diets is associated with increasing relative size and complexity of the caecum and proximal colon (Vorontsov, 1962) and all species that use fibrous diets to a substantial degree (e.g. voles, lemmings, cotton rats, wood rats) are caecotrophic or possess digesta separation mechanisms in the hindgut or have complex spiral structures in the colon that suggest a separation mechanism (Vorontsov, 1962; Snipes, 1979; Hörnicke and Björnhag, 1980; Kenagy and Hoyt, 1980; Ouellette and Heisinger, 1980; Björnhag, 1987).

Small rodents that separate digesta in the hindgut are able to use food of much lower digestibility than can species without separation mechanisms (Karasov, 1982; Batzli and Cole, 1979; Batzli, 1985; Hume *et al.*, 1993; Hume, Ch. 19). In Fig. 21.5*a*, I have compiled data on intake of total dry matter in relation to the digestibility of the diet for several microtine rodents that have complex hindguts and are known or expected to separate digesta (Hörnicke and Björnhag, 1980; Björnhag, 1987) and several species of small sciurids and murids that have relatively simple digestive tracts and apparently do not separate digesta (Cork and Kenagy, 1989b; Hume *et al.*, 1993). Apart from two exceptionally low values for microtines (due to extremely unpalatable diets; Batzli and Cole, 1979), intakes of dry matter by the sciurids and

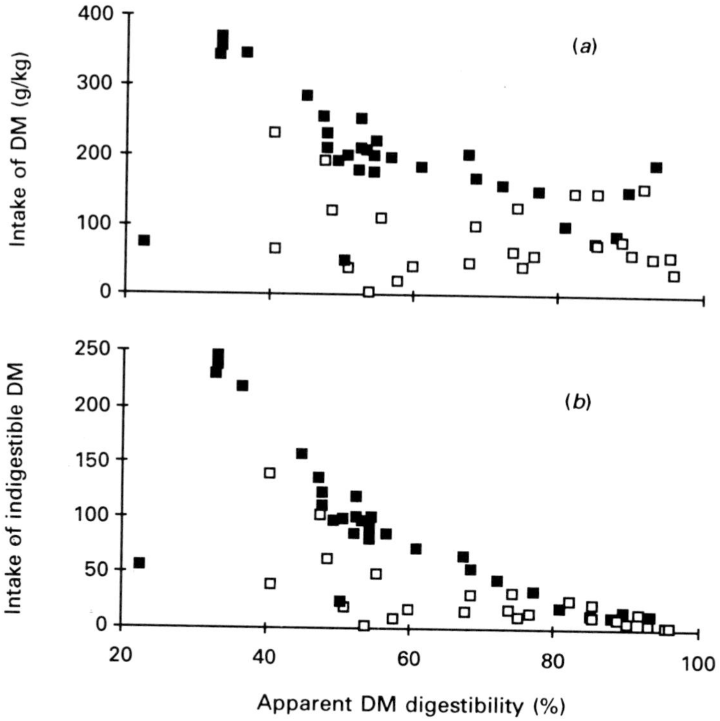

Fig. 21.5. Intake of (*a*) dry matter (DM) and (*b*) indigestible dry matter (g/kg body mass) by small rodents on various diets differing in digestibility. Closed squares are microtines known or expected to have digesta separation mechanisms, *Clethrionomys glareolus, Microtus arvalis* (Drozdz, 1968), *M. agrestis* (Hansson, 1971), *M. townsendii* (Hume *et al.*, 1993), *M. ochrogaster, M. californicus, Lemmus sibericus* (Batzli and Cole, 1979). Open squares are sciurids and murids known or expected to have no such mechanisms, *Ammospermophilus leucurus* (Karasov, 1982), *Spermophilus saturatus* (Cork and Kenagy, 1989a), *S. columbianus, Eutamias ameonus* (Hume *et al.*, 1993), *Apodemus agrarius, A. flavicollis* (Drozdz, 1968).

murids consistently are lower than those of microtines at digestibilities below about 70% (Fig. 21.5*a*). The two highest data points for sciurids/murids in Fig. 21.5*a* are for chipmunks from an experiment in which they failed to maintain body mass while voles on the same diet did (Hume *et al.*, 1993). The expected response of herbivorous mammals to declining digestibility (increasing dietary fibre concentration) is an increase in food intake (to maintain intake of digestible energy) until a 'critical digestibility' is reached below which food intake is limited by physical constraints on the capacity of the digestive tract for processing and eliminating dietary fibre (Baumgardt, 1970; van Soest, 1982). Figure 21.5*b* supports the notion of a physical limit to intake of fibre in the sciurids and murids because the amount of indigestible dry matter ingested appears to plateau below about 70% digestibility. In contrast, microtine rodents are able to increase intake of both total and indigestible dry matter down to a digestibility of about 30% (Fig. 21.5*a*,*b*). These trends reinforce the conclusion, drawn from the above discussion of arboreal folivores, that there appears to be a necessary association between utilisation of fibrous plant tissues and digesta separation with selective retention of solutes and/or fine particles.

Summary of trends: fibre-tolerance/intolerance among small mammals in relation to gut function

From the preceding discussion, some clear relationships emerge between digestive and metabolic adaptations and use of fibrous plant parts by small mammals (Table 21.2). Firstly, the highest levels of fibre-feeding are seen in species with little development of the foregut and extensive development of the caecum or caecum/colon. All evidence suggests that fermentation in the hindgut is not the primary source of digestible energy for these species. Within this group, moderately high levels of use of fibrous plant parts appear to require separation of digesta in the colon and selective retention of solutes and small particles in the caecum.

Among both hindgut fermenters and foregut fermenters, the only species that appear capable of eating high-fibre, high-lignin food exclusively (i.e. koalas and sloths) also have very low metabolic rates, suggesting that this is a necessary adaptation for folivory in this size range.

Among foregut-fermenting mammals less than 15 kg in body weight, other than three-toed sloths but including terrestrial as well as arboreal species, all species practise frugivory or granivory to a significant or substantial degree. All of these species have a partially tubiform forestomach that should allow higher passage rates and food intakes than are possible in most ruminants and might allow at least some digesta to escape fermentation in the sacciform region of the forestomach.

Issues: how much do we really understand?

The data reviewed here reveal clear trends, some of which can be reconciled with, if not predicted by, current theory but others are neither predicted nor explained by existing models. Hence, some challenging questions are raised that must be answered if current theory is to be relevant for small mammals.

The most fundamental question posed is whether or not the intakes of fibrous diet required to meet energy requirements in small mammals are attainable without exceptional adaptations of the digestive tract such as those reported in the fibre specialists. As discussed already, several current models show why digestive strategies that maximise retention and digestion of fibrous diets are untenable for small mammals. Of the two recent models that explore consequences of strategies that do not maximise retention of fibre, only one predicts where the likely limits to intake occur (Illius and Gordon, 1992).

The model by Illius and Gordon (1992) predicts, on the assumption that retention time of digesta scales with body mass$^{0.25}$, that small mammals of 1 kg or less should be able to meet their maintenance energy requirements from a diet of grass with as little as 35% cell contents or tree foliage with 40% cell contents (Fig. 21.6*a,b*). However, this is not consistent with considerable evidence that indicates most small mammals are *not* able to survive on such diets. For example, data on various small rodents with simple digestive tracts indicate or suggest inability to maintain a maintenance intake of herbage or artificial diets when dietary cell contents are less than about 60% (Baumgardt, 1970, 1974; Keys *et al.*, 1970; Hansson, 1971; Karasov, 1982; Cork and Kenagy, 1989a), and even voles do not survive on diets containing less than 45% cell contents (Keys and van Soest, 1970). Similarly, even mature foliage in broad-leaved tropical and temperate forests rarely contains less than 40% cell contents (Cork and Foley, 1991); yet no arboreal mammals less than 1 kg, and very few between 1 and 15 kg use it as primary food. Since the model by Illius and Gordon (1992) embodies most of what is known about size-related factors influencing digestion in mammals, it is instructive to use it as a framework for questioning the limitations of current knowledge.

Is the expectation of digesta retention scaling with body mass$^{0.25}$ realistic? This is a widely stated expectation in the literature and it arises because the duration of other time-related physiological processes scale in this way (Taylor, 1980; Calder, 1984), including the time between intestinal contractions (Clarke, 1927; Adolph, 1949). The few data that are available support the expectation directly and/or indicate that rate of food intake (retention time × gut volume) scales to body mass$^{0.75}$ (Calder, 1984; unpublished data in Batzli, 1985; Illius and Gordon, 1992). However, these data come mainly from studies of large mammals fed diets that were unlikely to push the animals near their limits. Furthermore, the few data available on small species come mostly from studies in which low-fibre diets were eaten. Hence, available data show that food intake and passage scale with the rate of gut contraction *when other factors are not limiting*, but they do not necessarily give

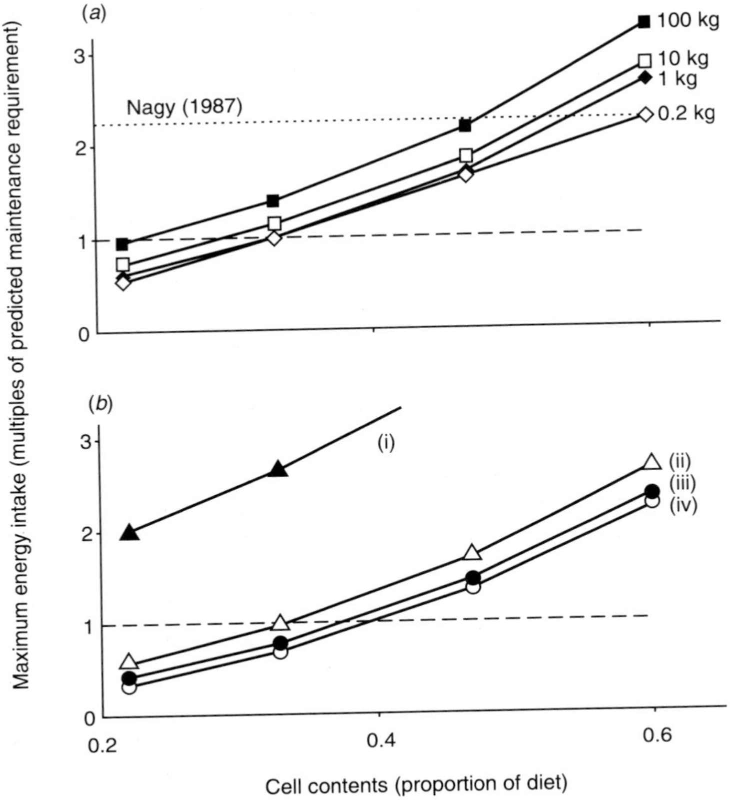

Fig. 21.6. The relationship between the proportion of cell contents in the diet and the maximum net energy intake. (*a*) Simulation (using the model of Illius and Gordon, 1991, 1992) of maximum net energy intake (multiples of predicted maintenance requirement) by hindgut-fermenting herbivores of various sizes in relation to the proportion of cell contents in their diet. The simulated diets vary in the ratio of cell contents to cell walls but have similar values for potential digestibility of cell walls (55%) and rate of cell-wall digestion (k_2 = 0.053/h); these values are characteristic of cell walls of mature *Phalaris arundinacea* grass (Illius and Gordon, 1991). Also shown is the intake of metabolisable energy required to meet predicted requirements for total daily energy expenditure in nature (Nagy, 1987). See text for further explanation. (*b*) The simulated relationship between proportion of cell contents in the diet and maximum net energy intake by a 1 kg hindgut-fermenting herbivore eating various diets differing in the digestion characteristics of the cell walls. The rate of digestion of cell walls in the large intestine (k_2, proportion per h) and the potential digestibility of cell walls (%) for each diet are: (i) 0.183, 90% (characteristic of cell walls in vegetative growth of young *Phalaris* grass; Illius and Gordon, 1991); (ii) 0.053, 55% (mature *Phalaris*); (iii) 0.0345, 36% (mature wheat straw); and (iv) 0.0185, 25% (*Eucalyptus punctata* cell walls eaten by hoalas (Cork *et al.*, 1983; Cork and Hume, 1983). All diets were assumed to have 5% ash.

realistic indications of the way that intake of high-fibre diets (e.g. mature grass or wheat straw) would scale if small mammals could be induced to eat them. Because passage rate can be influenced by a range of factors, including particle size, specific gravity and frequency of meals (Warner, 1981), it is important that passage rates of fibrous diets in small species are investigated directly rather than predicted from passage of low-fibre diets.

Are current data on digesta load valid for modelling purposes? Illius and Gordon (1992) assume that the prime determinant of cessation of feeding is digesta load (i.e. 'gut-fullness'). This assumption is supported by studies that show distension of various parts of the gut decreases appetite and sets upper limits to the amount eaten in humans, rats, ponies and pigs (Sharma, 1968; Houpt, 1984; Ralston, 1984; Forbes, 1985). Maximum digesta load is estimated empirically from data on mass of gut contents at slaughter and is assumed to scale more-or-less directly with body mass (Parra, 1978; Demment and van Soest, 1985; Illius and Gordon, 1992; Justice and Smith, 1992). Although this relationship is derived mostly from data on medium-sized to large species, with the few small species included not eating high-fibre diets, it is consistent with volumes of intestinal compartments calculated from linear dimensions over a wide range of mammals (Chivers and Hladik, 1980; Langer and Snipes, 1991; Snipes and Kriete, 1991). Recent studies have shown that small mammals are capable of considerable increases in gut capacity when switching from low-fibre to high-fibre diets (Hammond and Wunder, 1991; Batzli, Broussard and Oliver, Ch. 20), but whether or not this capability is comparable with that of larger species remains to be tested.

Are currently available estimates of maintenance energy requirements in mammals adequate? As pointed out already, the use of basal metabolic rate, or arbitrary multiples of it, as estimates of energy requirements in free-living mammals is inadequate and unnecessary since data have become available on actual expenditure of energy in natural environments (e.g. Nagy, 1987). Nevertheless, Justice and Smith (1992) use 2 × BMR as an estimate of maintenance requirements and Illius and Gordon (1991, 1992) use Brody's (1945) equation for net energy requirements of stock animals (net energy (MJ/d) = $0.4\ W(kg)^{0.73}$). In Fig. 21.6, Nagy's (1987) predicted metabolisable energy requirement of wild eutherian herbivores (FMR (kJ/d) = $5.94\ W(g)^{0.727}$) is applied to the Illius and Gordon (1992) model. We can expect metabolisable energy requirements to overestimate net energy requirements by about 15–19% (Brody, 1945; van Soest, 1982), so the Nagy line in Fig. 21.6 should be correspondingly lower. Even so, application of Nagy's data leads to predictions that are more consistent with observed patterns, suggesting that 1 kg mammals with average digestive and/or metabolic adaptations should be unable to meet maintenance energy requirements by eating mature grasses with more than about 50% fibre or by eating tree foliage with more than 45% fibre (Fig. 21.6*a*,*b*).

How does mastication of fibrous diets scale with body mass and how does it affect predictions of the limits to intake? This is an important consideration not addressed in any current models to my knowledge. Batzli (1985) suggested that finer grinding of food by the teeth in small compared with large species helps small mammals offset the disadvantages of rapid passage rates. Several studies indicate that small mammals obtain more from fermentation of fibre than predicted from current models (Karasov and Meyer, 1989; Justice and Smith, 1992; Batzli, Ch. 18). Although it would be expected that small particles of cell wall would be more rapidly degraded by microorganisms due to higher surface area/volume, few reliable data are available with which to model this factor (Illius and Gordon, 1991). Another important question to be considered is whether the time required for small species to grind fibrous foods imposes some limits.

Does the presence of tannins and other phenolics in tree foliage significantly affect limits to intake and digestion and can these effects be modelled? The foliage of woody dicotyledonous plants is well known for containing phenolic compounds, including tannins, at varying concentrations (see review by Cork and Foley, 1991). Considerable advances have been made recently in understanding the effects of tannins on intake and digestion of tree foliage by mammalian herbivores. Tannins have been shown to reduce the digestibility of cell contents in a range of browsing mammals (Cork *et al.*, 1983; Foley and Hume, 1987b; Robbins *et al.*, 1987) and of cell walls in both foregut- and hindgut-fermenting species that are not specialist browsers (Barry and Manley, 1984; Foley and Hume, 1987b; Robbins *et al.*, 1987, 1991; Hanley *et al.*, 1992). Data are becoming available on the different digestive consequences of different types of tannins (Hagerman *et al.*, 1992). Non-tannin phenolics that are absorbed also can influence intake and yield of nutrients from tree foliage due to their toxicity and the use of nutrients in their metabolic detoxification (Foley and Hume, 1987b; Foley, 1992; Hagerman *et al.*, 1992). There is evidence that ratios of tannins to other dietary components differ predictably between foliage selected versus avoided by small and moderately-small browsers (McKey *et al.*, 1981; Cork, 1992). The question of limits imposed by phenolics is given another dimension by the observations that some mammals, including small species, respond to tanniniferous diets by enhancing production of salivary proteins that have high affinity for tannins, which reduces the effects of tannins on other, more valuable, dietary and endogenous proteins (Mehansho *et al.*, 1987a,b). Therefore, consideration of the effects of phenolics on digestion and metabolism undoubtedly would improve the predictive ability of current models in relation to small browsers, but much more work is needed before highly accurate models can be built.

Do caecotrophy and/or digesta-separation mechanisms in the hindgut of small herbivores play a role in offsetting physiological limits? It has been suggested (e.g. Björnhag and Sperber, 1977; Cork and Warner, 1983; Cork and Foley, 1991; Foley and Cork, 1992) that a major function of digesta separation is the relatively rapid excretion of large particles of fibre compared with more-digestible components of the digesta, thus reducing the duration of the 'gut-filling' effect of dietary fibre and allowing higher food intakes than would have been possible otherwise. On the other hand, the small particles of cell walls selectively retained by this mechanism are likely to be more digestible than the large particles due to higher surface area/volume and lower lignin contents (Cork *et al.*, 1983; Bjorndal *et al.*, 1990), which should optimise yield from fermentative digestion within the limits set by rapid digesta passage. This aspect of digesta separation probably is more beneficial to small grazers than browsers because of the faster potential rate of degradation of the cell wall from grasses than from tree foliage.

Another important role of digesta separation and selective retention of fine particles and solutes is to minimise losses of cell solubles and/or endogenous and microbial energy and nitrogen in faeces. Such losses are expected to be high in mammals eating fibrous diets at high mass-specific rates of intake (Mason and Palmer, 1973; van Soest, 1982; Robbins, 1983). All the evidence available suggests that digesta separation and selective retention of solutes and fine particles does minimise losses of nitrogen in the faeces (Sperber, 1968; Hörnicke and Björnhag, 1980; Chilcott and Hume, 1984b; Cork, 1986; Foley and Hume, 1987c).

The benefits of selective retention of microorganisms in the hindgut are magnified by caecotrophy due to the return of high-quality microbial protein to the foregut where it can be digested and absorbed as intact amino acids (Hörnicke and Björnhag, 1980). All evidence to date suggests that ammonia rather than intact amino acids is the main product absorbed from the hindgut after digestion of the microbial protein therein (Robinson and Slade, 1974; Rérat, 1978); hence, the concentrating of microbial protein in the hindgut by selective retention of small particles is unlikely to help meet requirements for essential amino acids unless accompanied by reingestion. This poses the question as to why species like koalas and greater gliders do not practise caecotrophy, even though they have the physiological ability to separate digesta and concentrate microorganisms in the caecum. This remains a major challenge for future models of the role of the hindgut in small mammals (Foley and Cork, 1992).

This discussion suggests that, although there are constraints on use of fibrous diets by all small mammals, hindgut-fermenting species have more scope for evolving adaptations to offset these constraints than do foregut fermenters. Therefore, we need an explanation for why the main leaf-eaters in many tropical forests are foregut fermenters rather than hindgut fermenters. The first step in this explanation is to acknowledge that, contrary to the belief fostered in some literature, none of these arboreal foregut fermenters are fibre specialists; all use mixed fruit/seed and leaf diets and are found only in environments that provide that mix of food resources (Cork and Foley, 1991; DaSilva, 1992). Whereas retention and fermentation of all food would be an inefficient way to digest a strict diet of fruit or seeds and would be precluded by small body size as a way to use a diet of mature leaves, a strategy of foregut fermentation with the potential for bypass of the foregut by some digesta might be optimal for using a mixed fruit and leaf diet (Cork and Foley, 1991; see p. 350). This hypothesis offers a feasible explanation for foregut-fermenting arboreal primates and marsupials being found in habitats where both fruits (or seeds) and foliage are abundant (wet tropical forests) and absent from habitats like temperate eucalypt forests in which fruits and large seeds are rare (Cork and Foley, 1991). Another major question, for which I can offer no answer, is why so few hindgut-fermenting primates in tropical forests have evolved adaptations for folivory whereas several species of hindgut-fermenting marsupials in both tropical and temperate forests have.

Acknowledgements

I am extremely grateful to George Batzli, Bill Foley, Iain Gordon, Ian Hume, Andrew Illius, Bill Karasov and Carlos Martínez del Rio for their constructive comments on an earlier draft of this manuscript and for their inspirational research in this field. However, the ideas that emerged finally are entirely my own responsibility.

References

Adolph, E. F. (1949). Quantititive relations in the physiological constitutions of mammals. *Science*, **109**, 579–585.

Alexander, R. M. (1991). Optimization of gut structure and diet for higher vertebrate herbivores. *Philosophical Transactions of the Royal Society of London Series B, Biological Sciences*, **333**, 249–255.

Baker, R. H. (1971). Nutritional strategies of myomorph rodents in North American grasslands. *Journal of Mammalogy*, **52**, 800–805.

Baker, D. L. & Hobbs, N. T. (1987). Strategies of digestion: digestive efficiency and retention time of foraging diets in montane ungulates. *Canadian Journal of Zoology*, **65**, 1978–1984.

Barry, T. N. & Manley, T. R. (1984). The role of condensed tannins in the nutritional value of *Lotus pedunculatus* for sheep. *British Journal of Nutrition*, **51**, 493–504.

Batzli, G. O. (1985). Nutrition. In *Biology of New World Microtus*, ed. R. H. Tamarin, pp. 779–811. American Society of Mammologists.

Batzli, G. O. & Cole, F. R. (1979). Nutritional ecology of microtine rodents: digestibility of forage. *Journal of Mammalogy*, **60**, 740–750.

Bauchop, T. (1978). Digestion of leaves in vertebrate arboreal folivores. In *The Ecology of Arboreal Folivores*, ed. G. G. Montgomery, pp. 193–204. Washington DC: Smithsonian Institution Press.

Baumgardt, B. R. (1970). Regulation of feed intake and energy balance. In *Physiology of Digestion and Metabolism in the Ruminant*, ed. A. T. Phillipson, pp. 235–253. Newcastle-upon-Tyne: Oriel.

Baumgardt, B. R. (1974). Food intake, energy balance, and homeostasis. In *The Control of Metabolism*, ed. J. D. Sink, pp. 89–112. University Park: Pennsylvania State University.

Bell, R. H. V. (1969). The use of the herb layer by grazing ungulates in the Serengeti. In *Animal Populations in Relation to their Food Resources*, ed. A. Watson, pp. 111–123. Oxford: Blackwell.

Bjorndal, K. A., Bolten, A. B. & Moore, J. E. (1990). Digestive fermentation in herbivores: effect of food particle size. *Physiological Zoology*, **63**, 710–721.

Björnhag, G. (1972). Separation and delay of contents in the rabbit colon. *Swedish Journal of Agricultural Research*, **2**, 125–136.

Björnhag, G. (1987). Comparative aspects of digestion in the hindgut of mammals. The colonic separation mechanism (CSM)(a review). *Deutsche Tierärztliche Wochenschrift*, **94**, 33–36.

Björnhag, G. & Sperber, I. (1977). Transport of various food components through the digestive tract of turkeys, geese and guinea fowl. *Swedish Journal of Agricultural Research*, **7**, 57–66.

Bourlière, F. (1975). Mammals, small and large: the ecological implications of size. In *Small Mammals: Their Productivity and Population Dynamics*, ed. F. B. Golley, K. Petrusewicz & L. Ryszkowski, pp. 1–8. Cambridge: Cambridge University Press.

Brody, S. (1945). *Bioenergetics and Growth*, New York: Reinhold.

Calder, W. A. III (1984). *Size, Function and Life History*, Cambridge, MA: Harvard University Press.

Chilcott, M. J. & Hume, I. D. (1984a). Digestion of *Eucalyptus andrewsii* foliage by the common ringtail possum, *Pseudocheirus peregrinus. Australian Journal of Zoology*, **32**, 605–613.

Chilcott, M. J. & Hume, I. D. (1984b). Nitrogen and urea metabolism and nitrogen requirements of the common ringtail possum, *Pseudocheirus peregrinus*, fed *Eucalyptus andrewsii* foliage. *Australian Journal of Zoology*, **32**, 615–622.

Chilcott, M. J. & Hume, I. D. (1985). Coprophagy and the selective retention of fluid digesta: their role in the nutrition of the common ringtail possum (*Pseudocheirus peregrinus*) fed *Eucalyptus andrewsii* foliage. *Australian Journal of Zoology*, **33**, 1–15.

Chivers, D. J. & Hladik, C. M. (1980). Morphology of the gastrointestinal tract in primates: comparisons with other mammals in relation to diet. *Journal of Morphology*, **166**, 337–386.

Chivers, D. J. & Hladik, C. M. (1984). Diet and gut morphology in primates. In *Food Acquisition and Processing in Primates*, ed. D. J. Chivers, B. A. Wood & A. Bilsborough, pp. 213–230. New York: Plenum Press.

Clarke, A. J. (1927). *Comparative Physiology of the Heart*, Cambridge: Cambridge University Press.

Clutton-Brock, T. H. & Harvey, P. H. (1977). Species differences in feeding and ranging behaviour in primates. In *Primate Ecology*, ed. T. H. Clutton-Brock, pp. 557–579. London: Academic Press.

Cork, S. J. (1986). Foliage of *Eucalyptus punctata* and the maintenance nitrogen requirements of koalas, *Phascolarctos cinereus*. *Australian Journal of Zoology*, **34**, 17–23.

Cork, S. J. (1992). Polyphenols and the distribution of arboreal, folivorous marsupials in *Eucalyptus* forests of Australia. In *Plant Polyphenols: Synthesis, Properties, Significance*, ed. R. W. Hemingway & P. E. Laks, pp. 653–663. New York: Plenum Press.

Cork, S. J. & Foley, W. J. (1991). Digestive and metabolic strategies of arboreal mammalian folivores in relation to chemical defenses in temperate and tropical forests. In *Plant Defenses Against Mammalian Herbivory*, ed. R. T. Palo & C. T. Robbins, pp. 133–166. Boca Raton, FA: CRC Press.

Cork, S. J. & Hume, I. D. (1983). Microbial digestion in the koala (*Phascolarctos cinereus*, Marsupialia), an arboreal folivore. *Journal of Comparative Physiology*, **152**, 131–135.

Cork, S. J., Hume, I. D. & Dawson, T. J. (1983). Digestion and metabolism of a natural foliar diet (*Eucalyptus punctata*) by an arboreal marsupial, the koala (*Phascolarctos cinereus*). *Journal of Comparative Physiology*, **153**, 181–190.

Cork, S. J. & Kenagy, G. J. (1989a). Nutritional value of hypogeous fungus for a forest-dwelling ground squirrel. *Ecology*, **70**, 577–586.

Cork, S. J. & Kenagy, G. J. (1989b). Rates of gut passage and retention of hypogeous fungal spores in two forest-dwelling rodents. *Journal of Mammalogy*, **70**, 512–519.

Cork, S. J. & Warner, A. C. I. (1983). The passage of digesta markers through the gut of a folivorous marsupial, the koala *Phascolarctos cinereus*. *Journal of Comparative Physiology*, **152**, 43–51.

Cranford, J. A. & Johnson, E. O. (1989). Effects of coprophagy and diet quality on two microtine rodents (*Microtus pennsylvanicus* and *Microtus pinetorum*). *Journal of Mammalogy*, **70**, 494–502.

DaSilva, G. L. (1992). The western black-and-white colobus as a low-energy strategist – activity budgets, energy expenditure and energy intake. *Journal of Animal Ecology*, **61**, 79–91.

Dellow, D. W. (1982). Studies on the nutrition of macropodine marsupials III. The flow of digesta through the stomach and intestine of macropodines and sheep. *Australian Journal of Zoology*, **30**, 751–765.

Demment, M. W. & van Soest, P. J. (1985). A nutritional explanation for body-size patterns of ruminant and nonruminant herbivores. *American Naturalist*, **125**, 641–672.

Drozdz, A. (1968). Digestibility and assimilation of natural foods in small rodents. *Acta Theriologica*, **21**, 367–389.

Duncan, P., Foose, T. J., Gordon, I. J., Gakahu, C. G. & Lloyd, M. (1990). Comparative nutrient extraction from forages by grazing bovids and equids – a test of the nutritional model of equid bovid competition and coexistence. *Oecologia*, **84**, 411–418.

Eisenberg, J. F. (1978). The evolution of arboreal folivores in the class Mammalia.

In *The Ecology of Arboreal Folivores*, ed. G. G. Montgomery, pp. 135–152. Washington, DC: Smithsonian Institution Press.
Eisenberg, J. F. (1981). *The Mammalian Radiations*, Chicago: University of Chicago Press.
Elgar, M. A. & Harvey, P. H. (1987). Basal metabolic rates in mammals: allometry, phylogeny and ecology. *Functional Ecology*, **1**, 25–36.
Foley, W. J. (1987). Digestion and energy metabolism in a small arboreal marsupial, the greater glider, *Petauroides volans* fed high-terpene *Eucalyptus* foliage. *Journal of Comparative Physiology*, **157B**, 355–362.
Foley, W. J. (1992). Nitrogen and energy retention and acid-base status in the common ring-tailed possum (*Pseudocheirus peregrinus*): evidence of the effects of absorbed allelochemicals. *Physiological Zoology*, **65**, 403–422.
Foley, W. J. & Cork, S. J. (1992). Use of fibrous diets by small herbivores: How far can the rules be 'bent'? *Trends in Ecology and Evolution*, **7**, 159–162.
Foley, W. J. & Hume, I. D. (1987a). Passage of digesta markers in two species of arboreal folivorous marsupials – the greater glider (*Petauroides volans*) and the brushtail possum (*Trichosurus vulpecula*). *Physiological Zoology*, **60**, 103–113.
Foley, W. J. & Hume, I. D. (1987b). Digestion and metabolism of high-tannin *Eucalyptus* foliage by the brushtail possum (*Trichosurus vulpecula*) (Marsupialia: Phalangeridae). *Journal of Comparative Physiology*, **157B**, 67–76.
Foley, W. J. & Hume, I. D. (1987c). Nitrogen requirements and urea metabolism in two arboreal marsupials, the greater glider (*Petauroides volans*) and the brushtail possum (*Trichosurus vulpecula*) fed *Eucalyptus* foliage. *Physiological Zoology*, **60**, 241–250.
Foley, W. J., Hume, I. D. & Cork, S. J. (1989). Fermentation in the hindgut of the greater glider (*Petauroides volans*) and the brushtail possum (*Trichosurus vulpecula*). *Physiological Zoology*, **62**, 1126–1143.
Forbes, J. M. (1985). The importance of meals in the regulation of food intake. *Proceedings of the Nutrition Society of Australia,* **10**, 14–24.
Freudenberger, D. O., Wallis, I. R. & Hume, I. D. (1989). Digestive adaptations of kangaroos, wallabies and rat-kangaroos. In *Kangaroos, Wallabies and Rat-Kangaroos*, ed. G. Grigg, P. Jarman & I. Hume, pp. 179–187. Chipping Norton, NSW: Surrey Beatty.
Golley, F. B. (1960). Anatomy of the digestive tract of *Microtus*. *Journal of Mammalogy*, **41**, 89–99.
Hagerman, A. E., Robbins, C. T., Weerasuriya, Y., Wilson, T. C. & Mcarthur, C. (1992). Tannin chemistry in relation to digestion. *Journal of Range Management*, **45**, 57–62.
Hammond, K. A. & Wunder, B. A. (1991). The role of diet quality and energy need in the nutritional ecology of a small herbivore, *Microtus ochrogaster. Physiological Zoology*, **64**, 541–567.
Hanley, T. A., Robbins, C. T., Hagerman, A. E. & McArthur, C. (1992) Predicting digestible protein and digestible dry matter in tannin-containing forages consumed by ruminants. *Ecology*, **73**, 537–541.
Hansson, L. (1971). Habitat, food and population dynamics of the field vole *Microtus agrestis* (L.) in south Sweden. **Viltrevy**, **8**, 268–378.
Hansson, L. (1985). The food of bank voles, wood mice and yellow-necked mice. *Symposium of the Zoological Society of London*, **55**, 141–168.
Henry, S. R., Lee, A. K. & Smith, A. P. (1989). The trophic structure and species richness of assemblages of arboreal mammals in Australian forests. In

Patterns in the Structure of Mammalian Communities, ed. D. W. Morris, Z. Abramsky, B. J. Fox & M. R. Willig, pp. 229–240. Lubbock, TA:Texas Technical University Press.
Hladik, C. M., Charles-Dominique, P., Valdebouze, P., Delort-Laval, J. & Flanzy, J. (1971). La caecotrophie chez un primate phyllophage du genre *Lepilemur* et des corrèlations avec les particularitès de son appareil digestif. *Comptes Rendus Hebdomadaires des Seances de L'Academie des Sciences, Serie D – Sciences Naturelles*, **272**, 3191–3194.
Hobbs, N. T. (1990). Diet selection by generalist herbivores – a test of the linear programming model. In *Behavioural Mechanisms of Food Selection*, ed. R. N. Hughes, pp. 395–414. Berlin: Springer-Verlag.
Hofmann, R. R. (1989). Evolutionary steps of ecophysiological adaptation and diversification of ruminants: a comparative view of their digestive system. *Oecologia*, **78**, 443–457.
Hofmann, R. R. & Stewart, D. R. M. (1972). Grazer or browser: A classification based on the stomach structure and feeding habits of East African ruminants. *Mammalia*, **36**, 226–240.
Hörnicke, H. & Björnhag, G. (1980). Coprophagy and related strategies for digesta utilization. In *Digestive Physiology and Metabolism in Ruminants*, ed. Y. Ruckebusch & P. Thivend, pp. 707–730. Lancaster: MTP Press.
Houpt, T. R. (1984). Controls of feeding in pigs. *Journal of Animal Science*, **59**, 1345–1353.
Hume, I. D. (1982). *The Digestive Physiology and Nutrition of Marsupials*, London: Cambridge University Press.
Hume, I. D. (1984). Principal features of digestion in kangaroos. *Proceedings of the Nutrition Society of Australia*, **9**, 76–81.
Hume, I. D. (1989). Optimal digestive strategies in mammalian herbivores. *Physiological Zoology*, **62**, 1145–1163.
Hume, I. D. & Carlisle, C. H. (1985). Radiographic studies on the structure and function of the gastrointestinal tract of two species of potoroine marsupials. *Australian Journal of Zoology*, **33**, 641–654.
Hume, I. D., Carlisle, C. H., Reynolds, K. & Pass, M. A. (1988). Effects of fasting and sedation on gastrointestinal tract function on two potoroine marsupials. *Australian Journal of Zoology*, **36**, 411–420.
Hume, I. D., Morgan, K. R. & Kenagy, G. J. (1993). Digesta retention and digestive performance in sciurid and microtine rodents: effects of hindgut morphology and body size. *Physiological Zoology*, **66**, 396–411.
Illius, A. W. & Gordon, I. J. (1991). Prediction of intake and digestion in ruminants by a model of rumen kinetics integrating animal size and plant characteristics. *Journal of Agricultural Science*, **116**, 145–157.
Illius, A. W. & Gordon, I. J. (1992). Modelling the nutritional ecology of ungulate herbivores – evolution of body size and competitive interactions. *Oecologia*, **89**, 428–434.
Janis, C. (1976). The evolutionary strategy of the Equidae and the origins of rumen and cecal digestion. *Evolution*, **30**, 757–774.
Jarman, P. J. (1974). The social organization of antelope in relation to their ecology. *Behaviour*, **48**, 215–266.
Jarman, P. J. (1984). The dietary ecology of macropodid marsupials. *Proceedings of the Nutrition Society of Australia*, **9**, 82–87.
Justice, K. E. & Smith, F. A. (1992). A model of dietary fiber utilization by small mammalian herbivores, with empirical results for *Neotoma*. *American Naturalist*, **139**, 398–416.

Karasov, W. H. (1982). Energy assimilation, nitrogen requirement, and diet in free-living antelope ground squirrels *Ammospermophilus leucurus. Physiological Zoology*, **55**, 378–392.

Karasov, W. H. & Meyer, M. (1989). Digesta retention, nutrient absorption, and digestive efficiency in small herbivores. *Proceedings of the Fifth International Theriological Congress, Rome*, pp. 450.

Kenagy, G. J. & Hoyt, D. F. (1980). Reingestion of feces in rodents and its daily rhythmicity. *Oecologia*, **44**, 403–409.

Keys, J. E. & van Soest, P. J. (1970). Digestibility of forages by the meadow vole (*Microtus pennsylvanicus). Journal of Dairy Science*, **53**, 1502–1508.

Keys, J. E., van Soest, P. J. & Young, E. P. (1970). The effect of increasing dietary cell wall content on the digestibility of hemicellulose and cellulose in swine and rats. *Journal of Animal Science*, **31**, 1172.

Kinnear, J. E., Cockson, A., Christensen, P. & Main, A. R. (1979). The nutritional biology of the ruminants and ruminant-like mammals – a new approach. *Comparative Biochemistry and Physiology*, **64A**, 357–365.

Kleiber, M. (1975). *The Fire of Life*, New York: Wiley.

Landry, S. O. Jr (1970). The Rodentia as omnivores. *Quarterly Review of Biology*, **45**, 351–372.

Langer, P. (1974). Stomach evolution in the Artiodactyla. *Mammalia*, **38**, 295–314.

Langer, P. (1988). *The Mammalian Herbivore Stomach*, Stuttgart: Gustav Fischer.

Langer, P., Dellow, D. W. & Hume, I. D. (1980). Stomach structure and function in three species of macropodine marsupials. *Australian Journal of Zoology*, **28**, 1–18.

Langer, P. & Snipes, R. L. (1991). Adaptations of gut structure to function in herbivores. In *Physiological Aspects of Digestion and Metabolism in Ruminants*, ed. T. Tsuda, Y. Sasaki & R. Kawashima, pp. 349–384. San Diego: Academic Press.

Martin, R. D., Chivers, D. J., MacLarnon, A. M. & Hladik, C. M. (1985). Gastrointestinal allometry in primates and other mammals. In *Size and Scaling in Primate Biology*, ed. W. L. Jungers, pp. 61–89. New York: Plenum.

Mason, V. C. & Palmer, R. (1973). The influence of bacterial activity in the alimentary canal of rats on faecal nitrogen excretion. *Acta Agriculturae Scandinavica*, **23**, 141–150.

McKey, D. B., Gartlan, J. S., Waterman, P. G. & Choo, G. M. (1981). Food selection by black colobus monkeys (*Colobus satanus*) in relation to plant chemistry. *Biological Journal of the Linnean Society*, **16**, 115–146.

McNab, B. K. (1978). Energetics of arboreal folivores: physiological problems and ecological consequences of feeding on an ubiquitous food supply. In *The Ecology of Arboreal Folivores*, ed. G. G. Montgomery, pp. 153–162. Washington, D.C: Smithsonian Institution Press.

McNab, B. K. (1986). The influence of food habits on the energetics of eutherian mammals. *Ecological Monographs*, **56**, 1–19.

McNab, B. K. (1987). Basal rate and phylogeny. *Functional Ecology*, **1**, 159–167.

Mehansho, H., Ann, D. K., Butler, L. G., Rogler, J. C. & Carlson, D. M. (1987a). Induction of proline-rich proteins in hamster salivary glands by isoproterenol treatment and an unusual growth inhibition by tannins. *Journal of Biological Chemistry*, **262**, 12344–12350.

Mehansho, H., Butler, L. G. & Carlson, D. M. (1987b). Dietary tannins and salivary proline-rich proteins: interactions, induction and defense mechanisms. *Annual Review of Nutrition*, **7**, 423–440.

Montgomery, G. G. & Sunquist, M. E. (1978). Habitat selection and use by two-

toed and three-toed sloths. In *The Ecology of Arboreal Folivores*, ed. G. G. Montgomery, pp. 329–359. Washington, DC: Smithsonian Institution Press.
Müller, E. F., Kamau, J. M. Z. & Maloiy, G. M. O. (1983). A comparative study of basal metabolism and thermoregulation in a folivorous (*Colobus guereza*) and an omnivorous (*Cercopithecus mitis*) primate species. *Comparative Biochemistry and Physiology*, **74A**, 319–322.
Muul, I. & Liat, L. B. (1978). Comparative morphology, food habits, and ecology of some Malaysian arboreal rodents. In *The Ecology of Arboreal Folivores*, ed. G. G. Montgomery, pp. 361–370. Washington, DC: Smithsonian Institution Press.
Nagy, K. A. (1987). Field metabolic rate and food requirement scaling in mammals and birds. *Ecological Monographs*, **57**, 111–128.
Nagy, K. A. & Martin, R. (1985). Field metabolic rate, water flux, food consumption and time budget of koalas, *Phascolarctos cinereus* (Marsupialia: Phascolarctidae) in Victoria. *Australian Journal of Zoology*, **33**, 655–665.
Nagy, K. A. & Milton, K. (1979). Energy metabolism and food consumption by wild howler monkeys (*Alouatta palliata*). *Ecology*, **60**, 475–480.
Nagy, K. A. & Montgomery, G. G. (1980). Field metabolic rate, water flux, and food consumption in three-toed sloths (*Bradypus variegatus*). *Journal of Mammalogy*, **61**, 465–472.
Ouellette, D. E. & Heisinger, J. F. (1980). Reingestion of faeces by *Microtus pennsylvanicus*. *Journal of Mammalogy*, **61**, 366–368.
Parra, R. (1978). Comparison of foregut and hindgut fermentation in herbivores. In *The Ecology of Arboreal Folivores*, ed. G. G. Montgomery, pp. 205–230. Washington, D.C: Smithsonian Institution Press.
Penry, D. L. & Jumas, P. A. (1986). Chemical reactor analysis and optimal digestion. *Bioscience*, **36**, 310–315.
Penry, D. L. & Jumas, P. A. (1987). Modelling animal guts as chemical reactors. *American Naturalist*, **129**, 69–96.
Ralston, S. L. (1984). Controls of feeding in horses. *Journal of Animal Science*, **59**, 1354–1361.
Rérat, A. (1978). Digestion and absorption of carbohydrates and nitrogenous matters in the hindgut of omnivorous nonruminant animal. *Journal of Animal Science*, **6**, 1808–1837.
Robbins, C. T. (1983). *Wildlife Feeding and Nutrition,* New York: Academic Press.
Robbins, C. T., Hagerman, A. E., Austin, P. J., Mcarthur, C. & Hanley, T. A. (1991). Variation in mammalian physiological responses to a condensed tannin and its ecological implications. *Journal of Mammalogy*, **72**, 480–486.
Robbins, C. T., Mole, S., Hagerman, A. E. & Hanley, T. A. (1987). Role of tannins in defending plants against ruminants: reduction in dry matter digestion? *Ecology*, **68**, 1606–1615.
Robinson, D. W. & Slade, L. M. (1974). The current status of knowledge on the nutrition of equines. *Journal of Animal Science*, **39**, 1045–1066.
Sharma, K. N. (1968). Receptor mechanisms in the alimentary tract – their excitation and function. In *Handbook of Physiology – Alimentary Canal, Section 6, Vol. IV*, pp. 225. Washington: D.C: American Physiological Society.
Snipes, R. L. (1979). Anatomy of the cecum of the vole, *Microtus agrestis. Anatomy and Embryology*, **157**, 181–203.
Snipes, R. L. & Kriete, A. (1991). Quantitative investigation of the area and volume in different compartments of the intestine of 18 mammalian species.

Zeitschrift für Säugetierkunde – International Journal of Mammalian Biology, **56**, 225–244.

Sperber, I. (1968). Physiological mechanisms in herbivores for retention and utilization of nitrogenous compounds. In *Isotope Studies on the Nitrogen Chain*, pp. 209–219. Vienna: International Atomic Energy Agency.

Taylor, St C. S. (1980). Genetic size scaling rules in animal growth. *Animal Production*, **30**, 161–165.

van Soest, P. J. (1982). *Nutritional Ecology of the Ruminant*, Portland: Durham and Downey.

Vorontsov, N. N. (1962). The ways of food specialization and evolution of the alimentary system in Muroidea. *Symposium Theriologicum*, **1960**, 360–377.

Warner, A. C. I. (1981). Rate of passage of digesta through the gut of mammals and birds. *Nutrition Abstracts and Reviews, Series B*, **51**, 789–820.

22

The effects and costs of allelochemicals for mammalian herbivores: an ecological perspective

WILLIAM J. FOLEY and CLARE MCARTHUR

Mammalian herbivores encounter a diverse range of allelochemicals in their diets (Palo and Robbins, 1991). Clearly, theories of foraging behaviour and diet selection must focus on the effects of these substances as well as on nutrients and energy (Belovsky and Schmitz, 1991). This is particularly true for folivores and other browsers, because it is the leaves of trees and woody shrubs that contain the greatest concentration of allelochemicals (Bryant *et al.*, 1991; Cork and Foley, 1991; Meyer and Karasov, 1991).

Several studies have demonstrated that some allelochemicals are strongly deterrent to some herbivores (Clausen *et al.*, 1990, Bryant *et al.*, 1991, 1992), but many other species ingest significant quantities of, mainly carbon-based, allelochemicals (Cork and Foley, 1991; Meyer and Karasov, 1991; McArthur *et al.*, 1991). In most cases, the effect of these allelochemicals on food intake is unknown. This sometimes reflects problems in the analysis of allelochemicals in plants, but also reflects a lack of knowledge of the effects of ingested compounds on animals. Therefore, the occurrence, effects and costs of ingestion of the allelochemicals contained in different plants available to herbivores must form a central part of any theory of mammal–plant interactions.

The effects of allelochemicals on insect herbivores are much more widely documented (e.g. Bernays *et al.*, 1989); small size makes insects amenable to growth studies and the diets of many insects can be manipulated easily. In contrast, vertebrate herbivores are often difficult to maintain for detailed investigation of the effects of different diets on health and metabolism.

In order to understand the role of allelochemicals in mammal–plant interactions, we need to know the effects that they have on the animal and, in particular, what price the animal pays to ingest and excrete the compounds. Animals process allelochemicals in different ways and the effects and cost of excretion are not the same for all species. Understanding species differences should help us to understand separation of diets, food selection and nutritional ecology.

In this article we do not review the numerous studies that have sought correlations between diet selection and allelochemical content, nor do we discuss the occurrence of allelochemicals in different plants. There are recent reviews of these topics in Palo and Robbins (1991). Rather, we review, from an ecological perspective, what is known of the effects of ingested allelochemicals on mammalian herbivores and suggest some ways that the costs of ingestion and excretion can be measured.

Allelochemicals

Most theories of plant defence have divided allelochemicals into two functional groups. Originally, **toxins** were separated from **digestibility reducers** (Feeny, 1976; Rhoades and Cates, 1976). A more recent division separates **mobile** from **immobile** allelochemicals (Coley *et al.*, 1985), although both divisions produce similar groupings of chemicals. Toxic compounds are small molecules that are rapidly turned over in plants and are, therefore, mobile defences. On ingestion, they are usually absorbed from the gut and exert a specific toxic effect on the consumer. Typical examples are alkaloids, cyanogenic glycosides and non-protein amino acids. In contrast, digestibility reducers are larger molecules that are metabolically inactive in the plant and so are immobile defences. Their site of action in the consumer is supposed to be within the gut and they are thought to interfere with digestion of other nutrients. The best examples are tannins (Bernays *et al.*, 1989).

The effect of these particular allelochemicals on mammals, however, is conditional on the consumer (McArthur *et al.*, 1991). That is, some compounds act as toxins or as digestibility reducers depending on the way in which the consumer deals with them. Therefore, dietary separation may be explained by differences in the effects of allelochemicals in different herbivore species. This is because different effects result in different costs for ingesting the same group of compounds and, consequently, different feedbacks influence diet choice. Two examples illustrate this point.

Tannins have traditionally been regarded as digestibility reducers through their ability to form complexes with dietary and other proteins in the gut. There are many clear examples where ingested dietary tannin reduces protein digestibility (Glick and Joslyn, 1970; Lindroth and Batzli, 1984; Robbins *et al.*, 1991). Nevertheless, there is increasing evidence, both direct and indirect, that indicates that certain tannins sometimes are degraded and absorbed and thus act as toxins (O'Brien *et al.*, 1986; Mehansho *et al.*, 1987a, b; Clausen *et al.*, 1990; McArthur and Sanson, 1991; Hagerman *et al.*, 1992). How a tannin functions partly depends on its chemistry. For example, the condensed tannin quebracho reduces protein digestion in sheep and deer, whereas the low molecular weight gallotannin tannic acid does not (Hagerman *et al.*, 1992). Tannic acid is more easily degraded than quebracho, and it has been suggested that degradation occurs before strong protein–tannic acid complexes are formed. The resultant low molecular weight phenolics may be absorbed (Hagerman *et al.*, 1992). Consequently, in this example, tannic acid is functionally a toxin whereas quebracho is a digestibility reducer.

The effect of tannins also depends on the physiology of the consumer. Quebracho reduced the digestibility of protein in ruminants (Robbins *et al.*, 1987) and in some macropodoid marsupials (C. McArthur and G. D. Sanson, unpublished data), but it had no such effect in two marsupial possums (C. McArthur and G. D. Sanson, unpublished data). We believe this is because the possums deal with the tannin differently. Tannin–protein complexes appear to be broken down in the gut of common ringtail (*Pseudocheirus peregrinus*) and brushtail possums (*Trichosurus vulpecula*) because not all of the quebracho is recovered in the faeces. This suggests that the animal has absorbed a part of the quebracho phenolics. Again, quebracho tannin functions as either a digestibility reducer or a toxin, depending on the physiology of the consumer.

The second example concerns volatile monoterpenes which are a common component of many woody plants such as *Eucalyptus* ((Foley *et al.*, 1987) and *Artemesia* (sagebrush; Welch *et al.*, 1982). Monoterpenes are widely known for their anti-microbial effects, and early studies (Nagy and Tengerdy, 1968; Oh *et al.*, 1968) showed that some monoterpenes from sagebrush and Douglas fir (*Pseudotsuga menziesii*) could inhibit the cellulolytic activity of rumen microbes *in vitro*. Monoterpenes were consequently regarded as digestibility reducers (Nagy and Tengerdy, 1968, Connolly *et al.*, 1980). However, later studies *in vivo* found that monoterpenes were rapidly absorbed from the gut before exerting significant effects on microbes (Welch *et al.*, 1982; Foley *et al.*, 1987). Although absorption avoids possible digestibility-reducing effects, the absorbed compounds can now exert a toxic action: potentially disrupting cell membranes and causing liver damage (McLean *et al.*, 1993). The animal must rapidly excrete the compound from the body and this is presumed to involve some cost. These results suggest that the principal mode of action of monoterpenes is not an effect on digestibility but more likely a toxic action.

These examples illustrate the difficulty of trying to predict the effect of an allelochemical based solely on its chemistry. An understanding of both the chemistry of the allelochemical and the physiology of the consumer is essential for determining the cost of allelochemical ingestion.

Methods of dealing with allelochemicals

The effect of a particular allelochemical depends on the amount ingested and the rate and degree to which it can be neutralized or eliminated from the body. How quickly and to what extent this occurs depends on the method used to deal with it.

Salivary modification

Mammalian saliva plays a variety of roles including digestive, protective and buffering (Mandel, 1987). In human saliva, there is a group of 'proline-rich' proteins and peptides that influence mineral homeostasis (Bennick, 1982; Mandel, 1989; Madapallimattan and Bennick, 1990). Similar salivary proteins have been observed to interact strongly with dietary tannins and they are now more generally referred to as tannin-binding salivary proteins (TBSP) (Austin *et al.*, 1989).

Austin *et al.* (1989) postulated that TBSP form stable complexes with tannin in the gut and that these complexes are resistant to degradation in the gut. TBSP have a higher affinity for tannins than do many other proteins (including the major plant protein ribulose bisphosphate carboxylase/ oxygenase), because of their open conformation and enhanced hydrogen bond acceptor capacity (Hagerman and Butler, 1981). Consequently, tannin binds preferentially with TBSP even where there is an excess of other proteins. It has been assumed, but not explicitly demonstrated, that TBSP also bind more tannin per unit protein. This idea is supported by recent studies which showed that, in the presence of tannins, more protein is digested in those species that possess TBSP than in species which lack them (Robbins *et al.*, 1991).

The occurrence of TBSP has been examined more extensively among eutherians than marsupials. In eutherians, the production and effectiveness of TBSP varies and is loosely related to an animal's feeding niche. TBSP are constitutive and effective in browsers, e.g. deer, moose, beavers (Hagerman and Robbins, 1992), and some omnivores, e.g. humans (Mehansho *et al.*, 1987b) and bear (Hagerman and Robbins, 1992); they are inducible in other omnivores, e.g. rats (Mehansho *et al.*, 1983) and mice (Mehansho *et al.*, 1985); or uninducible by tannins and, therefore, ineffective in grazers, e.g. cattle (Austin *et al.*, 1989) and others, e.g. hamsters

(Mehansho *et al.*, 1987a). Some TBSP may have a high affinity for a variety of structural types of tannin, while others bind with a fairly restricted group of tannins which are apparently chemically similar (Hagerman and Robbins, 1993). There is some evidence that specialist browsers produce specific TBSP while generalist browsers produce non-specific TBSP (Hagerman and Robbins, 1992).

Preliminary evidence in marsupials (C. McArthur and G. D. Sanson, unpublished data; C. McArthur, A. M. Beal and G. D. Sanson, unpublished data) suggests that browsing macropodoids (e.g. Tasmanian pademelon *Thylogale billardieri*) produce TBSP but grazing macropodoids (e.g. red kangaroo *Macropus rufus*) do not. These marsupials, thus, show a similar pattern to ruminants. However, in folivorous hindgut-fermenting marsupials, (koala *Phascolarctos cinereus*, greater glider *Petauroides volans*, common ringtail possum, common brushtail possum) TBSP, although sometimes present, appear functionally unimportant. Proline-rich proteins from the parotid salivary glands of koalas and ringtail possums have an extremely low affinity for tannins compared with TBSP from other animals (Mole *et al.*, 1990). Furthermore, there appears to be little or no protein cost when these folivores consume a tannin-rich diet (Foley and Hume, 1986; McArthur and Sanson, 1991 and unpublished data). Interestingly, brushtail possums do produce non-specific TBSP (C. McArthur, A. M. Beal and G. D. Sanson, unpublished data) but the parotid gland responsible for their production and secretion is very small and flow rates are low (A. M. Beal personal communication). Consequently, we suggest that TBSP are functionally useless in brushtail possums. This observation emphasizes that TBSP not only have to be produced but also they have to be secreted in sufficient quantities to bind the tannin.

There are three advantages to producing TBSP. First, smaller amounts of protein are lost in the faeces of animals with TBSP than in animals which lack them. Secondly, there may be a saving of dietary essential amino acids when salivary protein–tannin complexes are excreted because proline is a 'cheap' non-essential amino acid. This may be particularly important for hindgut fermenters that have a greater reliance on dietary amino acids. Thirdly, by forming stable salivary protein–tannin complexes, tannin is no longer susceptible to degradation and absorption (Robbins *et al.*, 1991), so the risk of converting a digestibility reducer to a toxin is reduced. While the relative cost of producing specific or non-specific TBSP has yet to be investigated, there is presumably some advantage to specialist browsers in producing proteins which can bind specific tannins. An animal feeding on a diet with few tannin types may realize a greater protein saving if salivary proteins with high specificity bind a greater proportion of the tannin per unit protein than proteins that are non-specific. This argument, however, relies heavily on what might be a very small protein pay-off. It remains to be determined what, if any, are the benefits of specific TBSP. Competitive binding assays comparing specific versus non-specific salivary proteins may resolve this point, by indicating the relative capacity of each type for particular tannins.

In conclusion, salivary modification plays a role in some species in dealing with dietary tannins. In animals where TBSP are produced, toxic costs of tannins may be avoided and protein costs are reduced but still measurable. The effectiveness of TBSP may depend on having sufficiently high salivary flow rates and these may have evolved for other needs (for example buffering of forestomach contents in ruminants and macropodoids). The apparent lack of functional TBSP in folivorous, marsupial, hindgut fermenters may be due to a greater selective pressure in these animals to conserve protein. *Eucalyptus* foliage has a low nitrogen content and those species that feed on *Eucalyptus* may have offset the cost of excreting TBSP by evolving other mechanisms for dealing with tannin (McArthur and Sanson, 1991).

Microbial modification

Most mammalian herbivores house large microbial populations in expanded regions of their guts: the forestomach and/or the caecum–proximal colon. The microbial degradation of cellulose and associated carbohydrates is generally viewed as the primary role of these symbionts. However, it has also been suggested that microbial detoxification of ingested allelochemicals was a major evolutionary force in the development of such symbiotic relationships (Janzen, 1979).

Gut microorganisms have been shown to detoxify a range of allelochemicals including oxalates, alkaloids, cyanogenic glycosides and non-protein amino acids (O'Halloran, 1962; Galtier and Alvinerie, 1976; Smith, 1986). The potential importance of microbial detoxification is demonstrated by studies in Australian goats fed the shrub legume *Leucaena leucocephala*. The use of *Leucaena* as a stock food in Australia was limited by the presence of the toxic amino acid mimosine. In contrast, goats from Hawaii fed on *Leucaena* without problem. Experiments showed that the infusion of rumen fluid from Hawaiian goats into Australian goats negated the toxic effect and all treated animals started to eat the plant within hours (Jones and Megarrity, 1986). In this case, a specific microorganism existed that could detoxify mimosine. This spectacular result has led to a number of projects to engineer rumen microbes able to detoxify other allelochemicals (Smith, 1992).

It is possible that wild species harbour other microbes that may be equally effective in expanding an animal's feeding niche. Osawa (1992) has recently identified a biotype of *Streptococcus bovis* in the caecum of koalas which degrades protein–tannic acid complexes but it is not known whether the bacterium uses the protein or the tannic acid (or both) as substrate. Differences in microbial detoxification capacity are unlikely to be found between individuals within a population but it may be worthwhile comparing different populations within a species if significant differences in food choice are observed.

The role and importance of microbial detoxification depends in part on the nature of the ingested allelochemical and the digestive physiology of the animal. For example, some allelochemicals are actually made more toxic following microbial modification in the foregut (O'Hara and Fraser, 1975; Carrlson and Dickinson, 1978). Other compounds with antimicrobial actions, such as terpenes, may in fact be best dealt with by rapid absorption from a simple stomach rather than interaction with the microbial population. Furthermore, many ecologists have argued that herbivores with an extensive foregut fermentation may be better able to cope with diets rich in allelochemicals than hindgut fermenters because the foregut microbes act as a first-line of defence (Janzen, 1979; Langer, 1986). In contrast, hindgut fermenters are thought to be at a disadvantage for microbial detoxification. This notion has been very influential in the study of mammalian folivory (McKey *et al.*, 1981; Waterman *et al.*, 1988; DaSilva, 1992).

The most widely cited example of the potential usefulness of microbial detoxification in the foregut is the degradation of pyrrolizidine alkaloids in the rumen of sheep (Russell and Smith, 1968; Lanigan and Smith, 1970). However, studies by Cheeke (1984) showed that resistance to intoxication from pyrolizidine alkaloids was not simply a function of digestive anatomy. He found that while some foregut fermenters were resistant to the toxic effects, others were susceptible and a similar difference was observed in hindgut fermenters. For pyrrolizidine alkaloids at least, the potential toxic effect does not depend on the site of microbial action.

While the detoxifying ability of microbial symbionts is undoubtedly important in expanding the feeding niches of some species, it seems unwise to regard one system or the other as superior. Certainly, the evidence for a consistent advantage of foregut fermentation for detoxifying allelochemicals such as alkaloids is equivocal and cannot be used to explain the evolution of foregut fermentation in different groups.

Biotransformation of absorbed compounds

Allelochemicals that are absorbed from the gut usually undergo transformation prior to excretion. Compounds that are absorbed across the gut are usually lipid-soluble, non-polar compounds while those excreted in urine or bile are water soluble. To this extent, the general nature of the biotransformational processes are similar in all mammals (Caldwell, 1982). However, the particular enzymatic pathway and the level of enzymatic activity vary widely amongst species.

Biotransformation processes usually occur in two steps. The first step (often called a Phase I reaction) serves to introduce, or expose within the structure of the allelochemical, a functional group and this mostly (but not always) results in a less toxic product. The second step (Phase II) conjugates this modified product with a small molecule such as glucuronic acid, sulphate or glycine (Caldwell, 1982). The general nature of these processes is well known and was described in Freeland and Janzen's (1974) seminal work. Toxicologists have continued to study the mechanisms involved in great detail. However, given the potential importance of biotransformational processes in the interaction of mammals and woody plants, it is surprising that so few studies have been performed in wild species. In contrast, study of the role of biotransformational enzymes in insect–plant interactions is considerably more advanced. There is compelling evidence that differential toxicity of allelochemicals to insects is linked to differences in the pattern of activity of biotransformational enzymes (Lindroth, 1991; Brattsten, 1992).

The biotransformational systems of mammals, like those of insects, are highly complex. However, studying the mammalian systems from an ecological viewpoint is much more difficult. Some recent data on domestic animal species gathered and reviewed by Smith (1992) are relevant to the issues involved here.

1. The level of activity of biotransformational enzymes varies greatly amongst species and cannot be explained by phylogenetic or dietary groupings (e.g. carnivore/herbivore).
2. The inducibility of biotransformational enzymes makes it difficult to determine, from studies under standard conditions, what the effects of prior exposure to particular toxins will be (Smith, 1992). Almost all studies to date have been of domestic species and have used standard or model substrates rather than a relevant plant allelochemical. Little consideration has been given to the effects of prior exposure to the compound.
3. The site of biotransformational enzymes may be significant: the gut mucosa may be far more important in the biotransformation of ingested

allelochemicals than has been previously appreciated. For example, the activity of one biotransformational enzyme (UDP glucuronosyl transferase) was three times greater in the rumen wall than in the liver of sheep and cattle and many other enzymes in ileal tissue showed high activity. Given that the gut is the first organ exposed to ingested allelochemicals, these results may be very significant.

4. The action of biotransformational enzyme systems may sometimes result in toxication rather than detoxification, although this is relatively rare.

The possibility that biotransformational enzymes are induced by dietary constituents together with the effects of gender (e.g. Bergeron and Jodoin, 1991), developmental stage and nutritional status (Parke and Ioannides, 1981; Boyd and Campbell, 1983) suggests that it will be difficult to make specific predictions about the role of biotransformational enzymes in ecological interactions between plants and mammals. Nonetheless, given the probable significant interaction between nutritional state and biotransformational capacity, we urge that this aspect of the system be given attention. Factors such as body fat content, level of protein intake and mineral status can all affect the capacity of biotransformational enzymes. We know of no ecologically relevant example where these interactions have been studied.

Such a study will require us, initially, to accept the detailed mechanisms as a 'black-box' and to concentrate on the end-products of these processes. In effect, we need some index of the capacity of the biotransformational system in intact animals that can be measured under different nutritional states. Lindroth and Batzli (1983) have suggested a measure of urinary glucuronic acid excretion as a general index of detoxifying ability, but recent studies in arboreal marsupials (McLean *et al.*, 1993) have shown that the majority of terpenes and phenolics ingested as part of a diet of eucalypt leaves are excreted unconjugated. The measurement of total urinary organic acids may be a more appropriate parameter. This is much more likely to account for variable pathways (e.g. carboxylic acids, glycine conjugates) and the methods involved are relatively straightforward and accessible to nutritional ecologists. The development of indices of detoxification is an important research need in the study of interactions between mammals and woody plants.

Our coverage of the methods used by mammals to neutralize allelochemicals has, of necessity, been selective. However, there is little evidence that mammals use some of the other strategies adopted by insects for dealing with similar compounds. For example, many insects sequester allelochemicals, and, in others, high gut pH (Berenbaum, 1980) and the presence of surfactants (Martin and Martin, 1984) are important counters to allelochemicals. Genetic resistance through other, unknown mechanisms is certainly important in some cases (King *et al.*, 1978; Smith *et al.*, 1991), but presumably the conservative nature of mammalian physiology limits the adaptive options for dealing with allelochemicals.

Assessing the effects and costs of allelochemicals: what currency to use?

A single allelochemical can have different effects in different herbivores. Therefore, if we are to measure differences in the cost of ingestion and excretion of these compounds, we need to decide how to express the costs. Ingested allelochemicals affect animals on various levels, reflecting different types of costs with different units. We identify four major ways in which these costs can be assessed. First, there are effects related to site: both digestive (pre-absorptive) effects and metabolic (post-absorptive) effects. Secondly, there may be measurable changes in whole-animal nitrogen or energy needs that are attributable to allelochemicals. Thirdly, and more specifically, there may be costs associated with metabolic pathways and routes of excretion of allelochemicals. Finally, on a whole-animal scale, allelochemicals may affect fitness parameters such as growth and reproduction.

Effects of allelochemicals on digestion

The putative effect of tannins on digestion has probably been the most studied consequence of ingesting allelochemicals. Effects on digestibility are easily translated into a measurable cost (reduced protein digestion, reduced fibre digestion, reduced dry matter digestion) that, in theory, should allow simple, quantifiable comparisons between animals.

However, several recent reviews of tannins have failed to identify any clear effects on digestibility (Mole and Waterman, 1987; Bernays *et al.*, 1989). In those cases where effects on digestibility of dietary components have been identified, the mechanism of the effects is not well understood. Inhibitory effects of tannins on digestive enzymes have often been suggested as an important mechanism, but several studies have questioned this because of the apparent lack of free tannin available to bind digestive enzymes in the small intestine or caecum-colon (Foley and Hume, 1986; Mole and Waterman, 1987; Blytt *et al.*, 1988) or because of the apparent insensitivity of membrane-bound enzymes to condensed tannins (Blytt *et al.*, 1988). The limited effects of tannins on membrane-bound enzymes observed by Blytt *et al.* (1988) suggest that other membrane-associated processes such as nutrient absorption could also be little affected by tannins. There are few data to evaluate this possibility, but Karasov *et al.* (1992) showed that acute (but not sub-chronic) exposure to the hydrolysable tannin tannic acid inhibited carrier-mediated intestinal uptake of glucose and amino acids in mice. Although Karasov *et al.* (1992) stressed the complex nature of these processes, it may be worthwhile to use this approach in a range of species fed a range of different tannins.

However, digestive costs are often only part of the overall cost of ingesting tannins. There are two other aspects which must be considered. One relates to the fate of the tannin; the other relates to the overall effect of tannin on nutritional components of the diet. Unless it can be demonstrated that no part of the tannin fraction in a diet is degraded and absorbed, the cost is incomplete without considering post-absorptive or total metabolic effects. In some cases, it does appear that the tannin is fully recoverable in faeces, but this is by no means universal. Hydrolysable tannins in particular are often degraded. When tannins become functional toxins, part of the cost of ingesting them will be related to the processes leading to their excretion. As Robbins *et al.* (1991) pointed out, animals with tannin-binding salivary proteins may reduce the toxic cost of tannins. The measurement of toxic effects and costs is discussed in the next section.

Metabolic effects

The physiological effects of allelochemicals on mammalian browsers are poorly defined. Some allelochemicals are so toxic that they kill some herbivores very rapidly. However, this is probably relatively rare. Of greater ecological importance is the concept of toxicity at the sub-acute or chronic level; it has been widely argued that allelochemicals can restrict the amount of food consumed by a herbivore because animals have a limited capacity to detoxify allelochemicals (Freeland and Janzen, 1974; Freeland and Saladin, 1989).

Recently, two major physiological effects linked to the ingestion of allelochemicals have been identified. The first of these effects is a disturbance to the acid–base balance of marsupials fed *Eucalyptus* foliage and the second is a disturbance of sodium balance in lagomorphs fed woody shrubs.

Acidosis in folivorous marsupials

Studies of arboreal marsupials fed *Eucalyptus* foliage have shown that animals eating the leaves of some trees (e.g. *E. radiata, E. citriodora, E. dives*) excrete an acid urine whereas an alkaline urine is excreted following the ingestion of other species (e.g. *E. ovata*). Acid urines are characterized by a high level of ammonium and a low concentration of urea, a pattern characteristic of a metabolic acidosis. Foley (1992) argued that the acid load arises from the detoxification of terpenes and phenolics that have been absorbed from the diet. Subsequent chemical studies have confirmed this; the major urinary acids are derivatives of terpenes and phenolics found in the diet (McLean *et al.*, 1993; L. Johnson and W. J. Foley, unpublished data). The acid load is formed within hours of a change of diet and thus is clearly diet-related. This pattern has now been established in common ringtail possums, greater gliders and koalas (W. J. Foley, unpublished data) and suggests that diet-induced acidosis is a physiologically normal state in these animals.

Acid–base disturbances have wide-ranging effects on many organs and metabolic processes. For example, the conservation of nitrogen by urea cycling is diminished, muscles may be catabolized when nitrogen intakes are low and the animal is acidotic, and an animal's ability to concentrate urine may be lowered.

While there are certainly some differences between species in the metabolism of allelochemicals, the end-products are always strong organic acids. Therefore, Foley (1992) argued that effects similar to those seen in marsupials should be observed in other species consuming diets rich in allelochemicals.

Sodium wastage in rabbits

When lagomorphs are fed diets of browse or browse extracts, there is a marked increase in urinary sodium losses. For example, in mountain hares (*Lepus timidus*) fed a range of browse diets, urinary sodium losses resulted in a negative sodium balance (Pehrson, 1983). Similar results have been reported in mountain hares fed heather (G. Iason, unpublished data), mountain hares and European hares (*Lepus europaeus*) fed extracts of birch (Iason and Palo, 1991) and in snowshoe hares fed several browse species (Reichardt *et al.*, 1984).

What is surprising about all these results is that most of the sodium loss occurs via the urine. Normally, urinary losses of sodium are closely regulated by the renal system and the kidney would be expected to reabsorb sufficient filtered sodium to prevent negative sodium balance – unless that loss was obligatory. Pehrson (1983) suggested that negative sodium balance could occur through one of two mechanisms. Sodium imbalance could be due to the effects of a high dietary potassium intake. However, there is no evidence that the levels of potassium in any of these diets was notably high. Pehrson's second hypothesis was that many browse plants contain (unspecified) compounds that lead to sodium leakage. A third possibility is that sodium is lost as a result of an acidosis similar to that described in marsupials.

In contrast to most other species, lagomorphs appear to have a reduced capacity for ammoniagenesis during acidosis (Richardson *et al.*, 1978). We predict that all the lagomorph species in which sodium imbalance has been observed were excreting a load of organic acids from detoxified phenolics. Given the limited ability of lagomorphs to augment urinary ammonium, it may be that sodium is exchanged for hydrogen and excreted in the urine and that this is responsible for the negative sodium balance.

If sodium wastage in lagomorphs has a similar cause to the acidosis seen in folivorous marsupials, we will be better able to understand the effects of absorbed allelochemicals in mammals. Instead of having to understand the effects of the many different allelochemicals in the diet, we may be able to focus on the end-products of their metabolism. Since acid–base homeostasis is the most important regulatory necessity of any animal, processes that threaten it should be closely controlled. In particular, the speed at which acid metabolites are excreted is important. Acidic metabolites must not be allowed to accumulate and drive down systemic pH. In the next section, we briefly review the importance of kinetic studies of allelochemical excretion.

Rate of elimination of allelochemicals

The toxicity of an absorbed allelochemical depends on its concentration at the sites at which it causes damage. The rate at which it can be transformed or eliminated from the body is therefore an important factor in determining the degree of toxicity.

The speed of excretion of allelochemicals depends on a wide range of factors such as whether the allelochemical is metabolized before absorption, its lipid solubility and whether the metabolite is excreted in bile or urine or both. Details of these processes can be found in most recent textbooks of pharmacology (e.g. Klaasen and Rozman, 1991).

However, from an ecological perspective, the effect of body size on detoxification processes is particularly important. Walker (1978) has shown a relationship between body mass of mammals and the rate of metabolism of xenobiotics. Smaller species are capable of more rapid detoxification of xenobiotics than larger species. Freeland (1991) has used these data to argue that small species are more likely to evolve specialized food habits because they can maintain lower concentrations of allelochemicals in the plasma.

A slow rate of excretion of an allelochemical from the body increases the potential toxicity of the compound (Klaasen and Rozman, 1991), and therefore increases the cost. We suggest that examining the elimination characteristics of allelochemicals will be useful for estimating costs of consumption of simple or mixed diets. For example, explaining the limited intake of a single plant species may depend on establishing the intake of allelochemicals in that diet at which the detoxification or excretion pathways are saturated. Similarly, a common argument to explain the mixing of certain plant species in the diet of a herbivore is that it avoids overloading any one particular pathway for detoxification and elimination (Freeland and Janzen, 1974). Comparing elimination characteristics of allelochemicals from single plant species may reveal excretory differences related to pathway saturation. Intuitively, saturation of pathways at lower intakes implies greater costs for the consumer.

These ideas have been canvassed before (e.g. Freeland and Janzen, 1974; Freeland, 1991) but we are not aware of any studies that have attempted to explain the differential utilization of plants by mammals in these terms. We believe that this approach could be of great utility in studying the effects of allelochemicals on food selection and intake, and we urge its adoption.

Energy budgets and energy metabolism

Energy has been used as the currency for assessing the costs of allelochemicals in a number of studies (Cook *et al.*, 1952; Cork *et al.*, 1983; Foley, 1987). The advantages of doing so are that the energy intake and expenditure of the whole animal can be measured readily. Decreases in the energy available to the animal then represents an integrated cost of the ingestion and excretion of allelochemicals. The nutritional value of foods has traditionally been measured in terms of the metabolizable or net energy yield and so, in theory at least, it should be possible to express the costs of ingestion and excretion of allelochemicals in an ecologically meaningful way.

For example, Cook *et al.* (1952) and Foley (1987) showed that animals fed tree leaves rich in terpenes lost 40–50% of the digestible energy in the urine whereas on other diets the loss was only 10–15%. The difference was attributed to the excretion of metabolites of the ingested terpenes. On the surface, the cost of excreting terpenes appears to be enormous but it should be remembered that the gross energy of terpene-rich diets is significantly greater than other diets. Therefore, the net cost of excreting terpenes may not be nearly so great. Without detailed chemical examination of the urine, it is difficult to apportion the urinary energy loss to determine the 'energy increment' of the excretion of absorbed allelochemicals from these sort of data.

One disadvantage of measuring costs in terms of overall energy retention is that the results tell us little about the energy-demanding processes involved. Of more value are concurrent measures of heat production, but even then it is hard to attribute changes in whole animal heat production to the effects of a particular allelochemical.

This approach has been little used but one example illustrates both the utility and problems of such measures. It has been known for some time that the previous nutritional history of an animal can affect its basal metabolic rate (e.g. Marston, 1948). However, Thomas *et al.* (1988) observed that when voles (*Microtus pensylvannicus*) were maintained on diets containing allelochemicals, fasting or basal metabolic rate was significantly higher than in animals fed diets lacking allelochemicals. These data suggest that there is some carry-over effect of the allelochemicals that affects basal energy expenditure. It is possible that part of the reason is an increased protein synthesis for the production of biotransformational enzymes.

In future it will be necessary to partition the effects of allelochemicals far more carefully when measuring energy retention and heat production. For example, some allelochemicals can be excreted without metabolism because of their low pK_a and high water solubility (Scheline, 1978). Is there a measurable increment in heat production as a result? Manipulation of the routes of administration of allelochemicals (such as direct infusion into the bloodstream) may allow measurement of the costs of microbial detoxification. Careful choice of model allelochemicals may allow the costs of Phase I transformations to be measured separately from Phase II conjugations.

Specific nutrients and excretion of allelochemicals

The excretion of allelochemicals from the body may involve a cost in terms of specific nutrients. In mammals, allelochemicals are most often excreted conjugated to small molecules. The conjugating moiety is either a sulphate group (derived from sulphur-containing amino acids), glycine or glucuronic acid. The type of conjugate excreted depends on the nature of the allelochemical, the availability of the specific nutrient and the diet/digestive physiology of the animal.

It is possible to express the cost of excretion of allelochemicals in terms of the amount of conjugate present in the urine (or faeces). For example, Cork (1981) measured urinary glucuronic acid excretion in koalas fed an allelochemical-rich diet (*E. punctata* foliage) and calculated that it represented about 20% of the animals' fasting glucose production. Although it seems relatively simple to measure the quantity of conjugated product in this way and then determine how much nitrogen or carbohydrate it represents, in practice the approach is complicated by a number of factors. These include different routes of excretion and interconversion of the conjugating moieties within the body. For example, many glucuronide conjugates of terpenes are excreted partially in the urine and partially via the bile. Secondly, when loads of allelochemicals are high, the pathways of excretion may be different than at lower doses (Møller and Sheikh, 1983; Klaasen and Rozman, 1991).

Nonetheless, when excretory pathways are well known and when the sources of allelochemicals in the diet are relatively simple, this approach may have considerable utility. For example, Lowry and Sumpter (1987) fed sheep a range of tropical grasses containing varying amounts of phenolic acids (mainly *p*-coumaric and ferulic acids). Following absorption, these acids were excreted as the glycine-conjugated metabolite, hippuric acid. Only minor amounts of other metabolites were observed (e.g. benzoyl glucuronide). Lowry and Sumpter (1987) estimated that urinary excretion of hippurate represented up to 16% of the nitrogen ingested and a significantly greater proportion of the digested nitrogen. In cases where nitrogen availability is limited, this represents a significant cost.

Effects on growth

The ingestion of allelochemicals has been shown to have significant effects on the growth and survival of herbivores. For example, Jung and Batzli (1981) found that voles grew poorly and suffered significant mortality when fed extracts of a range of unpalatable arctic plants. Similarly, Lindroth and Batzli (1984) and Lindroth *et al.* (1986) observed reduced growth rates in voles fed a range of natural phenolics. Although these effects appear to be easily quantifiable, it is difficult to attribute them to any particular action of allelochemicals. This is illustrated by a recent study which showed that warfarin-resistant rats grew much more slowly than other rats but that the growth reduction was connected to differences in metabolic pathways rather than to the 'cost of detoxification' (Smith *et al.*, 1991).

Conclusions

Our understanding of the effects of allelochemicals on mammalian herbivores is still in its infancy. We have learnt that simple classification of plant allelochemicals as either toxins or digestibility reducers is no longer tenable and future theories of plant defence and foraging must accommodate a more dynamic view of the effects of allelochemicals. The effects of any allelochemical on a mammalian consumer will depend on the methods it uses to counteract and excrete the particular compound. Ultimately we need to understand both the chemistry of the allelochemicals and the physiology of the consumer.

Incorporating these ideas into foraging and plant defence theory may be difficult as we cannot yet make broad generalizations about the effects or costs of allelochemicals. Nonetheless, several of the areas identified in this review, such as the occurrence and utility of tannin-binding salivary proteins, the effects of absorbed allelochemicals on acid–base status and the possibility of differences in the kinetics of metabolism and excretion of allelochemicals between species, have the potential to provide this integrating framework. We urge that future studies of the interaction between mammalian herbivores and their food plants try to develop new ways to evaluate the costs of ingestion and excretion of allelochemicals.

References

Austin, P. J., Suchar, L. A., Robbins, C. T. & Hagerman, A. E. (1989). Tannin-binding proteins in saliva of deer and their absence in saliva of sheep and cattle. *Journal of Chemical Ecology*, **15**, 1335–1347.

Belovsky, G. E. & Schmitz, O. J. (1991). Mammalian herbivore optimal foraging and the role of plant defenses. In *Plant Defenses Against Mammalian Herbivory*, ed. R. T. Palo & C. T. Robins, pp. 1–28. Boca Raton, FL: CRC Press.

Bennick, A. (1982). Salivary proline-rich proteins. *Molecular and Cellular Biochemistry*, **45**, 83–99.

Berenbaum, M. (1980). Adaptive significance of midgut pH in larval lepidoptera. *American Naturalist*, **115**, 138–146.

Bergeron, J. M. & Jodoin, L. (1991). Costs of nutritional constraints on the vole (*Microtus pennsylvanicus*) along a time gradient. *Canadian Journal of Zoology*, **69**, 1496–1503.

Bernays, E. A., Cooper-Driver, G. & Bilgener, M. (1989). Herbivores and plant tannins. *Advances in Ecological Research*, **19**, 263–302.

Blytt, H. J., Guscar, T. J. & Butler, L. G. (1988). Antinutritional effects and ecological significance of dietary condensed tannins may not be due to binding and inhibiting digestive enzymes. *Journal of Chemical Ecology*, **14**, 1455–1465.

Boyd, J. N. & Campbell, T. C. (1983). Impact of nutrition on detoxication. In *Biological Basis of Detoxification*, ed. J. Caldwell & W. B. Jakoby, pp. 287–306. New York: Academic Press.

Brattsten, L. B. (1992). Metabolic defenses against plant allelochemicals. In *Herbivores: Their Interactions with Secondary Plant Metabolites*, 2nd edn.,

Vol. II: Ecological and Evolutionary Processes, ed. G. A. Rosenthal & M. R. Berenbaum, pp. 176–242. New York: Academic Press.

Bryant, J. P., Kuropat, P. J., Reichardt, P. B. & Clausen, T. P. (1991). Controls over the allocation of resources by woody plants to chemical antiherbivore defense. In *Plant Defenses Against Mammalian Herbivory*, ed. R. T. Palo & C. T. Robbins, pp. 83–102. Boca Raton FA: CRC Press.

Bryant, J. P., Reichardt, P. B., Clausen, T. P., Provenza, F. D. & Kuropat, P. J. (1992). Woody plant–mammal interactions. In *Herbivores: Their Interactions with Secondary Plant Metabolites*, 2nd edn., Vol. II: Ecological and Evolutionary Processes, ed. G. A. Rosenthal & M. R. Berenbaum, pp. 344–371. New York: Academic Press.

Caldwell, J. (1982). The conjugation reactions in foreign compound metabolism: definition, consequences and species variation. *Drug Metabolism Review*, **13**, 745–778.

Carrlson, J. R. & Dickinson, E. O. (1978). Tryptophan-induced pulmonary edema and emphysema in ruminants. In *Effects of Poisonous Plants on Livestock*, ed. R. F. Keeler, K. R. van Kampen & L. F. James pp. 261–269. New York: Academic Press.

Cheeke, P. R. (1984). Comparative toxicity and metabolism of pyrrolizidine alkaloids in ruminant and non-ruminant herbivores. *Canadian Journal of Animal Science*, **64**(S), 201–202.

Clausen, T. P., Provenza, F. D., Burritt, E. A., Reichardt, P. B. & Bryant, J. P. (1990). Ecological implication of condensed tannin structure: a case study. *Journal of Chemical Ecology*, **16**, 2381–2391.

Coley, P. D., Bryant, J. P. & Chapin, F. S. III (1985). Resource availability and plant antiherbivore defense. *Science*, **230**, 895–899.

Connolly, G. E., Ellison, B. O., Fleming, J. W. *et al.* (1980). Deer browsing of Douglas-fir trees in relation to volatile terpene composition and in vitro fermentability. *Forest Science*, **26**, 179–193.

Cook, C. W., Stoddart, L. A. & Harris, L. E. (1952). Determining the digestibility and metabolizable energy of winter range plants by sheep. *Journal of Animal Science*, **11**, 578–590.

Cork, S. J. (1981). *Digestion and Metabolism in the Koala (Phascolarctos cinereus Goldfuss): an Arboreal Folivore*. Sydney, NSW: University of New South Wales.

Cork, S. J. & Foley, W. J. (1991). Digestive and metabolic strategies of arboreal mammalian folivores in relation to chemical defences in temperate and tropical forests. In *Plant Defenses Against Mammalian Herbivory*, ed. R. T. Palo & C. T. Robbins, pp. 133–166. Boca Raton, FL: CRC Press.

Cork, S. J., Hume, I. D. & Dawson, T. J. (1983). Digestion and metabolism of a natural foliar diet (*Eucalyptus punctata*) by an arboreal marsupial, the koala (*Phascolarctos cinereus*). *Journal of Comparative Physiology*, **152**, 443–451.

DaSilva, G. L. (1992). The western black-and-white Colobus as a low energy strategist – activity budgets, energy expenditure and energy intake. *Journal of Animal Ecology*, **61**, 79–91.

Feeny, P. P. (1976). Plant apparency and chemical defense. *Recent Advances in Phytochemistry*, **10**, 1–40.

Foley, W. J., (1987). Digestion and metabolism in a small arboreal marsupial, the greater glider, *Petauroides volans* fed high terpene *Eucalyptus* foliage. *Journal of Comparative Physiology*, **157**, 355–362.

Foley, W. J. (1992). Nitrogen and energy retention and acid-base status in the common ringtail possum (*Pseudocheirus peregrinus*): evidence of the effects of absorbed allelochemicals. *Physiological Zoology*, **65**, 403–421.

Foley, W. J. & Hume, I. D. (1986). Digestion and metabolism of high-tannin *Eucalyptus* foliage by the brushtail possum (*Trichosurus vulpecula*) (Marsupialia: Phalangeridae). *Journal of Comparative Physiology*, **157**, 67–76.

Foley, W. J., Lassak, E. V. & Brophy, J. (1987). Digestion and absorption of *Eucalyptus* essential oils in Greater Glider (*Petauroides volans*) and Brushtail Possum (*Trichosurus vulpecula*). *Journal of Chemical Ecology*, **13**, 2115–2130.

Freeland, W. J. (1991). Plant secondary metabolites: biochemical coevolution with herbivores. In *Plant Defenses Against Mammalian Herbivory*, ed. R. T. Palo & C. T. Robbins, pp. 61–82. Boca Raton, FL: CRC Press.

Freeland, W. J. & Janzen, D. H. (1974). Strategies in herbivory by mammals: the role of plant secondary compounds. *American Naturalist*, **108**, 269–289.

Freeland, W. J. & Saladin, L. R. (1989). Choice of mixed diets by herbivores: the idiosyncratic effects of plant secondary compounds. *Biochemical Systematics and Ecology*, **17**, 493–497.

Galtier, P. & Alvinerie, M. (1976). In vitro transformation of ochratoxin A by animal microbial floras. *Annals de la Recherche Vétérinaire*, **7**, 91–98.

Glick, Z. & Joslyn, M. A. (1970). Effect of tannic acid and related compounds on the absorption and utilization of proteins in the rat. *Journal of Nutrition*, **100**, 516–520.

Hagerman, A. E. & Butler, L. G. (1981). The specificity of proanthocyanidin–protein interactions. *Journal of Biological Chemistry*, **256**, 4494–4497.

Hagerman, A. E. & Robbins, C. T. (1993). Specificity of tannin-binding salivary proteins relative to diet selection by mammals. *Canadian Journal of Zoology*, in press.

Hagerman, A. E., Robbins, C. T., Weerasuriya, Y., Wilson, T. C. & McArthur, C. (1992) Tannin chemistry in relation to digestion. Journal of Range Management, **45**, 57–62.

Iason, G. & Palo, R. T. (1991). Effects of birch phenolics on a grazing and a browsing mammal: a comparison of hares. *Journal of Chemical Ecology*, **17**, 1733–1743.

Janzen, D. H. (1979). New horizons in the biology of plant defenses. In *Herbivores: their Interactions with Secondary Plant Metabolites*, ed. G. A. Rosenthal & D. H. Janzen, pp. 331–351. New York: Academic Press.

Jones, R. J. & Megarrity, R. G. (1986). Successful transfer of DHP-degrading bacteria from Hawaiian goats to Australian ruminants to overcome the toxicity of *Leucaena*. *Australian Veterinary Journal*, **63**, 259–262.

Jung, H. J-G. & Batzli, G. O. (1981) Nutritional ecology of microtine rodents. Effect of plant extracts on the growth of Arctic microtines. *Journal of Mammalogy*, **62**, 386–392.

Karasov, W. H., Meyer, M. W. & Darken, B. W. (1992). Tannic acid inhibition of amino acid and sugar absorption by mouse and vole intestine: tests following acute and subchronic exposure. *Journal of Chemical Ecology*, **18**, 719–736.

King, D. R. Oliver, A. J. & Mead, R. J. (1978). The adaptation of some Western Australian mammals to food plants containing fluroacetates. *Australian Journal of Zoology*, **26**, 699–712.

Klaasen, C. D. & Rozman, K. (1991). Absorption, distribution, and excretion of toxicants. In *Toxicology. The Basic Science of Poisons*, ed. M. O. Amdur, J. Doull & C. D. Klassen, pp. 50–87. New York: Pergamon Press.

Langer, P. (1986). Large mammalian herbivores in tropical forests with either hindgut or forestomach fermentation. *Zeitschrift für Säugetierkunde*, **51**, 173–187.

Lanigan, G. W. & Smith, L. W. (1970). Metabolism of pyrrolizidine alkaloids in the ovine rumen. I. Formation of 7-alpha-hydroxy-alpha-methyl-8-alpha-pyrrolizidine from heliotropine and basicocarpine. *Australian Journal of Agricultural Research*, **21**, 493–500.

Lindroth, R. L. (1991). Differential toxicity of plant allelochemicals to insects; roles of enzymatic detoxication systems. In *Insect–Plant Interactions*, Vol. III, ed. E. Bernays, pp. 1–33. Boca Raton, FL: CRC Press.

Lindroth, R. L. & Batzli, G. (1983). Detoxication of some naturally occurring phenolics by prairie voles. A rapid assay of glucuronidation metabolism. *Biochemical Systematics and Ecology*, **11**, 405–409.

Lindroth, R. L. & Batzli, G. (1984). Plant phenolics as chemical defences: effects of natural phenolics on survival and growth of prairie voles (*Microtus ochragaster*). *Journal of Chemical Ecology*, **10**, 229–244.

Lindroth, R. L., Batzli, G. O. & Avildsen, S. I. (1986). *Lespedeza* phenolics and *Penstemon* alkaloids; effects on digestion efficiencies and growth of voles. *Journal of Chemical Ecology*, **12**, 713–728.

Lowry, J. B. & Sumpter, E. A. (1987). Hippuric acid output of sheep fed four tropical grasses. *Proceedings of the Australian Society for Animal Production*, **17**, 433.

Madapallimattan, G. & Bennick, A. (1990). Phosphopeptides derived from human salivary acidic proline-rich proteins. *Biochemical Journal*, **270**, 297–304.

Mandel, I. D. (1987). The functions of saliva. *Journal of Dental Research*, **66**, 623–627.

Mandel, I. D. (1989). The role of saliva in maintaining oral homeostasis. *Journal of the American Dental Association*, **119**, 298–304.

Marston, H. R. (1948). Energy transactions in the sheep. *Australian Journal of Scientific Research*, **B1**, 93–98.

Martin, M. M. & Martin, J. S. (1984). Surfactants: their role in preventing the precipitation of proteins by tannins in insect guts. *Oecologia*, **61**, 342–345.

McArthur, C., Hagerman, C. T. & Robbins, C. T. (1991). Physiological strategies of mammalian herbivores against plant defences. In *Plant Defenses Against Mammalian Herbivory*, ed. R. T. Palo & C. T. Robbins, pp. 103–114. Boca Raton, FL: CRC Press.

McArthur, C. & Sanson, G. D. (1991). Effects of tannins on digestion in the common ringtail possum (*Pseudocheirus peregrinus*), a specialized marsupial folivore. *Journal of Zoology* (London), **225**, 233–252.

McKey, D. B., Gartlan, J. S., Waterman, P. G. & Choo, G. M. (1981). Food selection by black colobus monkeys (*Colobus satanas*) in relation to plant chemistry. *Biological Journal of the Linnean Society*, **16**, 115–146.

McLean, S., Foley, W. J., Davies, N. W., Brandon, S., Duo, L. & Blackman, A. J. (1993). The metabolic fate of dietary terpenes from *Eucalyptus radiata* in the common ringtail possum (*Pseudocheirus peregrinus*). *Journal of Chemical Ecology*, in press.

Mehansho, H., Ann, D. K., Butler, L. G., Rogler, J. C. & Carlson, D. M. (1987a). Induction of proline-rich proteins in hamster salivary glands by isoproterenol treatment and an unusual growth inhibition by tannins. *Journal of Biological Chemistry*, **262**, 12344–12350.

Mehansho, H., Butler, L. G. & Carlson D. M. (1987b). Dietary tannins and salivary proline-rich proteins: interactions, induction, and defense mechanisms. *Annual Review of Nutrition*, **7**, 423–440.

Mehansho, H., Clements, S., Sheares, B. T., Smith, S. & Carlson, D. M. (1985). Induction of proline-rich glycoprotein synthesis in mouse salivary glands by isoproterenol and by tannins. *Journal of Biological Chemistry*, **260**, 4418–4423.

Mehansho, H., Hagerman, A. E., Clements, S., Butler, L., Rogler, J. & Carlson, D. M. (1983). Modulation of proline-rich protein biosynthesis in rat parotid glands by sorghums with high tannin levels. *Proceedings of the National Academy of Sciences, USA*, **80**, 3948–3952.

Meyer, M. & Karasov, W. H. (1991). Chemical aspects of herbivory in arid and semi-arid habitats. In *Plant Defenses Against Mammalian Herbivory*, ed. R. T. Palo & C. T. Robbins, pp. 167–188. Boca Raton, FL: CRC Press.

Mole, S., Butler, L. G. & Iason, G. (1990). Defense against dietary tannin in herbivores: a survey for proline rich salivary proteins in mammals. *Biochemical Systematics and Ecology*, **18**, 287–293.

Mole, S. & Waterman, P. (1987). Tannins as anti-feedants to mammalian herbivores – still an open question? In *Allelochemicals: Role in Agriculture and Forestry (A.C.S. Symposium 330)*, ed. G. R. Waller, pp. 572–587. Washington, DC: American Chemical Society.

Møller, J. V. & Sheikh, M. I. (1983). Renal organic anion transport system: pharmacological, physiological and biochemical aspects. *Pharmacology Reviews* **34**, 315–318.

Nagy, J. G. & Tengerdy, R. P. (1968). Antibacterial action of essential oils of Artemesia as an ecological factor. II Antibacterial action of the volatile oils of *Artemesia tridentata* (big sagebrush) on bacteria from the rumen of mule deer. *Applied Microbiology*, **16**, 441–444.

O'Brien, T. P., Lomdahl A. & Sanson, G. (1986). Preliminary microscopic investigations of the digesta derived from foliage of *Eucalyptus ovata* (Labill.) in the digestive tract of the common ringtail possum, *Pseudocheirus peregrinus* (Marsupialia). *Australian Journal of Zoology*, **34**, 157–176.

Oh, H. K., Jones, M. B. & Longhurst, W. M. (1968). Comparison of rumen microbial inhibition resulting from various essential oils isolated from relatively unpalatable plant species. *Applied Microbiology*, **16**, 39–44.

O'Halloran, M. W. (1962). The effect of oxalate on bacteria isolated from the rumen. *Proceedings of the Australian Society for Animal Production*, **4**, 18.

O'Hara, P. J. & Fraser, A. J. (1975). Nitrate poisoning in cattle grazing crops. *New Zealand Veterinary Journal*, **23**, 45–53.

Osawa, R. (1992). Tannin-protein complex degrading enterobacteria isolated from the alimentary tracts of koalas and a selective medium for their enumeration. *Applied and Environmental Microbiology*, **58**, 1754–1759.

Palo, R. T. & Robbins, C. T. (1991). *Plant Defenses Against Mammalian Herbivory*. Boca Raton, FL: CRC Press.

Parke, D. V. & Ioannides, C. (1981). The role of nutrition in toxicology. *Annual Review of Nutrition*, **1**, 207–219.

Pehrson, A. (1983). Digestibility and retention of food components in caged mountain hares *Lepus timidus* during the winter. *Holarctic Ecology*, **6**, 395–403.

Reichardt, P. B., Bryant, J. P., Clausen, T. P. & Wieland, G. D. (1984). Defense of winter-dormant Alaska paper birch against snowshoe hares. *Oecologia*, **65**, 58–69.

Rhoades, D. F. & Cates, R. G. (1976). Toward a general theory of plant antiherbivore chemistry. In *Recent Advances in Phytochemistry* ed. J. W. Wallace & R. L. Mansell, pp. 168–213. New York: Plenum.

Richardson, R. M. A., Goldstein, M. B., Stinebaugh, B. J. & Halperin, M. L. (1978). Influence of diet and metabolism on urinary acid excretion in the rat and rabbit. *Journal of Laboratory and Clinical Medicine*, **94**, 510–518.
Robbins, C. T., Hagerman, A. E., Austin, P. J., McArthur, C. & Hanley, T. A. (1991). Variation in mammalian physiological responses to a condensed tannin, and its ecological implications. *Journal of Mammalogy*, **72**, 480–486.
Robbins, C. T., Hanley, T. A., Hagerman, A. E. *et al.* (1987). Role of tannins in defending plants against ruminants: reduction in protein availability. *Ecology*, **68**, 98–107.
Russell, G. R. & Smith, R. M. (1968). Reduction of heliotrine by a rumen micro-organism. *Australian Journal of Biological Science*, **21**, 1277–1290.
Scheline, R. R. (1978). *Mammalian Metabolism of Plant Xenobiotics*. London: Academic Press.
Smith, G. S. (1986). Gastrointestinal toxifications and detoxifications in ruminants in relation to resource management. In *Gastrointestinal Toxicology*, ed. K. Rozman & O. Hanninen pp. 514–542. New York: Elsevier.
Smith, G. S. (1992). Toxification and detoxification of plant compounds by ruminants: an overview. *Journal of Range Management*, **45**, 25–29.
Smith, P. Townsend, M. G. & Smith, R. H. (1991). A cost of resistance in the brown rat? Reduced growth rate in warfarin-resistant lines. *Functional Ecology*, **5**, 441–447.
Thomas, D. W., Samson, C. & Bergeron, J. M. (1988). Metabolic costs associated with the ingestion of plant phenolics by *Microtus pennsylvanicus*. *Journal of Mammalogy*, **69**, 512–515.
Walker, C. H. (1978). Species differences in microsomal monooxygenase activity and their relationship to biological half lives. *Drug Metabolism Review*, **7**, 295–323.
Waterman, P. G., Ross, J. A. M., Bennett, E. L. & Davies, A. G. (1988). A comparison of the floristics and leaf chemistry of the tree flora in two Malaysian rain forests and the influence of leaf chemistry on populations of colobine monkeys in the Old World. *Biological Journal of the Linnean Society*, **34**, 1–16.
Welch, B. L., Narjisse, H. & McArthur, E. D. (1982). *Artemisia tridentata* monoterpenoid effect on ruminant digestion and forage selection. In *Aromatic Plants: Basic and Applied Aspects*, ed. N. Margaris, A. Koedam & D. Vokou. pp. 73–84. Boston: Martinus Nijhoff.

23

Short-chain fatty acids as a physiological signal from gut microbes

TAKASHI SAKATA

Many animal species depend on microbial activities in their digestive tract for a large part of their metabolic energy (Parra, 1978). Short-chain fatty acids (SCFA) such as acetic, propionic, butyric and valeric acids represent energy-carrying nutrients produced by gut fermentation (Bugaut, 1987).

These acids are major anions in the gut contents where fermentation takes place (Parra, 1978). The contribution of these acids to the host's energy economy varies among animals of different food habits. Nevertheless, all reptiles, birds and mammals possess some gut fermentation (Stevens, 1988). SCFA always exist at significant concentrations in gut fermentation chambers. Therefore, gut fermentation and the production of SCFA in the gut can be considered as part of the basic characteristics of land vertebrates.

Regulation of energy metabolism of 'classic' nutrients such as carbohydrates, proteins and lipids has been studied extensively. There are neural and humoral mechanisms that regulate the catabolism and anabolism of these nutrients to maintain their levels in the blood within an appropriate range. In contrast, relatively little is understood about regulatory mechanisms of energy metabolism involving SCFA, even though these acids are the main energy source in large herbivorous animals.

We humans, for instance, adjust our appetite according to blood sugar levels, which reflect the systemic energy balance. In other words, we have mechanisms to control the uptake of energy from classic nutrients. However, we do not know if we can control microbial activity based on the entry rate of SCFA into the host's body. In this regard, it may be plausible to ask how animals recognize bacterial activity?

It can be postulated that SCFA, more than gases (carbon dioxide, methane and hydrogen), pH and microbial cell biomass, reflect the level of bacterial activity in the digestive tract. If this is the case, then the host animal should be equipped with receptor mechanisms connected to transmission and signal-processing mechanisms for SCFA in order to assess and react to changes in microbial activity. In this chapter we discuss information on responses of animals to SCFA, sensory mechanisms of animals for SCFA, and transmission mechanisms for the resultant signals.

Effects of SCFA on motility of the ruminant digestive tract

In ruminants, SCFA generally inhibit the motility of the digestive tract when applied to the reticulorumen, abomasum, duodenum or caecum, especially at higher concentrations and at low pH (Table 23.1). Butyric acid has a stronger inhibitory action than that of acetic or propionic acid (Svendsen, 1972, 1974).

Earlier workers considered that these inhibitory effects were the result of direct local action of SCFA on smooth muscle cells (Svendson, 1974; Bolton *et al.*, 1976). However, there are dense networks of blood vessels in the lamina propria and sub-mucosa (Schnorr and Vollmerhaus, 1968) that should remove most of the SCFA absorbed from the lumen, leaving little to reach the underlying smooth muscle layers. Further, SCFA inhibit the motility of segments distant from the site of administration: SCFA administered into the duodenum inhibit the motility of the reticulorumen and the abomasum (Ehrlein and Hill, 1970). Therefore, the action of SCFA must be transmitted via a systemic mechanism. In addition, SCFA given into the rumen inhibit its motility in vagotomized sheep (Gregory, 1987a), indicating that SCFA can also inhibit motility via direct action on the local gastro-intestinal nerve system. These inhibitory actions in ruminants may explain the gastro-intestinal atony that results from the consumption of highly fermentable feed (Svendsen, 1975).

In spite of many studies that show inhibitory effects of SCFA, some workers demonstrated that SCFA stimulate phasic contraction of the rumen or caecum when SCFA are given into these organs. However, this stimulatory effect occurs only after the entire removal of the contents of the organ and washing of the site of application (Svendsen, 1972; Ruckebusch and Tomov, 1973). This difference in motility response could be explained by rapid adaptation of SCFA-receptors as found in the rat colon (Yajima, 1985).

The existence of a receptor mechanism for the presence of SCFA in the lumen and transmission of this information via the autonomic nervous system to the centre in the central nerve system that controls the gut motility has been proposed (Leek and Harding, 1975).

It is possible to estimate the location of this receptor mechanism from the time lag between the dose and the response and the diffusion rate of SCFA in the gut mucosa. Such a calculation gave an estimated distance of 150 μm

Table 23.1. *Effects of SCFA on gut motility of ruminants*

Application site	Acid	Dose (mM)	Effector site	Response	Reference
Rumen	Acetic	200–400	Reticulorumen (vagotomized)	Inhibition	Gregory (1987b)
	Propionic	400		Inhibition	
	n-Butyric	400		Inhibition	
Rumen	Acetic	75 or 300 (pH 3.5–5.0)	Rumen	Inhibition	Ruckebusch and Tomov (1973)
	Propionic	75 or 300 (pH 3.5–5.0)		Inhibition	
	n-Butyric	75 or 300 (pH 3.5–5.0)		Inhibition	
Rumen	Acetic	75 (pH 5.5–5.9)	Rumen	Stimulation	Ruckebusch and Tomov (1973)
	Propionic	75 (pH 5.5–5.9)		Stimulation	
	n-Butyric	75 (pH 5.5–5.9)		Stimulation	
Rumen	Acetic	70	Rumen	Lowered	Svendsen (1975)
	Propionic	70		Lowered	
	n-Butyric	70		Lowered	
Rumen	SCFA-mixture	> 5	Rumen	Inhibition	Svendsen (1975)
Abomasum	Acetic	300	Pyloric antrum	Lowered	Bolton *et al.* (1976)
	Propionic	300		Lowered	
	n-Butyric	300		Lowered	
Abomasum	SCFA-mixture	70 or 120	Abomasum	Lowered	Svendsen (1975)
Duodenum	SCFA-mixture		Forestomach and abomasum	Lowered	Ehrlein and Hill (1970)
Caecum	SCFA-mixture	0.1 1.0	Caecum	Stimulation, no change	Svendsen (1972)
Caecum	Acetic	10	Caecum	Inhibition	Svendsen (1972)
	Propionic	10		Inhibition	
	n-Butyric	10			

between the lumen surface and the receptor; therefore, the receptor should be at the level of the basal membrane of the epithelium in the rumen (Leek and Harding, 1975). The time lag between the dose in the abomasum and duodenum is similar to that in the rumen, indicating that the location of the SCFA receptor mechanism should be similar to that in the rumen. However, nobody has found any anatomical structure responsible for such a receptor function.

SCFA in the abomasum stimulate reticulorumen motility via a neural reflex. This accompanies the activation of the gastric centre in the brain (Leek and Harding, 1975) and the excitation of the efferent gastric branch of the vagus (Iggo and Leek, 1970). Thus, stimuli of SCFA are mediated via a vago-vagal reflex to be processed in the brain, very likely at its gastric centre.

Effects of SCFA on motility of the digestive tract in non-ruminant animals

SCFA also affect gut motility in non-ruminant mammals. Propionic or butyric acid (10 mM) infused into the cleaned distal colon stimulated propulsive motility of the distal colon *in vivo* in rats (Yajima and Sakata, 1987). This effect must be due not to lowered pH but to SCFA anions, because the stimulatory effect was observed at neutral pH (Yajima and Sakata, 1987). Ileal infusion of acetic, butyric, hexanoic and caprylic acids shorten the stomach-to-caecum transit time in rats (Richardson *et al.*, 1991). SCFA stimulate ileal motility in dogs (Kamath *et al.*, 1987) and humans (Kamath *et al.*, 1988) as well.

Such stimulatory effects of SCFA can be reproduced *in vitro* (Yajima, 1985). Propionic, butyric and valeric acids (above 0.1 mM) stimulate the longitudinal phasic contraction of isolated segments of middle or distal colon, but not of proximal colon. Acetic and lactic acids have no stimulatory effect. The contractile response to SCFA begins approximately 10 s after the dose and fades out about 60 s later. Such a stimulatory effect takes place only when SCFA are applied from the mucosal side. Serosal application or the exposure of isolated muscle layers to SCFA do not exert any contractile response (Yajima and Sakata, 1987); instead, relaxation of muscle layers is observed. These results indicate that the effect of SCFA on colonic motility is not directly on smooth muscle cells but is mediated via sensory and mediatory mechanisms.

Yajima and Sakata (1987) observed a typical desensitization in the contractile response of isolated rat colon to propionate and butyrate. The colon does not react to additional increases in the dose of SCFA when it is already stimulated by the same acid or another acid that evokes a contractile response. Even pre-treatment of the mucosal side of the segment with acetic or lactic acid, which have no stimulatory effect, abolishes the response of the tissue to stimulatory SCFA. Such desensitization may, at least in part, explain the insensitivity to SCFA observed in the caecum and proximal colon where active production of SCFA takes place. This hypothesis is supported by the finding that propionic and butyric acids stimulate the motility of the proximal colon of germ-free rats and that acetic and lactic acids also stimulates the colonic motility of germ-free rats (Yajima and Sakata, 1987). Further, infusion of SCFA into the large intestine of germ-free rats desensitizes the caecum and the proximal colon (Yajima and Sakata, 1987).

Further *in vitro* studies of Yajima (1985) clarified the mediatory mechanism for this effect. The short time lag between the dose and response indicates that the receptor site should exist within a few micrometres from the epithelium. The inhibition by atropine, luminal procaine and tetrodotoxin of the stimulatory effect of SCFA on *in vitro* motility indicate that the effect of SCFA should involve the enteric nerve system, especially sensory nerves and cholinergic nerves.

Physiological significance of the effects of SCFA on gut motility

It is still difficult to discuss the physiological significance of the effect of SCFA on gut motility. Generally speaking, SCFA either have no effect or inhibit the motility of gut segments, such as the reticulorumen, caecum and proximal colon, where active fermentation takes place. Inhibition of phasic contractile responses in such segments might be effective in increasing retention time of substrates for fermentation. On the other hand, stimulation of contractile responses in the distal colon may facilitate the caudal transport of faecal pellets; the stimulation of ileal contractions by SCFA may be a protective mechanism to remove refluxed fermentation contents from the ileum (Richardson *et al.*, 1991).

It is of prime importance to know how SCFA influence tonic contractions of the digestive tract and so define retention times of digesta and the capacity of gut segments.

Mucosal blood flow of the digestive tract

More than 80% of blood flow to the rumen of a domestic ruminant drains the mucosa (Dobson, 1984). The rumen mucosa has a dense blood vascular network (Schnorr and Vollmerhaus, 1968) which supports absorptive and metabolic activities of the mucosa. Postprandial blood flow rate in the sheep rumen mucosa is more than 3.1 ml/min per g, which is comparable to that of the kidney or thyroid and exceeds that of the heart. The level of postprandial mucosal blood flow in the rumen depends on concentrations of SCFA and carbon dioxide produced by the rumen fermentation (Dobson, 1984); SCFA increase the mucosal blood flow of the rumen.

Microscopically, this increase in blood flow is mainly due to the distension of venules, but not of capillaries (Sakata and Tamate, 1987a). This situation differs from the widening of capillaries, but not of venules, of the rumen by the administration of a hypertonic (1 l of 2.65% sodium chloride) solution into the sheep rumen (Sakata and Tamate, 1987a). Therefore, the effect of SCFA on the mucosal blood flow must be due to the effect of SCFA on subepithelial venules. SCFA also stimulate the blood flow to the dog large intestine (Kvietys and Granger, 1967). The effect of SCFA on mucosal blood flow is considered to be local (Kvietys and Granger, 1967) but this remains to be tested.

In general, an increase in gut mucosal blood flow results in an increase in absorption rate (Winne, 1978). In the rumen, increased blood flow rate due to SCFA stimulated absorption of water and electrolytes (Dobson, 1984). Thus, the stimulatory effect of SCFA on mucosal blood flow of the gut segment where fermentation takes place stabilizes the chemical environment of the fermentation chamber by increasing the removal of solutes produced or released by the gut fermentation. Further, this increased blood flow may support higher epithelial cell production rates caused by lumen SCFA (see p. 400) by supplying materials for epithelial cell production.

Effects of SCFA on endocrine pancreatic secretion in ruminants

Intravenous administration of propionic or *n*-butyric acid results in insulin release in sheep (Manns and Boda, 1967). Mineo *et al.* (1990a) compared effects of different SCFA and found that acetic acid had a very weak effect and that branched-chain SCFA, such as isobutyric or isovaleric acids, stimulated insulin release more strongly than *n*-butyric or *n*-valeric acids.

Propionic and *n*-butyric acids can stimulate insulin release from isolated pancreatic slice or isolated islets of Langerhans prepared from sheep (Jordan and Phillips, 1978). This indicates that SCFA can directly stimulate β-cells. However, a more recent study showed that systemic mediation via the sympathetic and parasympathetic nerve systems is more important than direct action on β-cells for stimulation of insulin secretion by SCFA *in vivo* (Bloom and Edwards, 1985). We do not know yet how, or where, SCFA stimulate the systemic mediatory system that leads to insulin secretion.

There is an argument that the effect of SCFA on insulin release in ruminants is not physiologically significant due to the physiologically irrelevant sites or doses of SCFA administration used in the above experiments (Stern *et al.*, 1970). However, some studies revealed that intraruminal (Trenkle, 1978) or portal (Manns *et al.*, 1967) SCFA infusion at physiological rates can stimulate insulin secretion from sheep pancreas. Such results support a physiological role for the stimulatory effect of SCFA on insulin release in ruminants (Brockmann, 1978).

Mineo *et al.* (1986) found that continuous intravenous infusion of acetic acid at a physiological rate in sheep does not stimulate insulin secretion but does activate the pancreatic β-cells to respond to glucose stimuli. Similar results were obtained in dogs and rats (H. Mineo and T. Sakata, unpublished data): continuous intravenous infusion of acetic acid increased the insulin response after intravenous glucose challenge.

There are many animals that depend on both auto- and alloenzymatic digestion for their energy supply. Such animals should have regulatory mechanisms that can control both glucose-based and SCFA-based aspects of their energy economy. The findings discussed suggest that continuous entry of acetic acid, although it may not make a significant caloric contribution, can affect glucose-based energy metabolism by increasing the insulin response to glucose load and by increasing the sensitivity of peripheral tissues to insulin.

SCFA stimulate the release of glucagon from the pancreas (Mineo *et al.*, 1990a,b,c). Branched-chain SCFA such as isobutyric or isovaleric acids had almost similar effects on glucagon secretion compared to their corresponding straight-chain SCFA (*n*-butyric and *n*-valeric acids) (Mineo *et al.*, 1990a), indicating that receptive mechanisms for insulin release and glucagon release should differ.

Effects of SCFA on endocrine pancreatic secretion in non-ruminant animals

SCFA have no stimulatory effects on insulin release in rats, rabbits and pigs (Horino *et al.*, 1968). However, ketone bodies such as β-hydroxybutyrate or acetoacetate, which are metabolites of SCFA, increase insulin release from isolated pancreatic islets (Goberna, 1974).

These studies used a pulse intravenous injection of SCFA. The effect of SCFA on pancreatic endocrine secretion in non-ruminant animals should be measured by administering SCFA in a more physiological way: by continuous infusion into either the gut lumen or the portal vein. Care should be taken with the preliminary feeding of animals in such studies in omnivores because the response to SCFA may be modified by the feeding history of the animal.

Effect of SCFA on pancreatic exocrine secretion

SCFA stimulate pancreatic exocrine secretion in ruminants (Kato *et al.*, 1991). Intraruminal and intraduodenal SCFA infusion stimulates fluid and amylase secretion from sheep pancreas (Magee, 1961; Taylor, 1962). Intravenous injection of SCFA also stimulates both fluid and amylase secretion, in a dose-dependent fashion. In sheep, butyric acid had a stronger effect than either acetic or propionic acids (Harada and Kato, 1983). The minimum dose of acetic acid for pancreatic amylase secretion in sheep (< 0.1 mM for isolated pancreatic tissue) is within the physiological range *in vivo* (Katoh and Tsuda, 1984).

Pancreatic exocrine responses to SCFA vary between animal species (Kato *et al.*, 1991). SCFA stimulate exocrine pancreatic secretion in sheep, goats, cattle, Japanese deer, guinea pigs, Japanese field voles and rats. On the other hand, SCFA do not stimulate exocrine pancreatic secretion in rabbits, mice, Syrian hamsters, pigs, cats and chickens.

In sheep and goats the structural requirement for SCFA as an agonist to stimulate pancreatic amylase release *in vitro* is the possession of a carboxyl group coupled with a hydrophobic group such as an aliphatic hydrocarbon chain, or a benzene or cyclohexane ring (Katoh and Yajima, 1989). The lack of a carboxyl group or the introduction of a hydrophilic component into the hydrophobic group diminishes the ability of SCFA derivatives to stimulate amylase release. Co-existing dicarboxylic acids (succinic and phthalic acids), which by themselves weakly stimulate amylase release *in vitro*, enhance the stimulatory effect of butyric acid (Katoh and Yajima, 1989).

SCFA seem to stimulate amylase release by activating intracellular processes in a way similar to acetyl choline (via Ca^{2+} release) but do not act via acetyl choline or CCK receptors in sheep pancreatic acinar cells (Kato *et al.*, 1991).

Gut epithelial cell proliferation

Cell proliferation in vitro

Generally, SCFA inhibit cell proliferation *in vitro* (for review: Sakata and Yajima, 1984), including that of primary cultures of rumen epithelial cells

(Galfi *et al.*, 1991) and of epithelial cells of acutely-isolated rat caecal tissue (Sakata, 1987). This inhibitory effect varies among acids and is dose-dependent, being reversible at lower doses but irreversible at higher doses (Ginsburg *et al.*, 1973; Kruh, 1982). Usually, SCFA alter gene expression (Kruh, 1991): the most critical effect seems to be the acetylation of histones by the reversible inhibition of histone deacetylase that arrests cells at the G_0-G_1 phase of the cell cycle (Kruh, 1982).

Stimulatory effect of SCFA on rumen epithelial cell proliferation

The rumen wall adapts to dietary changes mainly by changing its epithelial cell mass, which primarily determines the absorptive and metabolic capacity of the organ (Sakata and Yajima, 1984). Epithelial cell mass is the result of a dynamic equilibrium between cell gain and cell loss. There have been few studies on epithelial cell loss rate due to technical difficulties. Therefore, studies have mainly focussed on the rate of epithelial cell gain.

The extensive post-weaning development of the ruminant forestomach mucosa depends on SCFA (for review: Warner and Flatt, 1965). The effectiveness of different SCFA is in the order: butyrate > propionate > acetate. However, SCFA are not responsible for muscle growth of the ruminant forestomach during the post-weaning period (Tamate *et al.*, 1962, 1964). So far, nobody knows if the SCFA-dependent development of the rumen mucosa during the post-weaning period is the result of increased cell production rate or the result of decreased cell loss rate.

Intraruminal administration of sodium *n*-butyrate, sodium propionate or sodium acetate (18 mmol/kg body weight) increased rumen epithelial cell mitosis in fasted adult sheep (Sakata and Tamate, 1978b, 1979). Butyrate had a stronger effect than acetate or propionate. The effect of SCFA is transitional in spite of continued daily administration in fasted sheep. However, the stimulatory effect of sodium *n*-butyrate persists at least for 1 week with daily administration into the rumen of fed sheep (Galfi *et al.*, 1986). The stimulatory effect of butyrate depends on the administration rate (Sakata and Tamate, 1978b; Galfi *et al.*, 1986): rapid administration of sodium *n*-butyrate (18 mmol/kg body weight) within 10 s results in a marked increase in mitotic index of the rumen epithelium within a day, but the same dose has no effect when infused intraruminally over 20 h.

Demonstrations that SCFA inhibit rumen epithelial cell proliferation *in vitro* (Galfi *et al.*, 1981) have led to the suggestion that the stimulatory effect on rumen epithelial cells *in vivo* is indirect (Sakata *et al.*, 1980). Humoral mediation of the stimulation arising with SCFA was proposed; insulin was the first possible humoral mediator to be tested, since it was known that butyrate and propionate (but not acetate) stimulate insulin release from sheep pancreatic islets (Manns *et al.*, 1967; Manns and Boda, 1967).

Intravenous infusion of insulin (0.125 IU/kg per h plus 300 mg/kg glucose per h for 6 h) increased the epithelial mitotic index of fasted adult sheep within 3 h after the start of infusion, to return to initial levels within 42 h after the end of infusion (Sakata *et al.* 1980). This increase in mitotic index is larger than that caused by glucose infusion alone (300 mg/kg glucose per h for 6 h). Further, blood insulin levels in sheep given insulin plus glucose infusion resembled those after butyrate infusion (Sakata *et al.*, 1980). Therefore, it is clear that at least a part of the stimulatory effect of SCFA on rumen epithelial cell mitosis is mediated by insulin.

Such a mediation by insulin was further confirmed by *in vitro* studies (Galfi *et al.*, 1991). Insulin (1.6×10^{-6} or 1.6×10^{-7} mmol/l) stimulated the proliferation of a primary culture of rumen epithelial cells independent of the inhibitory action of coexisting butyric acid (2 or 10 mmol/l). Both the stimulatory effect of insulin and the inhibitory effect of butyrate were dose dependent. Glucagon, which is also released by SCFA (Mineo *et al.*, 1990a), stimulates the proliferation of primary cultures of rumen epithelial cells, but only in the absence of butyrate (Galfi *et al.*, 1991). Thus glucagon cannot be a physiological mediator for the tropic action of SCFA *in vivo*.

The stimulatory effect of SCFA *in vivo* may facilitate adaptation by the rumen mucosa to an increased intake of readily-fermentable feed, such as grains or fruits, for the forestomach fermenter. Readily fermentable substrates can be degraded rapidly in the forestomach to produce SCFA, resulting in an increase in epithelial cell mass as observed after the rapid intraruminal administration of sodium *n*-butyrate (Sakata and Tamate, 1978a). This should facilitate absorption and metabolism of SCFA produced in the forestomach. However, the increase in cell proliferation rate is usually accompanied by an increase in cell loss rate in non-cancerous tissues, leading to an increase in endogenous nitrogen loss via sloughed epithelial cells. This can be problematic when the forestomach fermenters eat readily fermentable but nitrogen-poor feed. In this regard, it would be worthwhile to study the influence of SCFA on rumen epithelial cell production under different nitrogen supply levels.

Effect of SCFA on epithelial cell proliferation in the digestive tract of non-ruminant animals

Conditions that reduce the production of SCFA depress epithelial cell proliferation not only of the large intestine but also of the small intestine. Thus, caeco-colonic bypass surgery (Sakata, 1988), parenteral nutrition (Jane *et al.*, 1977), feeding of substrate-free diet (Goodlad and Wright, 1983) and germ-free conditions (Komai *et al.*, 1982) depress mitotic activity of intestinal epithelia.

The stimulatory effect of SCFA on intestinal epithelial cell proliferation appears within a few days of daily administration and lasts for at least 2 weeks (Sakata, 1987). Daily administration of approximately 10% of the daily production level of SCFA (acetic, propionic and *n*-butyric acids, 100, 20 and 60 mM, respectively, pH 6.1; twice daily, 3 ml each) increased crypt cell production rate in the caecum and distal colon within 1–2 days (Sakata, 1987). SCFA increased crypt cell production rate in the jejunum and distal colon 3–4 fold without changing the patterns of their circadian fluctuations.

Acetic, propionic and *n*-butyric acids (except acetic acid in the jejunum) showed a dose-dependent stimulatory effect on epithelial cell production rates in the jejunum and distal colon (Sakata, 1987). The effectiveness was in the order: *n*-butyric > propionic > acetic acid.

Rombeau's group (Kripke *et al.*, 1989) found that continuous infusions of either butyric acid (20 to 150 mM) or a mixture of SCFA (acetic, propionic and *n*-butyric acids; 70, 35 and 20 mM, pH 6.1) increased mucosal DNA content of the jejunum and proximal colon of caecectomized rats. Their findings suggest that SCFA at physiological rates of administration can stimulate epithelial cell proliferation in the intestine (Sakata, 1991).

The trophic effect of SCFA does not need their further metabolism by bacteria, as the effect of SCFA occurs in germ-free rats (Sakata, 1987). It is not protons but SCFA anions that have the stimulatory effect because the pH of test solutions and control solutions was adjusted to 6.1 in the studies of Sakata (1987) and Kripke *et al.* (1989). However, this does not necessarily exclude the possibility of mitotic stimulation by low lumen pH. It might be possible that low lumen pH enhances the stimulatory effect of SCFA on intestinal epithelial cells.

It is unlikely that SCFA directly stimulate intestinal epithelial proliferation. First, it is hard to explain the effect of SCFA infused into the large intestine on jejunal epithelium (approximately 70 cm orad to the site of administration in the rat studies) by a direct mechanism (Sakata, 1987). Second, SCFA depress epithelial cell proliferation of short-term cultures of rat caecum

(Sakata, 1987); as these preparations contained histologically identifiable intestinal nerve system and enteroendocrine cells, the trophic action of SCFA must require a systemic mediating mechanism.

The receptors probably lie in or just below the mucosa near the site of administration (Sakata, 1991) because epithelial cells of the large intestine consume most of the *n*-butyric acid absorbed from the lumen, making its concentration in portal blood very low (Roediger, 1981).

The efferent transmission of the trophic effect of SCFA is very likely to be via the vascular system (Sakata, 1989). When a small segment of jejunum was autografted under the skin, with or without its mesenteric nerve connection maintained, disruption of the mesenteric nerve connection did not abolish the stimulatory effect of SCFA given into the caecum on the epithelial cell proliferation of the grafted jejunal segment. Enteroglucagon or intestinal peptide PYY may be the mediator for the trophic action of SCFA, because feeding of pectin or guar gum increased blood levels of these gut peptides in rats, and these peptides stimulate proliferation of intestinal epithelial cells both *in vivo* and *in vitro* (Goodlad *et al.*, 1987). Insulin cannot be a candidate for the efferent transmission of the trophic effect of SCFA in those non-ruminant species in which SCFA do not stimulate insulin release (Horino *et al.*, 1968).

Our recent studies using rats with microsurgical denervation of their caeca showed that the afferent transmission for the trophic effect of SCFA requires an autonomic nerve mechanism (Frankel *et al.*, 1991). These studies also suggested that a part of the trophic effect of SCFA on colonic epithelium is not systemically mediated.

Conclusions

The gastro-intestinal segment where fermentation takes place (forestomach and large intestine) seems to recognize changes in feeding and microbial conditions in terms of the production of SCFA. Yajima found a chemically specific receptor mechanism in the large intestine (Yajima, 1989, 1991), and similar receptors also exist in the rumen and duodenal mucosa (Leek, 1986).

Responses of gut motility, intestinal epithelial proliferation and pancreatic secretions to lumen SCFA and SCFA in the extracellular fluid are dose-dependent and vary between acids. Thus, production rates of each acid affect the above-mentioned responses of host animals. In other words, responses of the digestive tract to substrate fermentation can be influenced by the mode of fermentation, i.e. the rate of fermentation and the molar proportion of different acids. Thus, the chemical composition of the feed itself cannot be a good criterion to use to predict the responses of the digestive organs of the animal that eats the food. Other factors, such as particle size of the food entering into the fermentation chamber or entry rate of the substrate into the fermentation chamber, have considerable influences on both production rate and molar proportions of microbially produced SCFA.

Colonic receptors for SCFA are affected by their history. First, SCFA receptors, at least in the large intestine, are highly adaptable, which means that an additional dose of SCFA is not as effective as the first (Yajima, 1991). Second, long-term adaptation also occurs. Acetic acid by itself does not stimulate either colonic motility or colonic chloride secretion in conventional rats but it does in germ-free rats (Yajima, 1991). Further, SCFA do not stimulate the motility of the caecum or proximal colon of conventional rats where active fermentation takes place; however, they do so in germ-free rats (Yajima, 1991). These data indicate that SCFA receptors become highly sensitive when gut fermentation is very low and suggest that physiological (non-nutritional) responses of animal functions to fermentation products depend on what the animal ate previously.

Accordingly, evaluation of foods or food components as physiological stimuli cannot be made from simple chemical analysis of the food. Instead, it is very important to assess the production rate of each SCFA from the food in question in the fermentation chamber of the animal in question.

Acknowledgement

I am grateful for the valuable suggestions of Professor Ian D. Hume of University of Sydney and Professor Kazuo Katoh of Tohoku University.

References

Bloom, S. R. & Edwards, A. V. (1985). Pancreatic neuroendocrine responses to butyrate in conscious sheep. *Journal of Physiology*, **364**, 281–288.

Bolton, J. R., Merritt, A. M., Carlson, G. M. & Donawick, W. J. (1976). Normal abomasal electromyography and emptying in sheep and the effects of intraabomasal volatile fatty acid infusion. *American Journal of Veterinary Research*, **37**, 1387–1392.

Brockmann, R. P. (1978). Roles of glucagon and insulin in the regulation of metabolism in ruminants. *Canadian Journal of Veterinary Medicine*, **19**, 55–62.

Bugaut, M. (1987). Occurrence, absorption and metabolism of short chain fatty acids in the digestive tract of mammals. *Comparative Biochemistry and Physiology*, **86B**, 439–472.

Dobson, A. (1984). Blood flow and absorption from the rumen. *Quarterly Journal of Experimental Physiology*, **68**, 77–88.

Ehrlein, H. J. & Hill, H. (1970). Einflüsse des Labmagen- und Duodenalinhaltes auf die Motorik des Wiederkäuermagens. *Zentralblatt für Veterinärmedizin*, **A17**, 498–516.
Frankel, W., Zhang, W., Singh, A. *et al.* (1991). Trophic effects of short-chain fatty acids on rat jejunum are mediated by the autonomic nervous system. *American Gastroenterological Association*, May 10–13, San Francisco.
Galfi, P., Neogrady, S. & Kutas, F. (1986). Dissimilar ruminal epithelial response to short-term and continuous intraruminal infusion of sodium *n*-butyrate. *Journal of Veterinary Medicine*, **A33**, 47–52.
Galfi, P., Neogrady, S. & Sakata, T. (1991). Effects of volatile fatty acids on the epithelial cell proliferation of the digestive tract and its hormonal mediation. In *Physiological Aspects of Digestion and Metabolism in Ruminants*, ed. T. Tsuda, Y. Sasaki & R. Kawashima, pp. 50–59. San Diego: Academic Press.
Galfi, P., Veresegyhazy, T., Neogrady, S. & Kutas, F. (1981). Effect of sodium *n*-butyrate on primary ruminal epithelial cell culture. *Zentralblatt für Veterinärmedizin*, **A28**, 259–261.
Ginsburg, E., Salamon, D., Sreevalsan, T. & Freese, E. (1973). Growth inhibition and morphological changes caused by lipophilic acids in mammalian cells. *Proceedings of the National Academy of Sciences, USA*, **70**, 2457—2461.
Goberna, R. (1974). Action of β-hydroxybutyrate, acetoacetate and palmitate on the insulin release in the perfused isolated rat pancreas. *Hormone and Metabolism Research*, **6**, 256–260.
Goodlad, R. A., Lenton, W., Ghatei, M. A., Adrian, T. E., Bloom, S. R. & Wright, N. A. (1987). Proliferation effects of 'fibre' on the intestinal epithelium: relationship to gastrin, enteroglucagon and PYY. *Gut*, **28(S1)**, 221–226.
Goodlad, R. A. & Wright, N. A. (1983). Effects of addition of kaolin or cellulose to an elemental diet on intestinal cell proliferation in the rat. *British Journal of Nutrition* **50**, 91–98.
Gregory, P. C. (1987a). Inhibition of reticuloruminal motility in the vagotomized sheep. *Journal of Physiology*, **346**, 379–393.
Gregory, P. C. (1987b). Inhibition of reticuloruminal motility by volatile fatty acids and lactic acid in sheep. *Journal of Physiology*, **382**, 355–371.
Harada, E. & Kato, S. (1983). Effect of short-chain fatty acids on the secretory response of the ovine exocrine pancreas. *American Journal of Physiology*, **244**, G284–G290.
Horino, M. L., Machlin, J., Hertelendy, F. & Kipnis, D. M. (1968). Effects of short-chain fatty acids on plasma insulin in ruminant and nonruminant species. *Endocrinology*, **8**, 118–128.
Iggo, A. & Leek, B. F. (1970). Sensory receptors in the ruminant stomach and their reflex effects. In *Physiology of Digestion and Metabolism in the Ruminant*, ed. A. T. Phillipson, pp. 23–34. Newcastle upon Tyne: Oriel Press.
Jane, P., Carpentier, Y. & Willems, G. (1977). Colonic mucosal atrophy induced by a liquid elemental diet in rats. *American Journal of Digestive Diseases*, **22**, 808–812.
Jordan, H. N. & Phillips, R. W. (1978). Effects of fatty acids on isolated ovine pancreatic islets. *American Journal of Physiology*, **234**, E162–E167.
Kamath, P. S., Hoepfner, M. T. & Phillips, S. F. (1987). Short chain fatty acids stimulate motility of the canine ileum. *American Journal of Physiology*, **253**, G427–G433.
Kamath, P. S., Phillips, S. F. & Zinsmeister, A. R. (1988). Short chain fatty acids stimulate ileal motility in humans. *Gastroenterology*, **95**, 1496–1502.
Kato, S., Katoh, K. & Barej, W. (1991). Regulation of exocrine pancreatic

secretion in ruminants. In *Physiological Aspects of Digestion and Metabolism in Ruminants*, ed. T. Tsuda, Y. Sasaki & R. Kawashima, pp. 90–109. San Diego: Academic Press.

Katoh, K. & Tsuda, T. (1984). Effects of acetyl choline and short-chain fatty acids on acinar cells of the exocrine pancreas in sheep. *Journal of Physiology*, **356**, 479–489.

Katoh, K. & Yajima, T. (1989). Effects of butyric acid and analogues on amylase release from pancreatic segments of sheep and goats. *Pflügers Archiv für die gesamte Physiologie*, **413**, 256–260.

Komai, M., Takehisa, F. & Kimura, S. (1982). Effect of dietary fiber on intestinal epithelial cell kinetics of germ-free and conventional mice. *Nutritional Report International*, **26**, 255–261.

Kripke, S. A., Fox, A. D., Berman, J. M., Settle, R. G. & Rombeau, J. L. (1989). Stimulation of intestinal mucosal growth with intracolonic infusion of short-chain fatty acids. *Journal of Parenteral and Enteral Nutrition*, **13**, 109.

Kruh, J. (1982). Effect of sodium butyrate, a new pharmacological agent, on cells in culture. *Molecular and Cellular Biochemistry*, **42**, 65–82.

Kruh, J. (1991). Molecular and cellular effects of sodium butyrate. In *Short-Chain Fatty Acids: Metabolism, and Clinical Importance*, ed. J. H. Cummings, J. L. Rombeau & T. Sakata, pp. 45–50. Columbus: Ross Laboratories.

Kvietys, P. R. & Granger, D. N. (1967). Effects of volatile fatty acids on blood flow and oxygen uptake by the dog colon. *Gastroenterology*, **80**, 962–969.

Leek, B. F. (1986). Sensory receptors in the ruminant alimentary tract. In *Control of Digestion and Metabolism in Ruminants*. ed. L. P. Milligan, W. L. Grovum & A. Dobson, pp. 3–17. Englewood Cliffs, NJ: Prentice-Hall.

Leek, B. F. & Harding, R. H. (1975). Sensory nervous receptors in the ruminant stomach and the reflex control of reticuloruminal motility. In *Digestion and Metabolism in the Ruminant*, ed. I. W. McDonald & A. C. I. Warner, pp. 60–76. Armidale: University of New England.

Magee, D. F. (1961). An investigation into the external secretion of the pancreas in sheep. *Journal of Physiology*, **158**, 132–143.

Manns, J. G. & Boda, J. M. (1967). Insulin release by acetate, propionate, butyrate and glucose in lambs and adult sheep. *American Journal of Physiology*, **212**, 745–755.

Manns, J. G., Boda, J. M. & Wiles, R. F. (1967). Probable role of propionate and butyrate in control of insulin secretion in sheep. *American Journal of Physiology*, **212**, 756–764.

Mineo, H., Kanai, M., Kato, S. & Ushijima, J. (1990a). Effects of intravenous injection of butyrate, valerate and their isomers on endocrine pancreatic responses in conscious sheep (*Ovis aries*). *Comparative Biochemistry and Physiology*, **95A**, 411–416.

Mineo, H., Kitade, A. Kawakami, S., Kato, S. & Ushijima, J. (1990b). Effect of intravenous injection of acetate on the pancreas of sheep. *Research in Veterinary Science*, **48**, 310–313.

Mineo, H., Murao, R., Kato, S. & Ushijima, J. (1990c). Effect of intravenous injection of short chain fatty acids on glucagon secretion in sheep. *Japanese Journal of Zootechnical Science*, **61**, 349–353.

Mineo, H., Nishimura, M., Kato, S. & Ushijima, J. (1986). The effect of intravenous infusion of acetate on glucose-induced insulin secretion in sheep. *Japanese Journal of Zootechnical Sciences*, **57**, 765–769.

Parra, R. (1978). Comparison of foregut and hindgut fermentation in herbivores. In

The Ecology of Arboreal Folivores, ed. G. G. Montgomery, pp. 205–229. Washington, DC: Smithsonian Institution.

Richardson, A., Delbridge, A. T., Brown, N. J., Rumsey, R. D. E. & Read, N. W. (1991). Short chain fatty acids in the terminal ileum accelerate stomach to caecum transit time in the rat. *Gut*, **32**, 266–269.

Roediger, W. E. W. (1981). The effect of bacterial metabolites on nutrition and function of the colonic mucosa: symbiosis between man and bacteria. In *Colon and Nutrition*, ed. H. Kasper & H. Goebell, pp. 11–25. Lancaster: MTP Press.

Ruckebusch, Y. & Tomov, T. (1973). The sequential contraction of the rumen associated with eructation in sheep. *Journal of Physiology*, **235**, 447–458.

Sakata, T. (1987). Stimulatory effect of short-chain fatty acids on epithelial cell proliferation in the rat intestine: A possible explanation for trophic effects of fermentable fibre, gut microbes and luminal trophic factors. *British Journal of Nutrition*, **58**, 95–103.

Sakata, T. (1988). Depression of intestinal epithelial cell production rate by hindgut bypass in rats. *Scandinavian Journal of Gastroenterology*, **23**, 1200–1202.

Sakata, T. (1989). Stimulatory effect of short-chain fatty acids on epithelial cell proliferation of isolated and denervated jejunal segment of the rat. *Scandinavian Journal of Gastroenterology*, **24**, 886.

Sakata, T. (1991). Effects of short-chain fatty acids on epithelial cell proliferation and mucus release in the intestine. In *Short-Chain Fatty Acids: Metabolism and Clinical Importance,* ed. J. H. Cummings, J. L. Rombeau & T. Sakata, pp. 63–67. Columbus: Ross Laboratories.

Sakata, T., Hikosaka, K., Shiomura, Y. & Tamate, H. (1980). Stimulatory effect of insulin on ruminal epithelium cell mitosis in adult sheep. *British Journal of Nutrition*, **44**, 325–331.

Sakata, T. & Tamate, H. (1979). Rumen epithelial cell proliferation accelerated by propionate and acetate. *Journal of Dairy Science*, **62**, 49–52.

Sakata, T. & Tamate, H. (1987a). Influence of butyrate on microscopic structure of ruminal mucosa in adult sheep. *Japanese Journal of Zootechnical Sciences*, **49**, 687–696.

Sakata, T. & Tamate, H. (1987b). Rumen epithelial cell proliferation accelerated by rapid increase in intraruminal butyrate. *Journal of Dairy Science*, **61**, 1109–1113.

Sakata, T. & Yajima, T. (1984). Influence of short chain fatty acids on the epithelial cell division of digestive tract. *Quarterly Journal of Experimental Physiology*, **69**, 639–648.

Schnorr, R. & Vollmerhaus, B. (1968). Das Blutgefäßsystem des Pansenepithels von Ziege und Rind. *Zentralblatt für Veterinärmedizin*, **A15**, 799–828.

Stern, J. S., Baile, C. A. & Mayer, J. (1970). Are propionate and butyrate physiological regulators of plasma insulin in ruminants? *American Journal of Physiology*, **219**, 84–91.

Stevens, C. E. (1988). *Comparative Physiology of the Vertebrate Digestive System.* Cambridge: Cambridge University Press.

Svendsen, P. (1972). Inhibition of cecal motility in sheep by volatile fatty acids. *Nordisk Veterinaermedicin*, **24**, 393–396.

Svendsen, P. (1975). Experimental studies of gastrointestinal atony in ruminants. In *Digestion and Metabolism in the Ruminant*. ed. I. W. McDonald & A. C. I. Warner, pp. 563–75. Armidale: University of New England Publishing.

Tamate, H., McGilliard, A. D., Jacobson, N. L. & Getty, R. (1962). Effect of various diets on the anatomical development of the stomach in the calf. *Journal of Dairy Science*, **45**, 408–420.

Tamate, H., McGilliard, A. D., Jacobson, N. L. & Getty, R. (1964). The effect of various diets on the histological development of the stomach in the calf. *Tohoku Journal of Agricultural Research*, **14**, 171–193.

Taylor, R. B. (1962). Pancreatic secretion in the sheep. *Research in Veterinary Science*, **3**, 63–77.

Trenkle, A. J. (1978). *Journal of Dairy Science*, **61**, 281–293.

Warner, R. G. & Flatt, W. P. (1965). Anatomical development of the ruminant stomach. In *Physiology and Metabolism in Ruminants*. ed. R. W. Dougherty, pp. 24–38. Washington DC: Butterworth.

Winne, D. (1978). Blood flow in intestinal absorption rate. *Journal of Pharmacokinetics and Biopharmaceutics*, **6**, 55–78.

Yajima, T. (1985). Contractile effects of short-chain fatty acids on the isolated colon of the rat. *Journal of Physiology*, **368**, 667.

Yajima, T. (1989). Chemical specificity of short-chain fatty acid-induced electrical secretory response in the rat colonic mucosa. *Comparative Biochemistry and Physiology*, **93A**, 851.

Yajima, T. (1991). Sensory mechanisms of motor and secretory responses to short-chain fatty acids in the colon. In *Short-Chain Fatty Acids: Metabolism, and Clinical Importance*, ed. J. H: Cummings, J. L. Rombeau & T. Sakata, pp. 71–74. Columbus: Ross Laboratories.

Yajima, T. & Sakata, T. (1987). Influences of short-chain fatty acids on the digestive organs. *Bifidobacteria Microflora*, **6**, 7–14.

Part V

Synthesis and perspectives

24

Food, form and function: interrelationships and future needs

EDITORS and CONTRIBUTORS

We have considered the spectrum of factors involved in food, form and function, and the interrelationships between them. Thus, diets, gut morphology and digestion are each classified in detail, in terms of the nature of foods, the mechanics of their processing in the mouth and the chemistry of their digestion in the alimentary canal. Attempts have been made to relate them to each other and to indicate future directions for research. The aim in this concluding chapter is to synthesise the results of these deliberations as they have developed during the production of this book.

This must be done in an evolutionary context: the integrating framework for the diversity seen among mammals with respect to any body system and its interaction with the environment. This should be especially true for the digestive system and the wide range of foods exploited by mammals in the whole range of available environments. The biomes of the world are characterised by floral and faunal associations that provide trophic resources with changing profiles of potential diets, so that many configurations observed today may not be typical of the past. Nutritional niches tend to be narrow because of constraints on the digestive systems, but some mammals, especially primates, exhibit a wide range of diets because of a relatively unspecialised digestive system. Much can be gained by comparing the adaptations of primates with the specialisations of other animals for eating either animal matter or foliage. For such reasons, we do not restrict our attention to mammals but try to put them into a broader vertebrate perspective, especially by comparisons with the great radiation of birds.

It is over the last 50 million years, with the radiation of flowering plants, that environments have become so diverse and foods so varied and nutritious; hence the dramatic radiation of birds and mammals. The complexity is increased by changes between seasons and between years; this makes interpretation of the dietary adaptation of digestive systems even more problematic. Presumably, dietary flexibility is sustained by the needs to maximise nutritional intake at times of food abundance and of food scarcity. The ecological bottlenecks of the latter represent *acute* needs, but adaptations must be shaped equally by the chronic pressures operating most of the time.

Summary

We have reviewed this complex topic of the digestive system in mammals by focussing on food, form (anatomy) and function (physiology and energetics). The introductory section centred on scope and terminology, evolution and models. Mammals evolved in relation to the diversification of flowering plants (angiosperms) that established the great scope for the diversification of mammals and for the exploitation of a greater variety of plant parts, including foliage (Langer, Ch. 2). Faunivores have changed little in their anatomy and physiology during their evolution, but frugivores exploit caeco-colic fermentation with increasing body size and foliage in the diet. The original caeco-colic folivores either enhanced this system or moved over to forestomach fermentation, with some becoming ruminants.

Modelling provides a theoretical framework for digestive physiology, and Martínez del Rio, Cork and Karasov (Ch. 3) reviewed its development from non-reactor to reactor theory. McNeill Alexander (Ch. 4) considered in detail two different kinds of reactors, one good for fermentation (continuous-flow stirred tank) and one poor for fermentation but advantageous for rapid digestion (plug-flow); he considered whether richer diets are eaten (and processed) slowly or rapidly and how efficient they are. Such models are shown to fit well with the actual situation.

Concerning *Homo sapiens*, Hladik and Chivers (Ch. 5) remarked that the technical skills embedded in the socio-cultural traditions of most groups (or more recently developed in the industrial world) may partly replicate the various processes occurring inside the digestive tract of other species, such as fermentation, allowing more efficient use of several types of food. As for taste and perception, the cultural ability to transform food (outside the scope of this volume) can be viewed as an extension of the digestive system.

Langer and Chivers (Ch. 6) tried to produce a unifying classification of foods, based on an extensive review of the literature. Moir (Ch. 7) discussed the lack of key amino acids in the great grains of the world and how herbivores have to be carnivores, especially when young, to obtain adequate protein. Various myths, such as the specialised versus unspecialised dichotomy in birds, were exploded by Martínez del Rio (Ch. 8), with a detailed comparison of fruit-eating and flower-visiting birds and bats and an illuminating exploration of phenotypic and evolutionary plasticity of digestive functions.

Perrin (Ch. 9) reviewed critically the nutritional niche concept in relation to herbivores and their varied gastric adaptations and niche expansions; apparent overlap may not actually occur. Comparative niche analysis reveals problems of reasoning but provides hypotheses that now need testing. The role of taste, hitherto much neglected, in food selection and dietary differentiation was explored quantitatively by Simmen (Ch. 10) in relation to New World monkeys. Peters and O'Brien (Ch. 11) contrasted plant productivity in two areas of sub-tropical Africa – one semi-arid, the other sub-humid – from a hominid food perspective. While the harsher environment has less edible foods, a higher proportion of the woody flora is productive, even in the dry season, when increased amounts of coarser items pose special oral and digestive challenges.

Lucas (Ch. 13) considered the mechanical work done in the mouth, in relation to friction, flow and fracture and increasing access to food for enzyme action, as well as for swallowing; hence, particle size, shape, roughness and abrasiveness are relevant. Foods need to be analysed in terms of elastic moduli, fracture toughness and hardness, and related to tooth size and shape. Young Owl (Ch. 14) and Snipes (Ch. 15) refined old methods and developed new ones to measure intestinal surface areas, especially to incorporate the effects of microscopic surface features. These studies supply quantitative corroboration of the many old morphological observations and qualitative statements. Moreover, a new range of criteria based on microscopic morphometry allows categorisation according to food strategy.

Weaning time is shown to be particularly long in forestomach fermenters, even compared with caeco-colic fermenters, presumably because of the problems of acquiring an adequate microbial population (Langer, Ch. 16); gastric groove differentiation, not correlated with body size, is examined in detail in a wide range of mammalian stomachs. Björnhag (Ch. 17) showed the remarkable differentiation and functional diversity in the caecum and colon with regard to alloenzymatic digestion, absorption and, especially, separation mechanisms in relation to retention time and passage mechanisms.

Hume (Ch. 19) considered the variation in body size, gut morphology and digestive performance in rodents, affecting the teeth and guts in particular and proving to be more efficient in voles. Batzli *et al.* (Ch. 20) tackled the problem that microtine rodents violate the Bell–Jarman principle and maintain constant digestive energy intake over a wide range of diets with differing fibre content. They showed that this is done by a change in gut volume and processing variables (IPR, integrated processing response) in response to the change in energy demand of the dietary fibre content. Cork (Ch. 21) examined the constraints on small mammals that prevent them from feeding on fibrous foods, identifying the gaps in our knowledge that prevent explanations. Such foods have to be retained in elaborations of either the stomach or the caecum and/or colon (with separation and selective retention); such small animals may also have to be caecotrophic.

The effects and costs of allelochemicals on mammalian herbivores were discussed by Foley and McArthur (Ch. 22), emphasising the need to know both the chemistry of the compounds and the physiology of the consumers. Inability to generalise about the effects and costs of allelochemicals impedes the study of foraging and plant defence theory, but they identified four main ways in which costs can be assessed, which could provide the integrating framework for future progress. Sakata (Ch. 23) discussed the responses of animals to short-chain fatty acids and the sensory and transmission mechanisms, in terms of gut motility, mucosal blood flow, pancreatic secretions (endocrine and exocrine) and epithelial cell proliferation.

Perspective

Living matter – plant and animal – can be classified in terms of chemistry and structures; we have made a concerted effort to do this in a comprehensive manner (Figs. 24.1 and 24.2). The hierarchy of substances in cell contents and cell walls are recognised as is that in all the different parts of animals and plants; each such part can be said to be dominated by particular chemicals. The next step is to interrelate them, to fit them together; the pathway to this integration starts with the natural sub-division of plant parts according to predominance of fats, sugars, proteins, fibre and so forth. Some aspects of this are developed in Fig. 24.3.

A global view of sources of nitrogen (as percentage dry matter) and the water content of those sources in a variety of organisms that are potential food are presented in Fig. 24.3 (modified from Slansky and Scriber, 1985). (The nitrogen content of xylem and phloem fluids is on a moist base.) Nitrogen (N) values, rather than protein (conventionally $N \times 6.25$), are preferred as many sources, such as insects, contain considerable non-protein nitrogen, such as chitin. With few exceptions, actively metabolising tissues have a high water content (the poikilohydric lichens are an exception) coupled with a high nitrogen content. Both may be diluted by energy stores (sugar, starch, fat) and by plant structural components (cellulose, hemicellulose and lignin).

Low-fat animal bodies, such as iguana, crustaceans, sea-horse and fish, have very high nitrogen and water contents; whales and domestic ruminants,

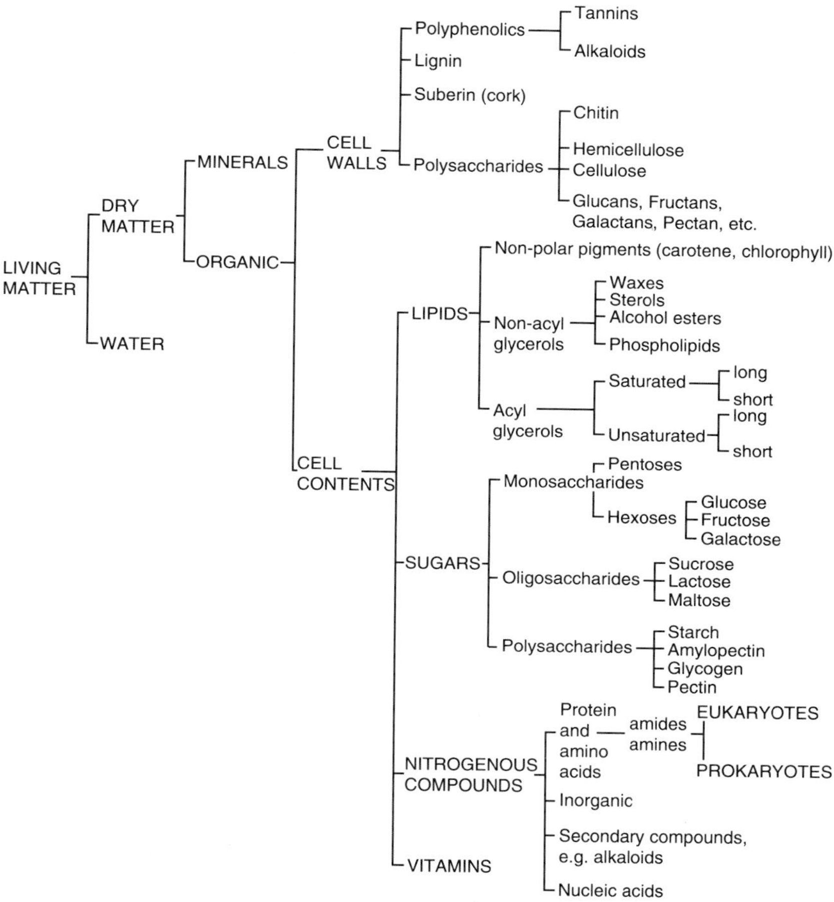

Fig. 24.1. Chemical composition of living matter.

flying birds and insects, pupae and larvae, with higher fat, have reduced water and, therefore, apparently reduced nitrogen in their dry matter (fat replaces water in wet mass and nitrogen in dry mass). Plant materials, with their non-protein structural fabric, have reduced levels of nitrogen in their dry matter but may still have high nitrogen contents in their seeds and leaves; lettuce has 95% water content, but still contains 4% nitrogen in its dry matter, as do milk and mushrooms. From moisture contents of 40% down to 15%, the materials available are low nitrogen, structural and senescent plant material. The non-metabolising plant storage seeds, grains and nuts have low (5–15%) moisture, 0.5 to 7% nitrogen and high carbohydrate or fat stores;

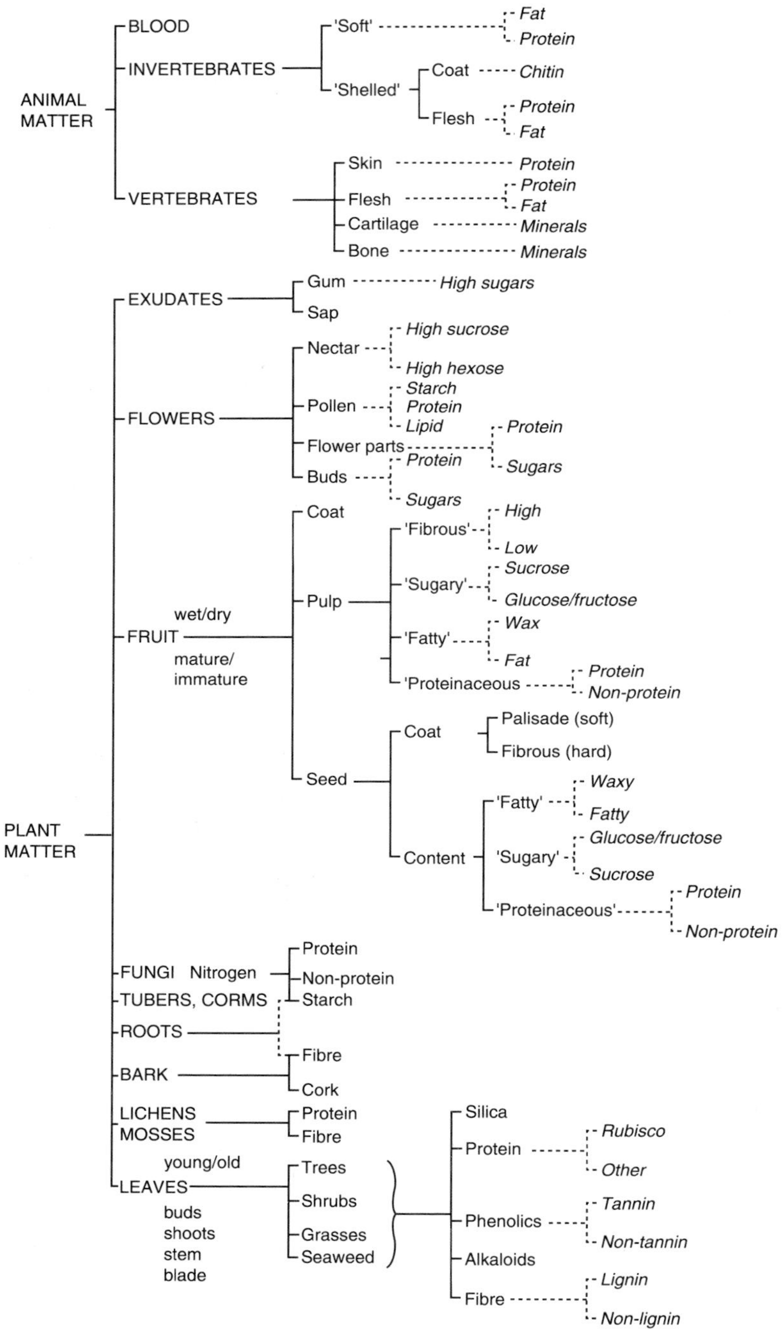

Fig. 24.2. Components of animal and plant matter.

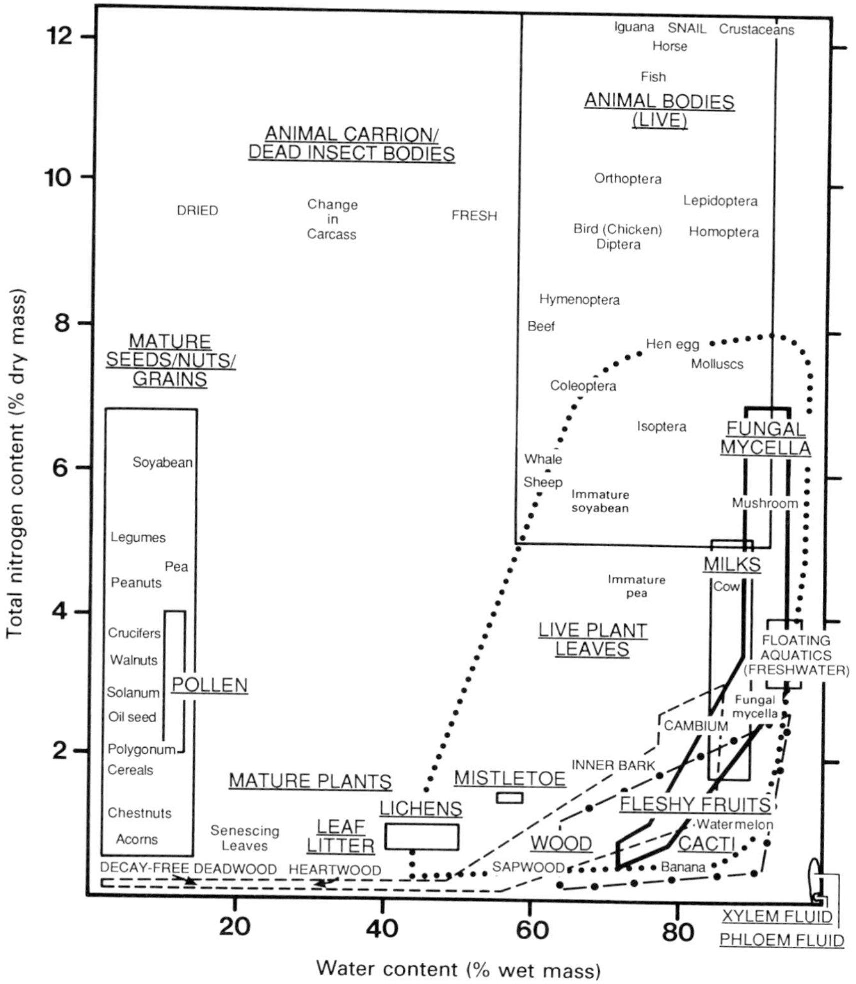

Fig. 24.3. Water and nitrogen contents of food. Nitrogen contents for phloem and xylem fluids are percentage of fresh weight; all other nitrogen contents are percentage of dry weight (from Slansky and Scriber, 1985; FAO, 1970–72).

dead and desiccated animal bodies have much higher nitrogen contents at low moisture levels. Many of these points are illustrated in Fig. 24.3.

The sequence of activities through the digestive system and the structural correlates are summarised in Table 24.1. The gastro-intestinal tract is subdivided into different regions, which constitute sections. These sections may have one or more functions and two or more sections may 'cooperate' to fulfill one or more functions. Structures or functions that are only found in

Table 24.1. *Functions and parts of anatomical regions in the gastro-intestinal tract of mammals*

Region	Parts	Function
Oral cavity	Lips, incisors/canines	Food acquisition
	Lips, cheeks, tongue	Control of food, Taste, Tension, Prehension, Transport regulation
	Premolars/molars	Mechanical processing
	Salivary glands	Secretion of lubricants, Buffering, Taste release?
	Cheeks, tongue, palate	Bolus formation
Pharynx		Bolus formation, Transport regulation
Oesophagus		Transport, Secretion of lubricants

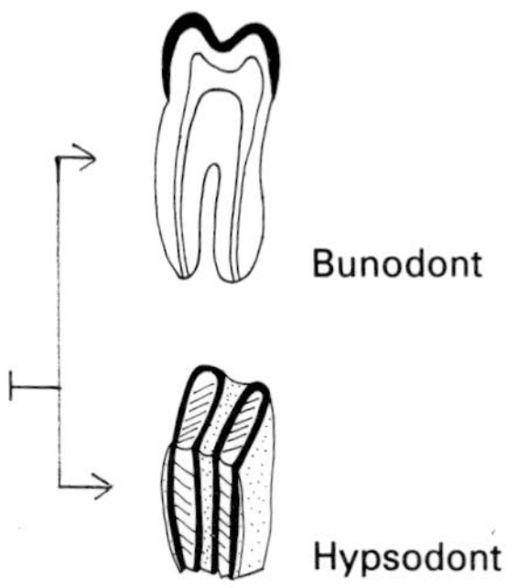

Stomach	Cardia Fornix/fundus corpus	Transport regulation Storage Sorting Mixing Autoenzymatic digestion
		(Forestomach formation) (Alloenzymatic digestion)
	Pars pylorica	Storage Sorting Mixing Autoenzymatic digestion
	Pylorus	Transport regulation
Small intestine	Duodenum Jejunum Ileum	Transport Secretion Absorption
Large intestine	Caecum and proximal colon	Transport Storage Mixing Secretion Absorption Separation Alloenzymatic digestion (Microbial digestion) (Formation of caecotrophs)

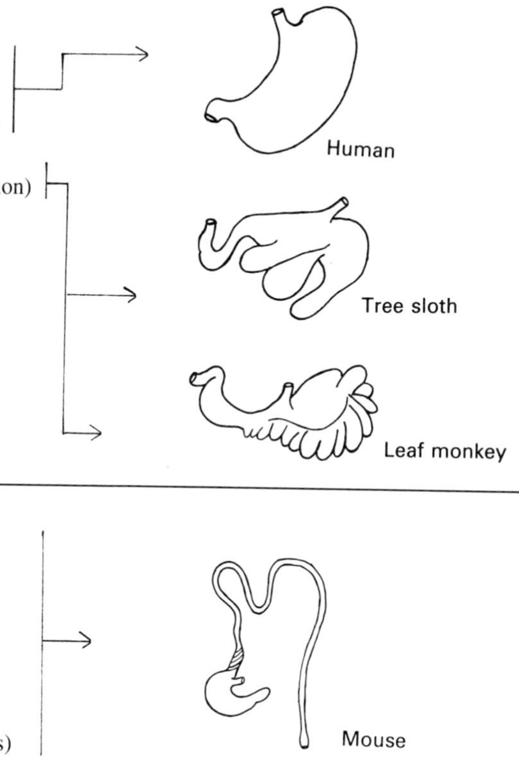

Table 24.1. Continued

Region	Parts	Function	
Large intestine	(Taeniae, haustra and plicae semi-lunares)	Transport regulation	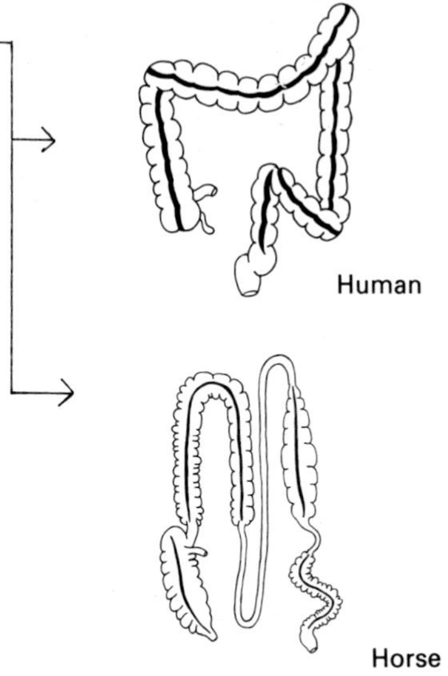
	Distal colon	Transport Secretion of lubricants Absorption (Faecal-pellet formation)	
	(Plicae semi-lunares)	Transport regulation	
	Rectum	Transport Storage Secretion of lubricants Skybala/scat formation	
Anus	Sphincter ani internus, externus	Transport regulation	

some species are given in brackets (Table 24.1). Complex carbohydrates are broken down by bacterial fermentation in the stomach and/or caecum and colon (where phenolics are most effectively neutralised) with the absorption of volatile fatty acids; simple sugars are processed mainly in the small intestine, along with simple proteins and fats.

The variety of digestive strategies in relation to body size and food type is summarised in Table 24.2. Hence, the largest mammals exhibit foregut fermentation – with or without rumination – or caeco-colic fermentation; medium-sized mammals have fermentation in the stomach or colon. In small mammals, there may be foregut fermentation with or without bypass features, or caeco-colic fermentation with or without separation of fine materials in the colon (with their return to the caecum). Some of the smaller mammals retain simple guts but may have storage features.

Lucas (Ch. 13) has argued that measurement of the mechanical properties of food is essential in order to make progress in relating the form of dentition to diet. A small fraction of the benefit derived from this can be shown graphically. Following Ashby (1989), the logarithms of the critical stress intensity factor (K_{Ic}) and the elastic modulus (E) are plotted for a variety of biological tissues in Fig. 24.4. All the tissues shown are either foods (e.g. parenchyma of apple flesh, seed cotyledon and seaweed) or casings which need to be broken in order to get access to food (e.g. seed and mollusc shells). Dental tissues are shown for comparison because these can also fracture. Where two points are joined by a line, this indicates either anisotropy, changes of properties with age or variation in measurement with different tests. However, detail is kept to a minimum here in order to make a general point.

The critical stress intensity factor (K_{Ic}) is a concept from linear elastic fracture mechanics and describes the stress field close to the tip of a crack. It is an alternative description of toughness to R, the work required to produce new unit surface area by fracture, that was used in Ch. 13. K_{Ic} is less appropriate than R if the food exhibits a non-linear stress–strain curve or if there is considerable ductility. However, to obtain a general picture, these restrictions have been waived here and the relation

$$K_{Ic}^2 = ER$$

has been assumed (some of the fracture toughness values in Fig. 24.4 have had to be converted from R to K_{Ic} using this formula). The thin lines crossing the graph obliquely represent constant K_{Ic}^2/E values and are, therefore, constant values of R. It is easy to see that biological tissues have a wide range of properties and, also, that the highest values of K_{Ic} tend to be associated

Table 24.2. *Body size and digestive characteristics versus diet among mammals. Unless indicated otherwise, solid bars mean that a substantial proportion of species in a size-gut category use a diet-type as a major component of food intake*

Body size	Digestive development	Examples	Diet types: High-lignin, dicot, leaf and twigs	Low-lignin, dicot, leaf	Low to medium fibre grasses	High-fibre grasses	Fruits	Seeds	Nectar and/or pollen	Fungi	Roots/ tubers	Animal tissue
Large[a] (>1200 kg)	Foregut (colon minor)											
	+ rumination	Few[a]										
	− rumination	Hippopotamus Whales[b]			■	■						?[b]
	Colon	Rhinoceros Elephant			■	■						
Medium (20–1200 kg)	Foregut (colon/caecum minor)	Ruminants Camelids Kangaroos	■	■	■							
	Colon	Horses Wombats			■	■						
	Caecum	Capybara			■	■						
Small[c] (1–20 kg)	Foregut (caecum/colon moderate)											

	+ bypass[d]	Some ruminants Colobid primates? Rat kangaroos			
	– bypass	Wallabies sloths	[e]		
	Caecum (± colon) + separation[f]	Koala Greater glider Lagomorphs			
	– separation	Many rodents Some primates Brushtail possum Some rodents (e.g. marmots) Hyrax	[g]		
Very small (<1 kg)	Caecum (± colon) + separation	Ringtail possums Sportive lemurs Many rodents	[h]		
	– separation	Many rodents (e.g. sciurids, some murids) Frugivorous and exudate-feeding marsupials Frugivorous and exudate-feeding primates			

Table 24.2. Continued

			Diet types									
Body size	Digestive development	Examples	High-lignin, dicot, leaf and twigs	Low-lignin, dicot, leaf	Low to medium fibre grasses	High-fibre grasses	Fruits	Seeds	Nectar and/or pollen	Fungi	Roots/ tubers	Animal tissue
Small/ medium	Gut simple (various degrees of development of stomach as storage organ)	Vampire bats Insectivores Carnivores Some frugivorous bats Primates and marsupials Nectarivores					■		■			■

[a] Demment and van Soest (1985) predicted that rumination and selective retention of fibre particles in the foregut is not advantageous in mammals above 600–1200 kg.
[b] Digestive systems and diet of whales are not well documented.
[c] Various models (see Parra, 1978). Demment and van Soest (1985) and Cork (Ch. 21) predict that heavy reliance on fermentation to meet energy requirements is precluded in mammals less than about 20 kg.
[d] Facility for some digesta to bypass fermentation in the forestomach. Not well documented in colobids. Occurs in wallabies, but to lesser extent than in rat kangaroos.
[e] Only sloths use high-lignin foliage, probably facilitated by their very low metabolic rate.
[f] Facility for selective retention of solutes and fine particles in the caecum ± proximal colon. Accompanied by caecotrophy in lagomorphs and many rodents (Björnhag, Ch. 17).
[g] Hyrax is the only grass-eater of which we know.
[h] Possibly facilitated by caecotrophy.

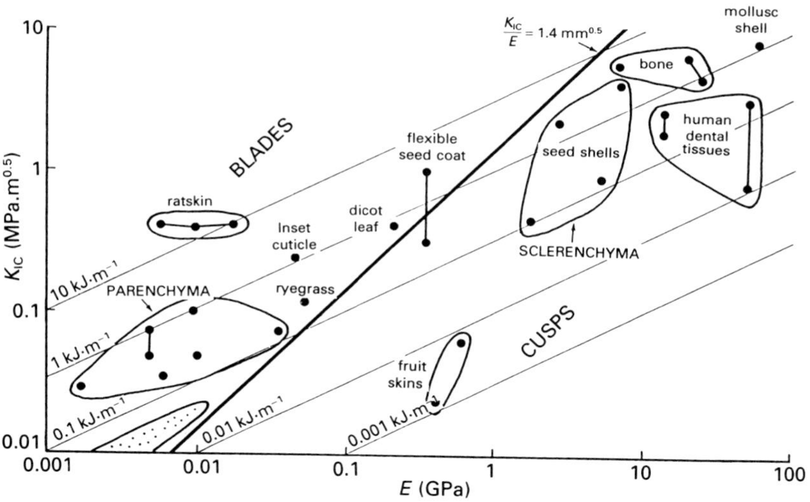

Fig. 24.4. The relationship between critical stress intensity factor (K_{Ic} units of MPa m$^{0.5}$) and elastic modulus (E, units of GPa) for a variety of biological tissues plotted on logarithmic scales. The thin lines show values of constant R (a measure of toughness) in kJ m^{-2}. The thick line represents $K_{Ic}/E = 1.4$ mm$^{0.5}$ and appears to distinguish two dental-dietary classes. The stippled area shown to the lower left of this line (and which probably extends to the right side) indicates solids that are too friable to pose any problem to the dentition. (From Lucas and Teaford, 1993.)

with foods of high E. However, K_{Ic} is a satisfactory measure of fracture resistance only in very stiff foods such as bone, seed and mollusc shells. Otherwise, resistance to cracking in lower modulus foods is determined not by K_{Ic} or R alone, but by the ratio K_{Ic}/E, which can also be expressed as $(ER)^{0.5}$ (Ashby, 1989). Rather than show a series of lines, only one, that at $K_{Ic}/E = 1.4$ mm$^{0.5}$, has been drawn (thick line in Fig. 24.4). It appears to distinguish between foods that crack easily ($K_{Ic}/E < 1.4$ mm$^{0.5}$) and those that do not ($K_{Ic}/E > 1.4$ mm$^{0.5}$). Virtually all mammals that eat foods to the left of the line (high K_{Ic}/E foods) used bladed post-canine teeth (or at least prominent ridges). These include carnivores, insectivores, graminivores and folivores. Mammals that eat foods to the right of the line (low K_{Ic}/E foods), such as those that break bone, mollusc shells or hard-shelled seeds in their mouths, use blunt-cusped post-canine teeth. Explanations for these dental shapes are discussed in Ch. 13.

Future directions

Food

Food classifications, such as those presented in Figs. 24.1 and 24.2, need to be rationalised and refined to be meaningful in relation to anatomy and physiology. We need to know more about potential foods in different types of environment, characterising them by the traditional food item classification but integrating mechanical and chemical properties. Developing a broad evolutionary perspective depends in large part on understanding such potentials, especially in the past.

Dietary flexibility and diversity need to be elaborated within mammalian groups (Table 24.2). A hierarchy of dietary components is needed for each mammalian taxon; a distinction between predominant and secondary food types would provide a starting point. This will help to develop a more ecological and evolutionary perspective.

In relation to mastication – to teeth and jaws – there is a need to develop laboratory tests for determining the mechanical properties of foods and field techniques for deriving the mechanical properties, since only the geometry of foods tends to survive storage.

Form

Ingestive behaviour needs to be described in detail, in relation to various facets of cranio-dental and post-cranial anatomy; that is, relating such behaviour to the use of the mouth, lips and tongue and of the limbs.

Features of tooth shape especially relevant to food properties and function, such as tooth sharpness and contiguity, need to be quantified. The importance of mammalian teeth in preparing food for digestion begs consideration of birds and reptiles. Herbivorous birds and reptiles may often feed on highly fibrous diets. The gizzard of some birds is presumed to be important in reducing the particle size of herbage but how gizzards actually work is poorly understood. Herbivorous lizards lack teeth and gizzards – how are they able to reduce the size of ingested plant material?

In relation to the gastro-intestinal tract, detailed knowledge of gross morphology needs to be supplemented with quantitative studies of microscopic anatomy, such as the epithelial lining and surface enlargements (e.g. villi), in relation to area and volume determinations. In addition, morphological studies should be extended to the ultrastructural level (transmission and scanning electron microscope). These can be correlated with morphometric studies of surface-enlargement factors at these levels, including, for example, microvilli.

It is also necessary to delineate clearly the different regions of the gastro-intestinal tract, as well as the sections of each region. The following criteria can be used for these studies: (a) innervation, (b) vascularisation (especially arterial supply), (c) relations of tract with mesenteries and (d) muscular architecture (especially for the stomach).

It also has to be determined whether gastro-intestinal regions of different mammalian taxa share derived characters, i.e. whether synapomorphies are differentiated. The taeniae in the stomach of Colobidae and Macropodidae, as well as in the large intestine of many phytophagous and polyphagous mammals, are an example of such synapomorphic structures. In contrast, the presence of oblique muscular fibres in the fornix/fundus and in the corpus of the mammalian stomach is a symplesiomorphic feature, i.e. it represents an ancestral character in the highly variable differentiations of the stomach in mammals. Knowledge of the details of anatomical structures will improve our understanding of the functional significance of important structural features, such as bypass or separating features for digesta of differing physical character.

A concerted effort should be made in future to bring together all morphological aspects at the different levels of observation (macroscopic to ultrastructural), some of which are presented in Part III, the Form chapters, into a comprehensive functional interpretation of the dietary strategy not only of individual species but also whole dietary groups.

Function

The flexibility of the digestive tract to changes in diet – both short term (even between meals) and long term (seasonal and climatic shifts) – is an area of great interest but little integrative information. These seasonal changes have actually been observed in the taste ability of some prosimian species (Simmen and Hladik, 1988). Indeed the taste ability of different species needs to be described, not only as the various thresholds for different soluble substances but also in terms of the supra-threshold responses occurring at the start of food processing in the first part of the gut.

Investigation is needed at both the integrated processing response level (intake, passage rate, volume, extent of digestion) and the metabolic level (uptake rates, metabolic adjustment, excretion) in order to identify bottlenecks to the flexibility of the gut.

In this area, there is a need to investigate the control of food intake in caeco-colic fermenters and the limits to using high-fibre diets (such as endogenous nitrogen losses, metabolic control of appetite, integrated processing response) in these animals. The costs of having a large and complex gut, in terms of large surface areas, high abrasion and energy and protein losses, need further definition. The roles of chitinase in animals (endogenous versus microbial) also demand attention, as do digestive organ adaptations for detoxification: (a) microbial (foregut fermenters), (b) salivary proteins and (c) post-absorptive mechanisms.

The importance of detoxification in the digestive tract is still uncertain. Certainly there is little evidence that foregut fermenters are necessarily better suited to toxic diets than caeco-colic fermenters. The role of salivary proteins is not as simple as first thought, and this suggests that the differences between grazers and browsers are more likely to be found in a number of factors, such as size of the digestive tract, kinetics of particle breakdown, than simply in detoxification ability.

The control of food intake in caeco-colic fermenters is only poorly understood. In particular, in small species it is unclear why passage rate is not increased on fibrous diets so that the intake of easily soluble constituents is maximised. Many small species maintain extensive fermentation areas and current measurements suggest that the contribution of fermentation to daily energy requirements is relatively small. We need to know whether this is the 'icing on the cake' or whether fermentation is necessary for other gut functions (e.g. water absorption).

Our understanding of the differences between foregut and 'hindgut' fermenters is much better than it was several years ago, but we still have a tendency to regard the foregut fermentation as a fibre-digesting process analogous to that in large animals such as kangaroos or sheep. We need to continue studies of small foregut fermenters since our current knowledge suggests that there is little benefit of this digestive pattern for small mammals. What other roles does fermentation serve in these species?

We also need to revisit a question which has been accepted as settled. 'Hindgut' fermenters are widely regarded as being unable to use microbial protein except by caecotrophy and this idea has underpinned our understanding of the differences between the two types of fermenters. Recent studies in herbivorous birds, however, have suggested an active transport of amino acids from the caecum (Foley and Cork, 1992), and we should look carefully in a range of mammals to see whether a similar pattern has been overlooked. Such a finding would have a large impact on our understanding of the nutritional ecology of caeco-colic fermenters.

Conclusions

We have added new information to old for a wide spectrum of topics concerned with feeding: the nature and variety of the foods of mammals and the structure and functioning of the digestive system. We have blended practical aspects to theory in this synthesis.

In classifying foods extensively and rigorously, a potentially unifying framework has been produced that will make further analysis and discussion of animal foods more productive. Progress has been made in defining the nature of niches and their temporal and spatial dynamics. Taste has been brought properly into the field of food selection. The variety of gut structure is better understood in relation to different diets and to digestive physiology, and methods have been refined. The functioning of the gut is better understood, albeit mainly among the herbivorous mammals – which themselves represent a dramatic range of diets and guts.

The essence of this volume is the application of very different lines of research to the digestive system: the development of models, the broadening of a focus on particular animal taxa or on particular parts of the gastro-intestinal tract by ecologists, anatomists, physiologists and biochemists and the integration of these different approaches. In particular, such syntheses will promote the development of a meaningful evolutionary framework.

Finally, we have a much clearer idea of where to go next, of which topics will be most productive in advancing our understanding of this central part of mammalian biology. Reproductive biology tends to obsess many biologists, but without enough of the right sort of food, no mammal can indulge effectively in the behaviour that is vital for perpetuating the species.

References

Ashby, M. F. (1989). Overview no. 80. On the engineering properties of materials. *Acta Metallurgica*, **37**, 1273–1293.

Demment, M. W. & van Soest, P. J. (1985). A nutritional explanation for body-sized patterns of ruminant and nonruminant herbivores. *American Nutritionist*, **15**, 641–672.

FAO (1970–72). *Amino-acid Content of Foods and Biological Data on Proteins.* Food Policy and Food Science Service, Nutrition Divison FAO. Rome: Food and Agricultural Organization of the United Nations.

Foley, W. J. & Cork, S. J. (1992). Use of fibrous diets by small herbivores: how far can the rules be 'bent'? *Tree*, **7**, 159–162.

Lucas, P. W. & Teaford, M. F. (1993). Functional morphology of colobine teeth. In *Colobine Monkeys: Their Evolutionary Ecology*, ed. A. G. S. Davies & J. F. Oates. Cambridge: Cambridge University Press, in press.

Parra, R. (1978). Comparison of foregut and hindgut fermentation in herbivores. In

Ecology of Arboreal Folivores, ed. G. G. Montgomery, pp. 205–229. Washington, DC: Smithsonian Institution.

Simmen, B. & Hladik, C. N. (1988). Seasonal variation of taste threshold for sucrose in a prosimian species, *Microcebus murinus*. *Folia primatologica*, **51**, 152–157.

Slansky, F. & Scriber, J. M. (1985). Food consumption and utilization. In *Comprehensive Insect Physiology, Biochemistry and Pharmacology*, Vol. 4: *Regulation: Digestion, Nutrition, Excretion*, ed. G. A. Kerkut & L. I. Gilbert, pp. 87–163. Oxford: Pergamon Press.

Index

Pages in *italics* indicate a figure; those in **bold** indicate a table.